BIOGRAPH[illegible]D [illegible]M IN PAIRED COMPARISONS

BIOGRAPHIES OF GENDER AND HERMAPHRODITISM IN PAIRED COMPARISONS

JOHN MONEY

The Johns Hopkins University and Hospital,
Baltimore, MD 21205, U.S.A.

Clinical Supplement to

THE HANDBOOK OF SEXOLOGY

Edited by J. Money and H. Musaph

1991

ELSEVIER

Amsterdam – New York – Oxford

This book is printed on acid-free paper.

ISBN 0-444-89129-3 (paperback)
ISBN 0-444-81403-5 (hardback)

Published by:
Elsevier Science Publishers B.V.
(Biomedical Division)
P.O. Box 211
1000 AE Amsterdam
The Netherlands

Sole distributors for the USA and Canada:
Elsevier Science Publishing Company, Inc.
655 Avenue of the Americas
New York, NY 10010
USA

Library of Congress Cataloging-in-Publication Data

Money, John, 1921–
Biographies of gender and hermaphroditism in paired comparisons : clinical supplement to the Handbook of sexology / John Money.
p. cm.
Includes bibliographical references.
ISBN 0-444-81403-5 (alk. paper)
1. Hermaphroditism--Case studies. I. Handbook of sexology. II. Title.
[DNLM: 1. Hermaphroditism--case studies. 2. Identification (Psychology)--case studies. 3. Sex Characteristics. 4. Sex Determination. WJ 712 M742b]
RC883.M66 1991
616.6′9409--dc20
DNLM/DLC
for Library of Congress

90-14095
CIP

Printed in The Netherlands

DEDICATION

This book is dedicated to those whose sexually remarkable lives have contributed so much knowledge to both humanistic sexology and sexological science, and to the welfare of those who succeed them.

Also by John Money

Hermaphroditism: An Inquiry into the Nature of a Human Paradox, 1952
The Psychologic Study of Man, 1957
A Standardized Road-Map Test of Direction Sense (with D. Alexander and H.T. Walker Jr.), 1965
Sex Errors of the Body: Dilemmas, Education, Counseling, 1968
Man and Woman, Boy and Girl: The Differentiation and Dimorphism of Gender Identity from Conception to Maturity (with A.A. Ehrhardt), 1972
Sexual Signatures (with Patricia Tucker), 1975
Love and Love Sickness: The Science of Sex, Gender Difference, and Pairbonding, 1980
The Destroying Angel: Sex, Fitness, and Food in the Legacy of Degeneracy Theory, Graham Crackers, Kellogg's Corn Flakes, and American Health History, 1985
Lovemaps: Clinical Concepts of Sexual/Erotic Health and Pathology, Paraphilia, and Gender Transposition in Childhood, Adolescence, and Maturity, 1986
Venuses Penuses: Sexology, Sexosophy, and Exigency Theory, 1986
Gay, Straight, and In-Between: The Sexology of Erotic Orientation, 1988
Vandalized Lovemaps: Paraphilic Outcome of Seven Cases in Pediatric Sexology (with M. Lamacz), 1989
Sexology of Genes, Genitals, Hormones, and Gender: Selected Readings, 1991

Edited by John Money

Reading Disability: Progress and Research Needs in Dyslexia, 1962
Sex Research: New Developments, 1965
The Disabled Reader: Education of the Dyslexic Child, 1966
Transsexualism and Sex Reassignment (with R. Green), 1969
Contemporary Sexual Behavior: Critical Issues in the 1970's (with J. Zubin), 1973
Developmental Human Behavior Genetics (with W.K. Schaie, E. Anderson and G. McClearn), 1975
Handbook of Sexology (with H. Musaph), 1977
Traumatic Abuse and Neglect of Children at Home (with G. Williams), 1980
Handbook of Human Sexuality (with B.B. Wolman), 1980
Handbook of Sexology, Vol. VI (with H. Musaph and J.M.A. Sitsen), 1988
Handbook of Sexology, Vol. VII (with H. Musaph and M.E. Perry), 1990

ACKNOWLEDGEMENTS

The matched pairs of biographies in four of the chapters in this volume are revised and expanded from preliminary publications as follows:

Chapter 6: J. Money and C. Lobato (1988) *Psychiatry: Interpersonal and Biological Processes*, 51:65–79.

Chapter 7: J. Money (1984) *International Journal of Family Psychiatry*, 5:317–381.

Chapter 9: J. Money (1968) *Clinical Pediatrics*, 7:331–339.

Chapter 10: J. Money and V.G. Lewis (1987) *Psychiatry: Interpersonal and Biological Processes*, 50:97–111.

The following people contributed to the tabulation of biographical data and the first drafts of their presentation:

Sandra Aamodt, Chapters 2, 3, 11;
Charles Annecillo, Chapter 3;
John T. Early III, Chapter 12;
James W. Frock, Chapter 10;
Daniel Herzog, Chapter 7;
Viola G. Lewis, Chapters 3, 9, 10;
Cecilia Lobato, Chapter 6;
Leonardo Tondo, Chapter 8.

From 1951 to the present many people, too numerous to name, who have worked in the Psychohormonal Research Unit contributed to the interviewing and testing of the patients who here constitute the matched pairs. Sandra Aamodt coordinated the typing, proofing and correcting of the book on the word processor.

CONTENTS

Foreword

At Harvard University, in 1948, a singular case of hermaphroditism was presented by George Gardner, M.D., to a small group of graduate students from the Psychological Clinic. This case (Money, 1952) became the sexological progenitor from which this book traces its descent. The case was one of male hermaphroditism with, according to today's diagnostics, the androgen-insensitivity syndrome. The parents had been professionally misguided into changing the announcement of sex from female to male in infancy, despite the absence of a penis.

The evidence of this case alone necessitated the sacrifice of the sacred cow of traditional sexological science, namely, that male and female, masculine and feminine, are absolute and eternal verities. Here was evidence that the different criteria or determinants of sex, although usually concordant among themselves, are not inevitably so. Although it was historically too soon for chromosomes to be actually counted, the evidence was that chromosomal sex, by itself alone, does not preordain the sex of the morphology of the genitalia, nor of the body at puberty. Mentally, here was a boy, the product of 15 years of male rearing, living in the body of a female, complete with large breasts.

As with chromosomal sex so also with gonadal sex. The fact that both gonads were testicles had been no guarantee that either the external genitalia or the body in general would differentiate as masculine. Enough had been learned in the preceding 30 years or so of experimental endocrinology to support the proposition that in fetal development, the body feminizes, if deprived of male hormones, or if unable to utilize them. Developmentally, Eve always takes precedence over Adam! Otherwise, knowledge of the endocrinology of this boy's syndrome as of all syndromes of hermaphroditism at that time, was rudimentary.

The methods of measuring hormone levels in the bloodstream were extremely crude, by today's standards. Nonetheless, it could reasonably be inferred that, with the body so completely feminized hormonally, the amount of male hormone in the bloodstream would have been developmentally insufficient for the brain to have been masculinized assuming, that is, that brain masculinization is exclusively dependent on hormones. The evidence of the boy's behavior and personal verbal report, however, did not confirm this inference. His brain had not been feminized. He was oriented indisputably toward not being a girl, but toward continuing to live as a boy, irrespective of having a body that failed to respond to male hormone and to virilize at puberty.

Independently of this hormonally feminized body, his mind had masculinized. His case demonstrated the part played by the encounters and experiences of postnatal social life in establishing masculinization (or feminization) of the mind and behavior. It was the first case among many whose biographies were investigated in person which, when added to the findings of a comprehensive literature survey, necessitated the coining of the term, gender role (Money, 1955). This term subsequently became divided into gender role and gender identity, both of which then became fused into G-I/R (gender-identity/role) (Money, 1986, Ch. 19). G-I/R is an umbrella term under which there is, as there must be, shelter for the sexuoerotic and genital aspects of masculine and feminine status, as well as for those aspects classified as educational, recreational, occupational, political, legal, economic, sartorial, cosmetic, and gestural. In the present case, the long-term sexuoerotic outcome was that the individual graduated as a professional health-care provider, established a sex life in marriage and, together with his wife, became a parent by adoption.

The discrepancies between the various criteria of sex recognized initially in this case led eventually to the publication (Money et al., 1955) of a list of the variables of sex that, although usually concordant with one another, are capable of being discordant. Updated, this list of the criteria or variables of sex is:

- chromosal sex
- TDF (testicular determing factor) (Page et al., 1987)
- H-Y antigen (Engel et al., 1980; Ohno, 1978)
- gonadal dimorphism
- prenatal hormones
- internal morphologic dimorphism
- external morphologic dimorphism
- brain dimorphism
- assigned sex
- socially stereotyped sex of rearing
- pubertal hormones
- G-I/R

The foregoing multiple variables of sex exert their influence not simultaneously, but sequentially in the development and differentiation of male and female. The graphic representation of multivariate and sequential determinism of G-I/R (Money and Ehrhardt, 1972) was devised originally for Money (1974), and modified in 1977 to include the newly discovered H-Y antigen (Fig. F.1) (Eicher et al., 1979; Koo et al., 1977; Silvers et al., 1977).

In my experience, Gardner's case from the Judge Baker clinic was the first, with many others subsequently, that pointed to the sex of assignment and social rearing as having a major role in the differentiation of G-I/R as male or female. In view of the analogy with native language (see below), this conception should not have been too surprising, but it was. Scientific dogma then, as now, attributed anything as fundamental as sex difference to 'real' science, namely biology. Whatever was left for social science was not considered fundamental but trivial and ephemeral. The antith-

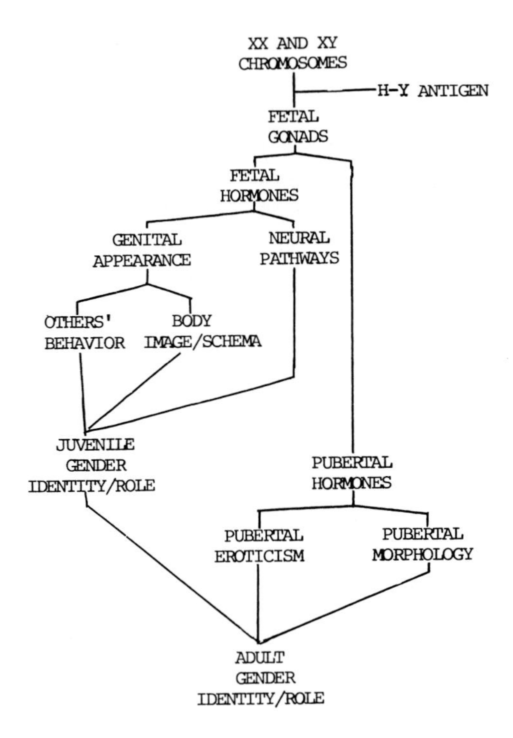

Fig. F.1. Gender-identity/role (G-I/R) is multivariately and sequentially determined and includes sexuo-erotic orientation.

esis between biology on the one hand, and social science and psychology on the other, derives from a failure to recognize that there is a biology, in the brain, of learning and remembering. It is an antithesis based also on the outmoded juxtaposition of nature versus nurture. This two-term proposition should be a three-term one: nature/critical-period/nurture (Money, 1986, Ch. 7). When nature and nurture interact at a critical developmental period, the resultant outcome may be permanent — just as one's native language is permanent.

In any science, the function of a single case is not to prove a general law, but to alert to the possibility that one might exist and be waiting to be discovered. Its discovery will be contingent on the availability of more cases. In sexology, hermaphroditism and related conditions constitute a window through which to examine the determinants of masculine and feminine differentiation. Etiologically, there are many syndromes of hermaphroditism, some more rare, and some more prevalent than others. Even when it is possible to assemble a group homogeneous for diagnosis, it will almost certainly not be possible to match them on all other pertinent variables. That is why, when two individuals do happen to be concordantly matched except for a major discordance – for example concordant for diagnostic history but discordant for sex of rearing – it is of value to the future of the science of sexology to keep a record of the respective outcomes.

Like a single case, a lone matched pair does not prove a general rule. As the pairs mount up, however, each pair diagnostically different from every other pair, so also does the evidence of a general principle at work. In this book, the general principle is that the prenatal precursors of the male/female differentiation of G-I/R in the brain are hormone dependent. Postnatally, G-I/R is dependent on input that reaches the brain through the skin senses, the eyes, and the ears, and is contingent, in general, on the sex of assignment. When the G-I/R diverges from the assigned sex, the factors involved may be either prenatal or postnatal in origin, or both. They have not yet been quantifiably distinguished from one another.

The existence of a prenatal, hormonally mediated phase in the differentiation of G-I/R allows for the possibility that a hormonal anomaly may induce an anomaly of the G-I/R. It may do so either directly or indirectly. The indirect effect would constitute a vulnerability or predisposition that would materialize only in the presence of the necessary postnatal conditions. Thus, contingent on postnatal conditions in the sex of rearing, the outcome of a bisexual predisposition could be a bisexual G-I/R, or an exclusively heterosexual or homosexual one.

Postnatal flexibility in the differentiation of G-I/R in the human species has its counterpart, but to a lesser degree, in subhuman primates. Subprimate mammals, however, are more like 'hormonal robots,' in that their mating behavior as male or female is ordained by the hormonalization of the brain, prenatally, more than by postnatal variations or manipulations of the social environment.

There is a hypothesis, still in the undocumented realm of science-fiction hypotheses, that explains the lack of prenatal hormonal fixation of the G-I/R in the human brain. According to this hypothesis, the phylogenetic heritage of speech and logical thinking in the human species required a high degree of developmental brain plasticity and release from preordained functional fixation. Hence the release of the sexual brain from its earlier robot-like functioning, and the necessity of sensory input from the social environment for the completion of the postnatal phase of G-I/R differentiation.

For the development of language in individual members of the human species, the brain in prenatal life must be correctly formed and healthy. Postnatally, language development is contingent on signals that typically enter the brain through the ears and that are progressively matched by vocal responses. Early in life, the native language is established as a brain/mind template which becomes virtually immutable.

By analogy, the G-I/R develops in a similar fashion. Postnatally, its signals are relayed into the brain through all the senses. The masculinity/feminity ratio of the brain/mind template differentiates from early infancy according to the principles of identification of oneself with those of one's own natal sex, and complementation of one's self to those of the other natal sex. By age 5, the precursors of adult masculinity and femininity are so well established as to have become virtually immutable.

The stereotypes of masculine and feminine apply not only to work, play, schooling, dress, ornament, budget, and legal rights, but also to romantic, erotic, and genital sex roles. In cases of hermaphroditism and related birth defects of the sex organs, a boy may grow up to have a masculine G-I/R, except that his sex role, in the specific anatomical and genital sense, is impaired. Correspondingly a girl may be similarly affected. Nonetheless, except for the anatomical defect, the boy may be romantically and erotically masculine, and the girl feminine. The romantic and erotic orientation of such a boy or girl belongs to an aspect of the G-I/R for which recently the name *lovemap* was coined (Money, 1986).

A lovemap develops as a pattern or template in the brain and the mind. Its development begins in childhood. The years around age 8 are critical in the shaping of its development as well-formed or malformed.

A lovemap registers the ideas and images of the idealized lover, the idealized love affair, and the idealized activities and practices that induce sexual arousal and lead to sexual climax. It appears spontaneously in explicit dreams, romantic fantasies, and masturbation fantasies, with or without external stimuli. These dreams and fantasies serve as rehearsal for real-life performances.

Disclosure of a lovemap in public is tabooed by those people who define all erotically explicit stories or pictures as pornographic. For this reason, the specific investigation of lovemaps has seldom been undertaken in sexological research.

In the biographies presented in this book, each case has been selected because of the research significance of the lovemap. The plan of presentation is the same, case by case. First comes the diagnostic and clinical biography which establishes the specific sexological significance of the case for lovemap formation as masculine or feminine. Then follows that part of the social biography which is nonerotically gender coded, and which sets the stage for the sexual and erotic explicitness of the lovemap biography.

Each biography is explicit insofar as it employs excerpts from the patient's own writings and verbatim quotations from the transcripts of audiotaped interviews. For economy of space, the patient's own words are given precedence over those of the interviewer. The actual dialogue is reproduced only when essential for the meaning.

Since they are arranged in matched pairs, each pair is followed by an exposition of their sexological significance.

BIBLIOGRAPHY

Eicher, W., Spoljar, M., Cleve, H., Murken, J.-D., Richter, K. and Stangel-Rutkowski, S. (1979) H-Y antigen in trans-sexuality. *Lancet*, ii: 1137–1138.

Engel, W., Pfäfflin, F. and Wiedeking, C. (1980) H-Y antigen in transsexuality, and how to explain testis differentiation in H-Y antigen-negative males and ovary differentiation in H-Y antigen-positive females. *Hum. Genet.*, 55: 315–319.

Koo, G.C., Wachtel, S.S., Krupen-Brown, K., Mittl, L.R., Genel, M., Breg, W.R., Rosenthal, I.M., Borgaonkar, D.S., Miller, D.A., Tantravahi, R., Schreck, R.R., Erlanger, B.F. and Miller, O.J. (1977) Mapping the locus of the H-Y gene on the human Y chromosome. *Science*, 198: 940–942.

Money, J. (1952) *Hermaphroditism: An Inquiry into the Nature of a Human Paradox*. Doctoral Dissertation, Harvard University Library. University Microfilms Library Services, Xerox Corporation, Ann Arbor, MI 48106, 1967.

Money, J. (1955) Hermaphroditism, gender and precocity in hyperadrenocorticism: Psychologic findings. *Bull. Johns Hopkins Hosp.*, 96: 253–264.

Money, J. (1974) Intersexual and transexual behavior and syndromes. In: *American Handbook of Psychiatry*, Vol. III, 2nd edn., revised (S. Arieti and E.B. Brady, Eds.). New York, Basic Books.

Money, J. (1986) *Venuses Penuses: Sexology, Sexosophy, and Exigency Theory*. Buffalo, Prometheus Books.

Money, J. (1988) *Lovemaps: Clinical Concepts of Sexual/Erotic Health and Pathology, Paraphilia, and Gender Transposition in Childhood, Adolescence and Maturity*. Buffalo, Prometheus Books.

Money, J. and Ehrhardt, A.A. (1972) *Man and Woman, Boy and Girl: Differentiation and Dimorphism of Gender Identity from Conception to Maturity*. Baltimore, Johns Hopkins University Press.

Money, J., Hampson, J.G. and Hampson, J.L. (1955) An examination of some basic sexual concepts: The evidence of human hermaphroditism. *Bull. Johns Hopkins Hosp.*, 97: 301–319.

Ohno, S. (1978) The role of H-Y antigen in primary sex determination. *J. Am. Med. Ass.*, 239: 217–220.

Page, D.C., Mosher, R., Simpson, E.M., Fisher, E.M.C., Mardon, G., Pollack, J., McGillivray, B., de la Chapelle, A. and Brown, L.G. (1987) The sex determining region of the human Y chromosome encodes a finger protein. *Cell*, 51: 1091–1104.

Silvers, W.K. and Wachtel, S.S. (1977) H-Y antigen: Behavior and function. *Science*, 195: 956–960.

CHAPTER 1

Introduction
The method of matched pairs

STATISTICS AND THE EXTREME CASE

When the Polish gynecologist, Neugebauer (1908), published his great turn of the century work on hermaphroditism, he warned darkly, in the moral fashion of the day, of the peril in a girls' school of letting loose a wolf in sheep's clothing. His wolf was a male hermaphrodite, a child who, though living as a girl, underwent a masculinizing puberty. Since diagnosis was pretty hit-and-miss in those days, the hapless child may well have been a female hermaphrodite with the virilizing adrenogenital syndrome, and not as feared, a male hermaphrodite, at all. Little did that matter. A wolf was a person with masculine passions, and he could not be trusted to resist the temptation of indiscriminate molestation when put down in a veritable sheepfold of innocent young virgins.

This was the age in which the moral medical tone was set by such books as William Acton's volume on *The Functions and Disorders of the Reproductive Organs in Childhood, Youth, Adult Age, and Advanced Life, Considered in Their Physiological, Social and Moral Relations* (1875). 'The majority of women,' Acton wrote, 'happily for themselves and for society, are not very much troubled with sexual feeling of any kind. ... As a general rule, a modest woman seldom desires any sexual gratification for herself. She submits to her husband, but only to please him; and, but for the desire of maternity, would far rather be relieved from his attentions.'

Neugebauer may have been concerned about an epidemic of lupine pregnancies, but unnecessarily so, because most of his wolves in sheep's clothing would have been sterile. He was as much concerned, of course, about sexual behavior. In that proscriptive and sexologically still prescientific era, he accepted the common sense of the barnyard that testicles are responsible for male sexual behavior in all its ungovernable lechery.

By the middle of the 20th century, there had been some dilution of such crude sexism. It was this diluted version that physicians lived by in those mid-century days when courses in sexological medicine were equated with pornography (as they still are in some medical institutions), when grand rounds in hermaphroditsm were not far removed from sex-segregated freak-shows, and when the gonads were the sine qua non of a hermaphroditic sex. The gonads gave the promise, albeit in vain, of fertility; but above all they gave what was believed to be a prophecy of whether the baby's sexual instincts would be male or female when awakened at puberty. The wolf would thus have normal wolfish instincts, ostensibly, in or out of the sheepfold.

The historical fact is that many people have sexually conceptualized the human species as becoming at puberty a type of instinct-driven robot. This concept is still alive and well. The robot is merchandized in two models, one with a male, and one with a female instinct or drive, the male aggressive and the female submissive. In the prestige journal, *Science*, as recently as 1974 Imperato-McGinley and associates wrote that 'male sex drive appears to be testosterone related and not dihydrotestosterone related. ...' She based her conjecture on a pedigree of male hermaphrodites with a hormonal anomaly based on a deficiency of the enzyme, 5α-reductase, which changes testosterone into dihydrotestosterone (Imperato-McGinley and Peterson, 1976; Peterson et al., 1977), and has maintained it by ignoring contrary evidence (Saenger et al., 1978; Rubin et al., 1981).

Those who use the concept of a sexual drive or instinct, and who think of its sexual dimorphism as being gonadally preordained, take it as a matter of great moment, one may even say fanaticism, to ensure that a hermaphroditic baby is assigned to what they believe to be the correct gonadal sex. They argue against the wisdom of assigning a baby to the morphologic sex in which it will be enabled, with proper clinical management, to belong sexologically in romance, eroticism, love, pair-bonding, and coition. They base their argument on their acceptance of the traditional ideology that sexological function is contingent exclusively on gonadal structure, and they defend their ideology with the statistics of the normal curve. The normal curve, however, reveals only contiguities in time or space, not cause and effect. Cases that deviate from the normal curve, or are at the extreme of the normal distribution, are very often the ones from which the secrets of causality may be unraveled. Cases of intersexuality are a prime example. To use normal-curve statistics based on the general population to predict the sexological future of an intersex is to lose the opportunity provided by a so-called experiment of nature to differentiate the variables of masculinity and femininity, and to probe for cause and effect.

Mathematicians themselves caution against the misuse of their science. 'The entire mathematical arsenal that our modern sages command cannot establish facts,' wrote the Russian mathematician, Evgrafov, as quoted by Murphy (1980). 'Practical people should always keep this in mind when they ask mathematicians for help.' Theoretical people should keep in mind that mathematics can legitimate concepts and principles, but it cannot create or invent them. Only theoreticians can do that. Murphy, a mathematical geneticist, cautioned against the misuse of mathematics in the case of schizophrenia which is, he wrote, 'totally irreducible to a natural scale of measurement. ... The courses open to those studying such a disease are of two broad types. First, one may invent an arbitrary scale of measurement with the accompanying hornet's nest of establishing secure properties and sound meaning for it. ... The alternative way of dealing with the problem is to divorce the notion of schizophrenia from the realm of quantitation: to argue that schizophrenia is an entity that does not depend ontologically on measurement. It thus bears a similarity to the idea of person, or of beauty, or of creativeness. This metaphysical kind of trait is the antithesis of a trait such as short stature that would scarcely have any meaning at all if it were deprived of its quantitative status. It may prove much easier to grasp the meaning of the characteristic itself than to grasp the meaning of the measurement devised to study it; and it is much less apt to come to pieces in our hands.'

It is not sufficient to say that one grasps the concept of schizophrenia by intuition, clinical osmosis, mystical apperception, or any other inarticulate process. For any concept to have scientific utility, it need not be subdivisible into units measurable on a scale that, like a meter stick, has a fixed zero, and equal incremental units. But it must be subdivisible into component characteristics or categories (in schizophrenia they are symptoms) that can be counted and summated. Until such subdivision is achieved, a concept remains too amorphous for science to deal with — just as the concept of moon-walking used to be too amorphous, and as the concepts of muse, unicorn, firebird, angel, demon, sixth sense, flying saucer, and Bermuda triangle still are too amorphous.

Even when a concept can be broken down into component characteristics, that is no guarantee that the concept will yield good returns scientifically. Obedience to methodology (what Thomas Kuhn (1970) calls normal science) is no substitute for astute concept formation and analysis. That is why in the development of any science there are phases when new concepts need to be enunciated. It is not possible to enunciate a concept and to prove it at the same time. Sometimes there are strategic and tactical limitations that impose a time delay between the enunciation of a concept and its orthodox experimental and statistical support or proof. At such a juncture, there may be great scientific value in nonstatistical evidence — especially the evidence of the extreme case that disobeys established doctrine. For example, in the annals of hermaphroditism, one single case of a chromosomal, 46,XX female whose postnatal biography unremarkably conforms to that of a stereotypic 46,XY male (Money, 1952; Money et al., 1955) shows beyond doubt that the prenatal chromosomal constitution as XX does not automatically and robot-like govern the postnatal gender-identity/role (G-I/R) as that of a female. It shows also that there are other variables, the roles of which remain to be uncovered.

Since the existence of such a case cannot be contrived experimentally, it must exist as an experiment of nature and be sought out and made use of on the basis of availability, in much the same way as celestial events, paleontological specimens, or ethnographically unique communities are made use of when they occur or are discovered.

There is a special place in sexology at this present phase of its history for the detailed single case report that exemplifies a particular, new concept. It leads to the formulation of a new hypothesis. More cases accumulate, or new technology leads to the contrived, experimental investigation of this new hypothesis, so that eventually it can be put to statistical test.

The outstanding single case stands out when compared with the accepted norm which it challenges. The matched pair of outstanding cases may each stand out when compared with the norm. But that is more or less coincidental to the matched-pair method which has its own methodological significance, namely that one case serves as a comparison or contrast for the other. There is some similarity with a before-and-after study of change in which the one case serves as its own comparison over time.

In the matched-pairs method, there are two individuals concordant on the basis of one criterion or set of criteria, but discordant on the basis of another. In the present instance, the two sets of criteria are prenatal and postnatal. Thus two individuals with a birth defect of the sex organs may be matched as concordant for prenatal developmental history and diagnostic status, but postnatally and in maturity discordant for G-I/R (Chapters 2 and 5–9).

An alternative possibility is that two individuals may be divergent for prenatal developmental history, but postnatally convergent on the criterion of G-I/R differentiation (Chapters 3, 4, 10, 11).

In the best of all possible clinical worlds, there would be many more than one pair of patients who could be matched on the basis of the same criterion. Ideally, the size of the sample of matched pairs would be large enough to warrant their comparison not only one pair at a time, but as two contrasting groups. Then the differences between the two groups could be evaluated statistically in the usual way. In the health-care world that actually exists, however, some syndromes are so rare that it is not possible to assemble a large, clinically homogeneous series of cases in one medical center, even in the course of an entire professional lifetime. A collaborative project in which many medical centers pool their resources would require a degree of concinnity presently incompatible with doctrinal and fiscal professional rivalries.

Rivalry impedes not only collaborative effort, but also the very continuity of longitudinal follow-up. It was politically militant doctrinal rivalry that forced the termination at Harvard Medical School of the long-term follow-up of children born with an anomalous number of chromosomes (Culliton, 1974). At Johns Hopkins, it was a rivalrous political and moral doctrine that terminated the follow-up program of surgical sex reassignment in transexualism (Meyer and Reter, 1979). On the international academic scene, doctrinal rivalry regarding the origins of gender identity led to an alliance with an unscrupulous media (Diamond, 1982) that prematurely terminated a unique longitudinal study of identical twins (Money, 1975a). A BBC crew of television sleuths, incited by the prospect of airing a doctrinal dispute, traced the whereabouts of the twins and their family and unethically invaded their privacy for programing purposes.

If they survive doctrinal dispute, all long-term outcome studies in sexology, whether of single cases or matched pairs, have the value of scarcity. That justifies restoration of the tradition, widely accepted a century ago, of publishing long and thorough case studies. They are the foundation of the developing theoretical opus of a new science — and sexology is a new science.

METHODOLOGY OF MATCHED-PAIR SEXOLOGICAL OUTCOME STUDIES

In laboratory studies of animals or tissue specimens, it is possible to get subjects or specimens on schedule. In human clinical studies, there is no fixed schedule for the intake of new subjects. One works in the clinic and awaits the admission of new patients. Only rarely do they appear in matched pairs. Sometimes it may take months or years to find the perfect match for a matched-pair comparison and outcome study. In cases of birth defect of the sex organs, it is necessary to continue follow-up until maturity in order to ascertain the sexological outcome. Thus, the very first requirement of the longitudinal outcome studies recorded in this book was an act of either faith or foolhardiness that what was begun in 1951 would continue to be financed by extramural grant support for 40 years and more. Alas, times have not changed. That is one reason why the chapters of this book are unique.

Over the years, I have had by mail and by phone a continuous flow of requests for what I consider to be pretentiously called the 'instruments' or 'tools' of sexological measurement. I call them what they are: tests, questionnaires, and interviews, so as to not have a false sense of mensurational accuracy. Accurate or not, there are very few tests, questionnaires, or interviews designed expressly to standardize the ascertainment of sexological data either longitudinally or in cross section.

There are three categories of test used in clinical psychoendocrinology and sexology. In category one is the test as a standardized sampling of a psychological function demonstrated in the here and now — for example, the intelligence test. It tests intelligence by requiring responses to questions and problems that necessitate intelligence for their solution; and it presumes that the score is predictive of the level at which intelligence can be expected to function in general.

In category two is the test as a standardized scale of opinion, attitude, or self-evaluation — for example the personality inventory or questionnaire. Each item requires a forced-choice response of the either/or, multiple-choice, or numerical-rating type, thus limiting the range of information a respondent may provide. It is presumed that the items test not only verbalization but also the deeds or actions which that verbalization purports to represent.

In category three is the test as an interview with a standardized agenda or schedule of inquiry. It is also known as a structured interview, to differentiate it from a nondirective interview. The standardization in this type of test applies not to the fixed wording of both the questions and the fixed-choice responses, as in a standardized scale, but to the agenda and to the scoring criteria used by the professionals who do the scoring.

In an interview in which the agenda is embodied in a schedule of inquiry, each topic is introduced with an open-ended request for information so as to avoid the stiltedness of fixed-word questions. Open-endedness allows the respondent free rein to define the topic in terms of what it means to him or her, personally. Subsequently the inquiry may become progressively less open-ended until eventually each particular topic may be closed with an interrogatory, employing forced-choice questions. Such questions are suitable for eliciting exact details of vital statistics, chronology, prevalence, location, frequency, dimension, and so on, for which exact answers may legitimately be expected to exist. Even an apparently straightforward statistic, such as marital status and number of children, however, fails to give information about composition of the household, for it does not give information about cohabitation, single parenthood, foster children, or relatives in residence. An open-ended request for information about household composition would allow such information to be retrieved.

An open-ended inquiry may be as uncomplicated as announcing the next topic on the standardized schedule of inquiry. It is a good idea to have a copy of the schedule in view, with an extra copy provided for the convenience of the subject. In this way, the subject knows that the topics are, indeed, standard, and that the interview is not an intimate intrusion on personal inviolacy. This is a significant issue in taking a sex history insofar as requesting intimate information is popularly recognized as one way of 'coming on' to someone.

Another safeguard against intruding on personal inviolacy is to request information about topics that, though sensitive in their personal application, are also in the public domain. For example, as an open-ended starter, one may say: 'There's been a lot of television coverage of gay activists these days — gay pride, gay parades, and gay victims of AIDS. Have you been following any of that?' This request does not require the disclosure of confidential information that might be self-incriminating or self-stigmatizing, but it does legitimate conversation pertaining to being gay. It also lets the listener know that the interviewer can be entrusted with information on the topic. Thus it becomes feasible to lead into a request for more personal information.

Personal disclosure of being gay is in the public domain only for those who have come out of the closet and are known in public as gay, but not for those who have not. Especially in teenage, admitting gay or bisexual propensities, even to oneself, may be experienced as mortifying. It is in this context that the Catch-22 query may be of value. A Catch-22 situation is one in which you're damned if you do, and also damned if you don't. The Empire State Building Test is a gay Catch-22. It is probably the meanest question on earth, I tell a teenaged respondent, but I ask it anyway: You are on top of the building and a crazed sex terrorist with a gun forces you to the edge of the parapet and says, "Okay, suck my dick or you go over." Which would you do? (The sex of the terrorist may be male or female, homosexual or heterosexual, relative to the sex of the respondent.) Whatever the response, it permits a distinction to be drawn between a homosexual act and a homosexual orientation or status.

Another technique for legitimating a potentially stigmatizing and self-incriminating topic that is taboo-ridden and not in the public domain is the Parable Technique, as I named it after first using it as recorded in Case #1 of Chapter 7. The parable is a story based on either a single or a composite case, and is selected because of the likelihood that it may have personal application to the respondent. For example, in counseling a teenaged or young adult woman with a history of treatment for the syndrome of congenital virilizing adrenal hyperplasia, I may quote from a case that illustrates my own research findings regarding the relative prevalence of bisexual fantasy and/or lesbian experience in the syndrome. The quote would draw attention to the fact that, as a teenager, the girl in that case was afflicted with extreme reticence and could not say anything about her own romantic, sexual, and erotic development until she had sufficient evidence that I could be entrusted with what she had to reveal. For the listener, the parable of this case serves as a narrative projective test, designed to elicit her own response, without imposing a forced-choice answer. At one extreme, she may respond with elective mutism, or with flippant evasiveness. At the other extreme, she may unlock and tell of the solitary torment she has been going through because of an unrequited and secret love affair with a female friend. Without the parable, she would not have been able to unlock, for she would have lacked evidence of my compassion and my competence to be nonjudgmental in hearing her self-disclosure. She needed evidence that my ears would not explode when they heard her story!

The Parable Technique is a special example of enhancing the flow of information by application of the maxim, 'Don't teach, preach, or judge, just tell.' The speaker always tells whatever is being told in the first person singular. Thus he or she tells something about his or her own experiences, encounters, activities, reactions, memories, thought, imagery, dreams, proposals, expectancies, or such like. The listener is not being interrogated, and has the option of not responding as well as of responding and telling something about himself.

Without a preliminary sense that they will not be judged, people adopt the all-pervasive assumption of the moral, religious, and legal institutions of our society that they are voluntarily responsible for the behavior by which they are gauged as good or bad, guilty or not guilty, criminal or not criminal. Acting on this assumption in a sexological interview, they respond as if from the position of publicly exonerating themselves by enlisting the approval of the interviewer as a potential advocate of their moral rectitude. Their success is dependent on the degree of their semantic capability of being evasive by talking in platitudes. For example, a parent who self-righteously accuses a child of being rebellious and disobedient will impart an unqualified expectation that the listener will agree without demanding detailed substantiation of where, why, or how.

The way not to be sucked into platitudes of any variety, accusatory or otherwise, is to use the sportscaster technique. This technique requires, in the place of platitudinous and diffusely global generalities, a play-by-play, detailed account of the specifics of what actually happened in a particular experience or encounter. Otherwise the raw data of the sportscaster's account become semantically slanted in favor of inferential attributes of blame, motivation, intent, and responsibility which serve to support, as a hidden agenda, the expediency of conscripting the listener as a friendly ally. Hearing the adversary's version, also on the basis of the sportscaster's play-by-play technique, increases the degree to which the history of what actually happened can be accurately reconstructed.

The scope and number of items on the agenda of a schedule of inquiry is governed by the range and extent of the concepts and hypotheses under scrutiny. When a research design is cross-sectional, the agenda may be more limited than when it is longitudinal. In longitudinal studies, however, it is unwise to lose data by not collecting them. Limitations are set later, and they apply to what will be excerpted from the written record. Initially, the net is cast widely in what is tagged pejoratively by its detractors as a fishing expedition. Sometimes it is only with the knowledge of hindsight that data preserved in the record will prove essential to the formulation of hypotheses and explanations. For example, when I first began research in the sexology of birth defects of the sex organs, I did not have a hypothesis concerning the function of sexual rehearsal play — nor even a name for it. However, there was a place under sex education in the schedule of inquiry for 'playing doctor' and for other sexual games, so that relevant data did get preserved.

The surest way to guarantee that information will be retrievable as given is to record it, verbatim. Videotaping preserves the image of the speaker along with the spoken word, but there is no system for transcribing the visual content in the same way as the auditory. With a schedule of inquiry, an interview can be recorded so that, with each change of topic, the words that indicate the new topic are simultaneously spoken and written verbatim on a sheet of paper, in capitals if spoken by the interviewer, and in lower case if by the respondent. These words then constitute subheadings. They are seen by the typist, and listened for. When they are heard, they are set in the transcript as subheadings. They facilitate the subsequent location and retrieval of information in the cumulative record, as well as the formation of a table of contents and an index.

There are some interviews, especially those with great emotional momentum, that need to be recorded from beginning to end. There are others that become too long-winded, digressive, and redundant. To record them in full would be uneconomic in both transcription cost and the time required of those who subsequently read them. In such instances, the interviewer keeps a list of topics covered, and concludes by enlisting the respondent to share the task of making a summary interview. The interviewer abridges for the summary those parts of the interview that were too wordy, and the respondent recapitulates those parts in which his/her own words are essential. The instances are rare in which a respondent is not able to say for the second time what he said the first. The proceedings of an hour can thus be reduced to 15 or 20 minutes. With the microphone live, the summary interview may be extended to introduce additional topics for which the spontaneity of the first response is required.

Those who use a tape recorder as an electronic extension of their own memory use the same style of talking whether the microphone is on or off. The respondent follows suit, and only rarely elects to have the microphone off. Listening to, or reading the transcripts of one's own interviews serves also as a valuable self-corrective.

The biographies in this book were excerpted from the patients' consolidated psychohormonal files. Each file begins with such information as can be retrieved about the patient's birth, and the parents' own report of the delivery and what happened to them in the next few days, weeks, and months.

The records from the prenatal and neonatal period are the beginning of what will become, in the long term, the consolidated, developmental history. The components of this consolidated history comprise the intramural clinical and laboratory history from each division of the hospital in which the individual has been seen: the surgical history; the history from extramural clinics and hospitals; the history from social service, penal, academic, and other institutions in which the person has been known; and the psychohormonal unit's history.

The psychohormonal history, like the rest of the clinical history, is cumulative over time and changes of personnel. The potentially adverse effects of personnel changes, which are inevitable in a longitudinal study, are minimized by having all interviews subdivided or shared among two or more interviewers. In this way, the patient and family establish a sense of belonging in the psychohormonal unit, and not just to a single person. They also have a chance to reveal themselves differently to different interviewers — male and female, black and white, old and young. Cohesiveness and unity among interviewers is achieved by means of a joint session of summing up, in which interviewers, patient, and family all participate. This joint session circumvents problems of secrecy and evasiveness, although respecting the requirements of each person's confidentiality.

The joint interview allows a child, as well as an older patient, to be an active collaborator in his/her case management, and not a victim of what may be experienced as nosocomial stress and abuse. It allows diagnostic and prognostic information to be given incrementally and in such a way that it does not traumatize, but has a therapeutic and preventive effect. It prevents the possibility of traumatic disclosure later in life when, under the freedom of information act, any patient has legal access to his/her own medical files. Even earlier in life, a young person can be traumatized by inadvertently overhearing personal clinical information, or by seeing his/her own clinical files. The young are also likely to read or to see on television documentaries of their own syndrome, or to study it in biology class.

There is a more traumatic and a less traumatic way of presenting even bad news. For example, it is traumatizing to tell a girl that she is a male with male chromosomes, but not traumatizing to explain that the nature creates some XY girls and some XX boys among many other varieties of karyotype, and that she is an XY girl.

When the gonads are incongruous, as they are in the androgen-insensitivity syndrome and other cases of birth defect of the sex organs, the nontraumatizing way of presenting the information, first of all to the parents, and later to the patient, is with the help of diagrams of prenatal development (Money, 1968, 1975b) which demonstrate that testicles and ovaries grow from the same gonadal cells and are both called gonads. Under the microscope, a girl's gonads may look testicular in structure, and a boy's ovarian, or either may have a combined appearance. For self-reference, regardless of medical terminology, they are either gonads or sex-glands.

Professionals may avoid the entrapment of pejorative, stigmatizing, and traumatizing idiom, which is derived from the search for pathology, by writing all reports in the expectation that a copy will eventually be read by the patient. This criterion is automatically satisfied when the summary of an interview is made in the patient's presence with him/her collaborating.

The consolidated history, not the patient, is the data base for outcome research. Each specific outcome study begins with identifying those records that qualify for inclusion, according to the criteria established with respect to diagnosis, treatment history, present age, geographical availability, and such like. The pages of each record are sequenced chronologically and according to the classification of a standard Table of Contents (for example, Endocrine History, Surgical History, Hospitalization History, School History, Social Agency History, Psychological Testing, Psychohormonal Unit Interviews, and so on). The pages are then numbered so that the history can be indexed in the way that a book is indexed.

The index is a conceptual one, not a terminological one. The conceptual categories are decided upon in relationship to the purposes and hypotheses of each particular investigation. They are discussed, critiqued, and revised in detail, ahead of time. They derive from the basic theoretical issue or issues involved. For example, the conceptual categories in one investigation may be: gender transpostition, neonatal dissension regarding assigned sex, marital-sexual discord, and family psychopathology; whereas in another investigation they may be romantic dating, age of onset of puberty, academic progress, and history of sexual learning.

After the history is indexed, the next step is to tabulate the data revelant to each conceptual category, and then to reduce the content of the tabulation to a number. The simplest reduction is binary, that is a rating of present or absent with respect to a given characteristic. The next degree of complexity is a three-point rating scale, such as, strong/medium/weak, or severe/moderate/mild. A rating scale of more than three points is not excluded, but it is typically avoided because the degree of precision it implies turns out to be spurious. If there is a problem of agreement between two independent raters, then another rater, or more than one, is coopted to form a jury. If the jury remains divided, then the rating is put into a special category, namely undecided.

The aforesaid method of reducing data has an advantage over subjects' (patients') own ratings, for they each have their own subjective criterion standards. When the same two or three trained personnel do the ratings, they bring the same criterion standards to each case. This is the objective type of scoring that is routine in, for example, the Wechsler Intelligence Scales.

The objective scoring method is eminently suited to longitudinal studies based on clinical histories. The developmental history of each individual is unique, despite the shared commonality of the same diagnosis, prognosis, and treatment. The clinical history of each individual differs, especially with respect to chronology, frequency, and duration of contacts. It cannot be otherwise when the subjects are human beings who cannot be experimentally controlled and regimented like experimental animals.

Each clinical history of each individual differs also with respect to the informant and the source of information from which a given rating is derived. For example, regarding parenting behavior manifested in juvenile play, the source may be direct observation of the child, the written report of a case worker, the taped and transcribed interview with a parent, or the retrospecive recall (taped and transcribed) of the patient in later years. Either singly or in combination, these various sources can be accommodated to when the aforesaid method of rating is employed.

The final step is the consolidation of tabulated scores from each case into a master table which shows the number of subjects (or responses) in each category. These data can then be subjected to statistical evaluation. They lend themselves typically to simple statistical procedures, chiefly chi-square, and *t*-test, in accordance with the standard methodology of science, though multivariate analysis is sometimes applicable.

The scientific merit of this book lies not in the method of tabulating data from multiple cases, in order to compare groups of individuals matched to be homogenous for diagnosis. Instead, this book's merit lies in tabulating data from cases matched in pairs for concordance on one set of criteria and discordance on another, so that their contrasting clinical, social, and lovemap biographies can be narrated and compared. Both methods have their proper place in science. This book is the place to which one turns in order to comprehend the total biographical complexity of what it means to be a person whose entire existence is, for the advancement of sexological science, the equivalent of what the first moonwalk was for the advancement of zero-gravity biology.

TELEOLOGY, REDUCTIONISM, AND EXIGENCY THEORY

In terms of the philosophy of science, there are two theoretical premises that suffuse sexological writing and practice: one is teleological determinism, and the other is reductionistic determinism. Teleological determinism is formulated as motivational theory. Motivation is attributed not only to motives, conscious or unconscious, but also to urges, instincts, drives, needs, wants, wishes, dreams, desires, preferences, and choices. There is also the teleological theory of evolutionary determinism, the sexological application of which is in comparative sexology, and also in the sociobiological theory of sex.

Reductionistic determinism in sexology is formulated as stimulus-response, behavior-modification theory. Modification of behavior is attributable to modification of either the stimulus, or of the response, or both. The response is modified positively by reinforcement, or negatively by aversion. Reductionism exists also as genetic and physiological determinism, the sexological application of which is in hormonal and neurobiological sexology.

'Don't bother me with the evidence; my mind has already decided.' Another version of this epigram is: 'Don't distract me with your findings; my theory already explains them.' The jeopardy inherent in all theories of determinism is that they steer one's attention selectively, not only toward recognizing some evidence at the expense of what is neglected, but also to giving it a name and a meaning.

In order to avoid this hazard, I advocate not being committed to explanatory principles in sexology that are derived exclusively from either teleological or reductionistic theories of determinism. Eclecticism is not a viable alternative, for an array of theories without a superordinate theory of when to use them is chaotic. My escape from this chaos has been to formulate a theory, exigency theory (Money, 1986a,b, Ch. 8) which is not itself deterministic, but is capable of accomodating both teleological and reductionistic determinism.

Whereas exigency theory does not itself provide causal explanations, it does provide a classificatory template, the five universal exigencies of existence, within which causal explanations may be catalogued as in need of being discovered, or as having been discovered. There is, of course, no guarantee that causal explanations will be discovered, nor of the order of their discovery, but the five universal exigencies do offer a guarantee against inadvertent omission of what to look for. The five are named: pairbondage, troopbondage, abidance, ycleptance, and foredoomance (Money, 1984, 1986b, Ch. 9).

Pairbondage (or pairbondship) means being bonded together in pairs, as in the parent-child pairbond, or the pairbond of those who are lovers or breeding partners. In everyday usage, bondage implies servitude or enforced submission. Though pairbondage is defined so as not to exclude this restrictive connotation, it has a larger meaning that encompasses also mutual dependency and cooperation, and affectional attachment. Pairbondage has a twofold phyletic origin in mammals. One is mutual attachment between a nursing mother and her feeding baby, without which the young fail to survive. The other is mutual attraction between males and females, and their accommodation to one another in mating, without which a diecious species fails to reproduce itself.

Male-female pairbonding is species specific and individually variable with respect to its duration and the proximity of the pair. In human beings, the two extremes are represented by anonymous donor fertilization versus lifelong allegiance and copulatory fidelity.

Troopbondage (or troopbondship) means bondedness together among individuals so that they become members of a family or troop that continues its long-term existence despite the loss or departure of any one member. Human troopbondage has its primate phyletic origin in the fact that members of the troop breed not in unison but asynchronously, with transgenerational overlap, and with age-related interdependency. In newborn mammals, the troopbonding of a baby begins with its pairbonding with its mother as the phyletically ordained minimum unit for its survival and health. After weaning, it is also phyletically ordained for herding and troopbonding species that isolation and deprivation of the company of other members of the species or their surrogate replacements is incompatible with health and survival. Nonhuman primate species are, in the majority of instances, troopbonders like ourselves.

Abidance means continuing to remain, be sustained, or survive in the same condition or circumstances of living or dwelling. It is a noun formed from the verb, to abide (from the Anglo-Saxon root, *bidan*, to bide). There are three forms of the past participle, abode, abided, and abidden.

In its present usage, abidance means to be sustained in one's ecological niche or dwelling place in inanimate nature in cooperation or competition with others of one's own species, amongst other species of fauna and flora. Abidance has its phyletic origin in the fact that human primates are mammalian omnivores ecologically dependent on air, water, earth, and fire, and on the products of these four, particularly in the forms of nourishment, shelter, and clothing, for survival. Human troops or individuals with an impoverished ecological niche that fails to provide sufficient food, water, shelter, and clothing do not survive.

Yclept is an Elizabethan word, one form of the past participle of *to clepe*, meaning to name, to call, or to style. Ycleped and cleped are two alternative past participles. Ycleptance means the condition or experience of being classified, branded, labeled, or typecast. It has its phyletic basis in likeness and unlikeness between individual and group attributes. Human beings have named and typecast one another since before recorded time. The terms range from the haphazard informality of nicknames that recognize personal idiosyncracies, to the highly organized formality of scientific classifications or medical diagnosis that prognosticate our futures. The categories of ycleptance are many and diverse: sex, age, family, clan, language, race, region, religion, politics, wealth, occupation, health, physique, looks, temperament, and so on. We all live typecast under the imprimatur of our fellow human beings. We are either stigmatized or idolized by the brand names or labels under which we are yclept. They shape our destinies.

Doom, in Anglo-Saxon and Middle English, meant what is laid down, a judgment or decree. In today's usage it also means destiny or fate, especially if the predicted outcome is adverse, as in being doomed to suffer harm, sickness or death. A foredoom is a doom ordained beforehand. Foredoomance is the collective noun that, as here defined, denotes the condition of being preordained to die, and to being vulnerable to injury, defect, and disease. Foredoomance has its phyletic origins in the principle of infirmity and the mortality of all life forms. Some individuals are at greater risk than others because of imperfections or errors in their genetic code. Some are at greater risk by reason of exposure to more dangerous places or things. All, however, are exposed to the risk, phyletically ordained, that all life forms, from viruses and bacteria to insects and vertebrates, are subject to being displaced by, and preyed upon by other life forms. Foredoomance applies to each one of us at first hand, in a primary way, and also in a derivative way insofar as it applies also to those we know. Their suffering grieves us; their dying is our bereavement.

These five universal exigencies apply not only to our existence as a species, but also to the personal existence of each one of us. Thus, they constitute a template against which to analyze the progress of any biography, and to trace both the continuity and the contiguity of its component elements, clinical and nonclinical. They provide a place for all of the variables that may affect the development of an individual human life. In clinical evaluation and history-taking, they are a safeguard against inadvertent omissions in the agenda of the schedule of inquiry. Their influence is present in each of the biographies in this book.

Within the five universals, attributions of continuity and contiguity to variables of the biography, past and present, may be ascertained, for the system is a dynamic, not a static one. Dynamic interconnectivity does not, however, presume a cause-and-effect relationship. In sexology, as more generally in psychology, genuinely authenticated cause and effect is an extremely rare commodity. Most causal explanations are semantic artifacts produced by loosely applying terms like because to two events that are sequentially related, temporally, but not causally.

Within the confines of motivation theory, causal statements may sound good, but they have no more explanatory value than a tautological verbalism. Thus: 'He is a masochist because he rents masochistic videotapes' has the same explanatory value as, 'He rents masochistic videotapes because he is a masochist.'

Attributing causality to motivation jeopardizes professional nonjudgmentalism and impartiality by insidiously attributing responsibility for being motivated or not motivated to the person being adressed. It is inherent in the idiom of the language that terms like preference, choice, urge, want, wish, and desire put the onus of responsibility on the one who does the preferring, choosing, et cetera. The not so hidden implication is that motivation to change is a voluntary responsibility. If change does not take place – for example, if the masochist fails to quit renting videotapes – then the insidious implication of blame aligns his would-be therapist into an agonistic relationship with the masochist. The proper alignment should be that of the professional and the patient together as allies against the symptom or the syndrome.

Attacking the patient instead of the disease is a legacy of the theory of demon possession in which the patient was held responsible for the sin of allowing the devil to take possession of his soul. He should, therefore, deserve punishment, not relief.

Motivational explanations are the property of the law which cannot exist without the postulates of personal responsibility, blame and guilt. The clinic is not a courtroom. The doctor's job is not to pass judgment, but to be impartial and provide treatment. Judgmentalism is inherent in motivational theory, no matter how insidiously it may be under cover. Nonjudgmentalism is inherent in exigency theory, in open alliance with humane impartiality, compassion for human frailty, and the restoration of well-being.

BIBLIOGRAPHY

Acton, W. (1875) *The Functions and Disorders of the Reproductive Organs in Childhood, Youth, Adult Age, and Advanced Life, Considered in Their Physiological, Social and Moral Relations* (6th edn.). Philadelphia, Presley Blakiston.

Culliton, B.J. (1974) Patients' rights: Harvard is site of battle over X and Y chromosomes. *Science*, 186:715–717.

Diamond, M. (1982) Sexual identity, monozygotic twins reared in discordant sex roles and a BBC follow-up. *Arch. Sex. Behav.*, 11:181–186.

Imperato-McGinley, J. and Peterson, R.E. (1976) Male pseudohermaphroditism: The complexities of male phenotypic development. *Am. J. Med.*, 61:251–272.

Imperato-McGinley, J., Guerrero, L., Gautier, T. and Peterson, R.E. (1974) Steroid 5α-reductase deficiency in man: An inherited form of male pseudohermaphroditism. *Science*, 186:1213–1215.

Kuhn, T.S. (1970) *The Structure of Scientific Revolutions*. Chicago, University of Chicago Press.

Meyer, J.K. and Reter, D.J. (1979) Sex reassignment: Follow-up. *Arch. Gen. Psychiatry*, 36:1010–1015.

Money, J. (1952) *Hermaphroditism: An Inquiry into the Nature of a Human Paradox*. Doctoral Dissertation, Harvard University Library. University Microfilms Library Services, Xerox Corporation, Ann Arbor, MI 48106, 1967.

Money, J. (1968) *Sex Errors of the Body: Dilemmas, Education and Counseling*. Baltimore, Johns Hopkins University Press.

Money, J. (1975a) Ablatio penis: Normal male infant sex-reassigned as a girl. *Arch. Sex. Behav.*, 4:65–71.

Money, J. (1975b) Psychologic counseling: Hermaphroditism. In: *Endocrine and Genetic Diseases of Childhood and Adolescence*, 2nd edn. (L.I. Gardner, Ed.). Philadelphia, W.B. Saunders.

Money, J. (1984) Five universal exigencies, indicatrons and sexological theory. *J. Sex Marital Ther.*, 10:229–238.

Money, J. (1986a) Longitudinal studies in clinical psychoendocrinology and sexology: Methodology. *J. Dev. Behav. Pediatr.*, 7:31–34.

Money, J. (1986b) *Venuses Penuses: Sexology, Sexosophy, and Exigency Theory*. Buffalo, Prometheus Books.

Money, J., Hampson, J.G. and Hampson, J.L. (1955) An examination of some basic sexual concepts: The evidence of human hermaproditism. *Bull. Johns Hopkins Hosp.*, 97:301–319.

Murphy, E.A. (1980) Quantitative genetics: A critique. *Soc. Biol.*, 26:126–141.

Neugebauer, F.L. von (1908) *Hermaphroditismus beim Menschen*. Leipzig, Klinkhardt.

Peterson, R.E., Imperato-McGinley, J., Gautier, T. and Sturla, E (1977) Male pseudohermaphroditism due to steroid 5α-reductase deficiency. *Am. J. Med.*, 62:170–191.

Rubin, R.T., Reinisch, J.M. and Haskett, R.F. (1981) Postnatal gonadal steroid effects on human behavior. *Science*, 211:1318–1324.

Saenger, P., Goldman, A.S., Levine, L.S., Korth-Schutz, S., Muecke, E.C., Katsumata, M., Doberne, Y. and New, M.I. (1978) Prepubertal diagnosis of steroid 5α-reductase deficiency. *J. Clin. Endocrinol. Metab.*, 46:627–634.

CHAPTER 2

Matched pair of cases diagnostically concordant for 46,XY androgen-insensitivity syndrome (AIS) and discordant for sex of rearing

SYNOPSIS

The two cases presented in this chapter were diagnostically concordant for the complete form of the androgen-insensitivity syndrome (AIS). An X-chromosome bearing egg from the mother had joined with the Y-chromosome bearing sperm from the father and formed an XY zygote which, according to the actuarial statistics of genetics, should have differentiated into a male baby. But it did not. The X-chromosome bearing egg carried also an X-linked recessive defect that ordained that each cell of the developing embryo and fetus would be unable to process the molecules of male sex hormone that, after the earliest phase of sexual differentiation, would be prerequisite to the complete differentiation of a male. The earliest stage of differentiation began as that of a male. The bilateral clusters of undifferentiated gonadal cells formed into testes. Each testis secreted its antimullerian hormone which signaled the bilateral mullerian ducts, the precursors of the uterus and fallopian tubes, to quit growing and become vestigial. Then each testis secreted its quota of male sex hormone which should have signaled the bilateral wolffian ducts, the precursors of the internal male reproductive organs, to enlarge and differentiate. The signal failed, however, insofar as the wolffian ducts were powerless to respond to male sex hormone. The precursors of the external male reproductive organs also failed to enlarge and differentiate as male. Instead they fell back on the female template of differentiation, which did not require sex hormones. The baby was born with the genital appearance of a girl.

Close inspection, which is not done routinely in the delivery room, would have revealed the vagina to be a blind pouch or dimple, lacking a connection with a cervix, uterus, and fallopian tubes. Palpation of the labia majora and the groins might have led to the discovery of lumps, the testicles, in the inguinal canals, but such was not the case. The condition would have been recognized by chromosome testing, but the newborn are not routinely tested chromosomally. Although the chromosomal karyotype would have been 46,XY, it would have confirmed that the baby should be assigned and habilitated not as a boy, but as a 46,XY girl.

Looking ahead to puberty, it could have been predicted that the body in both cases would develop as that of a female. That is exactly what happened. Being unable to respond to male hormone secreted by the testicles, all cells of the body, at puberty and thereafter, responded exclusively to the female hormone normally secreted in the male by the testicles, and normally overpowered by the greater quantity of male hormone. Because there was no uterus, there was no menstruation and, of course, no possibility of intrauterine pregnancy by in vitro fertilization. Technologically however, an extrauterine pregnancy by in vitro fertilization is a future possibility. The testicles, which must be able to use male hormone in order to produce sperms, were sterile.

At birth, each baby of this pair was initially declared to be a girl. In Case #1, that decision was never questioned. In Case #2, it was, and the child was, at age 18 months, erroneously redeclared as a boy. Thereafter, the two children were discordant for the sex in which they were reared and in which they lived.

In the files of the Johns Hopkins Psychohormonal Research Unit, there is only one case, the present one, of a child with the complete form of the androgen-insensitivity syndrome reared as a boy. The matching case is drawn from a sample of 36 similar cases reared as girls. Both cases meet the criteria of having an extensive amount of follow-up data on file. Both demonstrate the difficulty of coping with a genital and procreative anomaly. The girl's case demonstrates that there was at puberty no gulf between the sex declared by the body and by the mind. Both were female. The boy's case demonstrates the opposite. There was at puberty an unbridgeable gap between the hormonal femininity declared by the body and the masculinity declared by the mind. In young adulthood, the boy's life ended in suicide by hanging. The girl's life continued into mid-life, as a wife and mother. She surmounted the hurdles of wifehood and adoptive motherhood, even though they were challenging and difficult.

CASE #1: 46,XY AIS, REARED FEMALE

Diagnostic and clinical biography

When she was 12, nearly four years in advance of being formally diagnosed, the girl whose case is here presented first realized that other girls did not have swellings in their groins like she did. She wondered 'if there was something wrong' with her. She assumed she was like her sister, three years older, who had been told at age 13, by the family doctor, that she had hernias.

The older sister's pubertal development was not remarkable, except that by age

17 she had not menstruated. This was the age when, in the surgical outpatient clinic, she had had an evaluation for bilateral hernias. In accordance with the terminology of the day, male hermaphroditism of the testicular feminizing type was suspected, and the patient was referred to the specialist in that syndrome, namely, the pediatric endocrinologist (see Wilkins, 1965). The diagnosis was confirmed, but the terminology of male hermaphroditism was avoided in the presence of the patient and family, so as not to be unnecessarily traumatizing. Androgen-insensitivity syndrome is today the preferred name of this particular birth defect.

The older sister did not hide her knowledge of her condition and its prognosis from her younger sister who assumed that her own diagnosis and prognosis would be the same. Thus she knew she would 'not be able to have children, and I can't have my monthly like a normal girl would, and I think there isn't anything that can be done about it. ... The glands are misplaced, and they're infantile, and an infantile womb'. Official confirmation came a year later when, at age 15:8, she agreed to being examined and to having a skin biopsy procedure to provide tissue for a sex chromatin determination (the future technique for counting the chromosomes was four years away). On the same occasion she was accompanied by her younger sister, age 6:9, who proved to have the syndrome also. A still younger sister, who had died of pneumonia at age 18 months, was presumed not to have the syndrome, insofar as the mother had not noticed swellings in the groins, as she had in the other three. There were two other unaffected siblings, a brother who was the oldest of the five living children, and the next oldest, a sister, who had menstruated regularly from the age of 16 onward.

At the time of the patient's first physical examination at age 15:8, the height, 5 ft. 4 in. (161.8 cm) was slightly above average for age. The weight, 91½ lb (41 kg), was about 21 lb (9.5 kg) below average for both height and age. The body was described as 'set on a light skeletal frame with somewhat slender hips, and the musculature fairly well-developed with rounded, feminine contours' (Fig. 2.1). The diameter of the breasts was 14 cm (5½ in.) and of the areolae, 3.5 cm (1½ in.). They were described as 'well-developed, adolescent in type, and with considerable glandular tissue'. The bra size was 34B. The voice was alto and normal for a female. There was no facial or body hair. Instead of the total absence of axillary and pubic hair

Fig. 2.1. Case #1. Appearance of the body habitus at age 20:3. Pubertal feminization had required no treatment.

Fig. 2.2. Case #1. Female appearance of the vulva at age 20:3, immediately prior to internal surgical lengthening of the vagina.

Fig. 2.3. Case #2. Appearance of the external genitalia at age 11:11, after completion of surgical masculinization.

Fig. 2.4. Case #2. Appearance of the body habitus at age 11:11, prior to surgical breast removal.

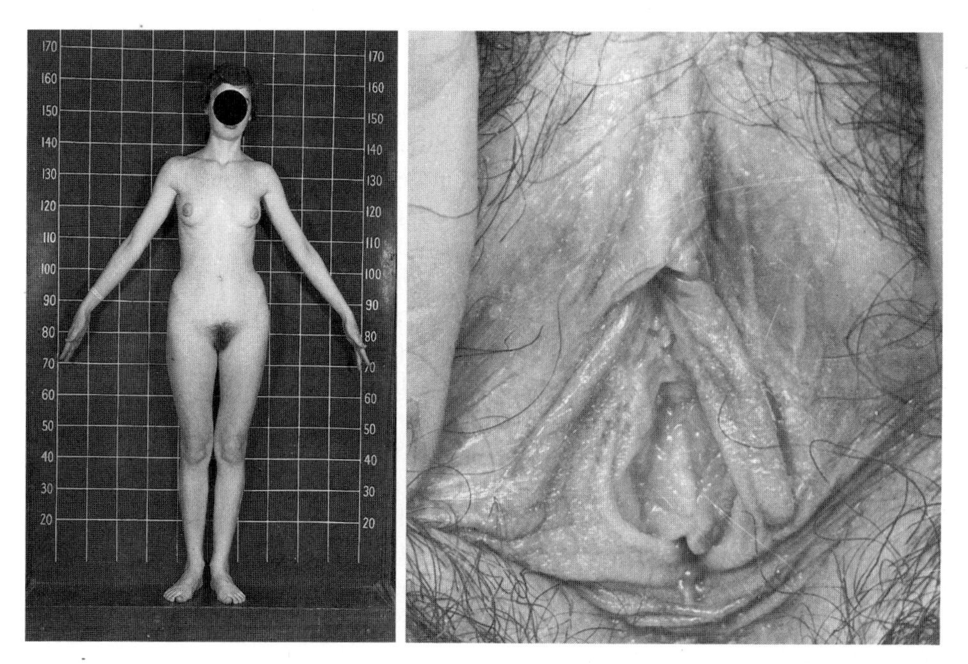

Fig. 2.1

Fig. 2.2

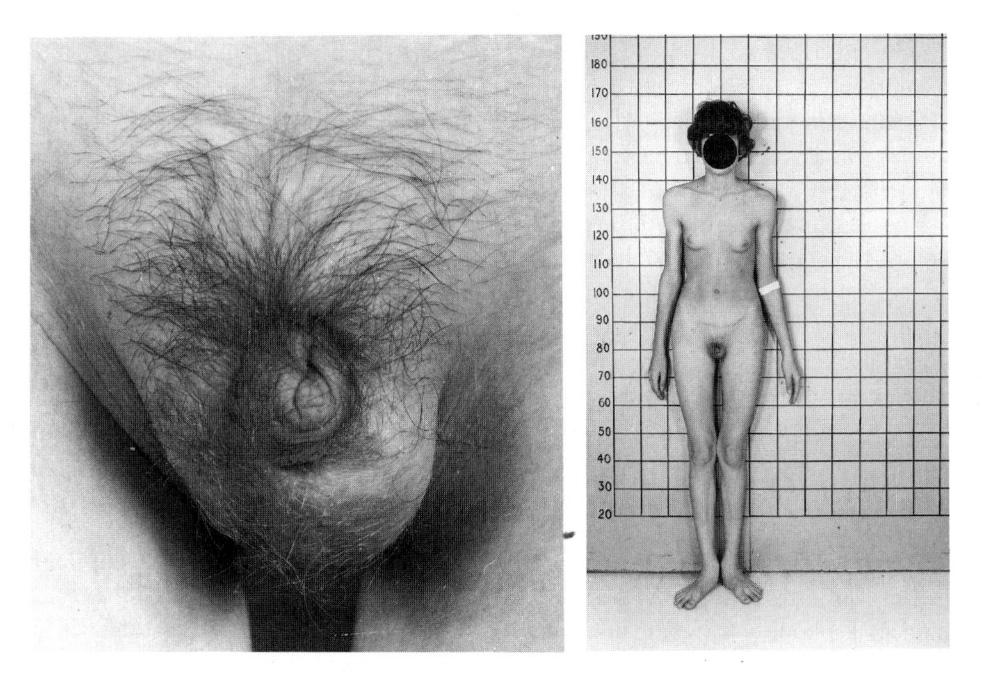

Fig. 2.3

Fig. 2.4

characteristic of some cases of androgen insensitivity, there was a moderate growth in both areas.

The configuration of the external genitalia was normally feminine (Fig. 2.2). The clitoris was small and not easily seen. The vaginal smear showed estrinization. The vagina was measured $2^1/_2$ in. (6 cm) in depth and ended blindly. There was no uterus palpable by rectal examination. The gonads were palpable in the inguinal canals and could be moved into the labia majora. They measured approximately 3.5×2.0 cm.

By age 19 the meaurements of height and weight were 5 ft. 5 in. (165 cm) and 101 lb (46 kg), respectively. Urinary ketosteroids and estrogens were measured in connection with a study of their response to gonadotropin administration. The hormonal levels were found to be within the range expected for the syndrome.

A subsequent hormonal investigation was done when the patient was aged 31 (Brown et al., 1982; see also Amrhein et al., 1976). It was part of a larger study in which AIS patients volunteered to give a blood sample for a study designed to test, by means of the radioimmunoassay technique, the androgen-binding capability of androgen-insensitive cells. Skin cells from this patient and her two AIS sisters, were found to be capable of binding the androgen molecule within the cell nucleus, whereas they were deficient in utilizing it. In other AIS cases and pedigrees, it is intracellular binding itself that is deficient.

The patient's age was 20:3 when she consented to, and underwent, surgery for gonadectomy and vaginoplasty. The clinic's policy at that time was to recommend surgical vaginoplasty, insofar as the efficacy of vaginal dilation had not then been rediscovered (Frank, 1938; Fordney, 1978). The policy, then as today, was to recommend gonadectomy also on the criterion of an increased eventual risk of testicular cancer in undescended testes, especially those that are anatomically and/or functionally imperfect, as is the case in the androgen-insensitivity syndrome (Morris and Mahesh, 1963; Manuel et al., 1976).

The elective timing of both surgical procedures had, in accordance with the clinic's policy, been made known to the patient since her initial evaluation. Because the family was not secretive, she knew when she was aged 18 that her older sister's surgical admission had been for the procedures that eventually she herself would presumably undergo. When, at age 20, her own surgery was done, vaginoplasty and gonadectomy were performed sequentially in one operation. The vaginoplasty was done according to the method of McIndoe, whereby the vault of the vaginal pouch was incised, extended, and lined with a split-thickness skin graft taken from an area of the right hip that would be covered by a bikini bathing suit. The postsurgical recovery was uneventful.

Insofar as feminization in the androgen-insensitivity syndrome is contingent upon the level of estrogen normally secreted by the testes, surgical removal of the testes induces a premature menopause, replete with acute menopausal symptoms. It is, therefore, necessary to give female hormone replacement therapy. For this reason, the patient was put on a cyclical maintenance dose of Premarin, 1.25 mg/day, for three weeks of each month. There has been no problem with compliance over the ensuing years.

Apart from the AIS, the patient's clinical history had relatively infrequent entries. There was a chronic problem of partial hearing loss in both ears which the patient attributed, in retrospect, to ear infections at age 4, treated with daily irrigation which the mother erroneously continued for nine years, until age 13. In the fourth decade of life, there had been, in the course of several months, recurrent outbreaks of hives, which spontaneously remitted when intrafamilial and marital difficulties eased up. Under the influence of sexological counseling, concurrent connubial problems of mutual sexuoerotic nonresponsiveness also went into remission.

Gender-coded social biography

Throughout the developmental years of her infancy and childhood, no one had any clue whatsoever that this girl would eventually prove to be different from daughters, sisters, or schoolgirls in general. She herself did not suspect anything different about herself until, at age 12, she realized that something might be wrong when she saw swellings in her groins that girls, other than her older sister, did not have. She knew that her sister had been diagnosed only as having hernias, and assuming the same would apply to herself, went on with living the life of a girl. She overcame an initial apprehension of being seen naked under the shower, after gym class at school, when she realized other girls were not paying sufficient attention to single her out as being different. Thus she remained secure in her status as a girl as she went through the premenarcheal changes of pubertal feminization. Subsequently, she remained equally secure in her status as a female when, as a sequel to her older sister's diagnosis, she had to face up to having the same diagnosis herself, and with it the prognosis of never menstruating and never carrying a pregnancy. Thus, despite the anomaly of the genetic and gonadal sex in this case, the social biography throughout childhood and teenage, as well as subsequently, has been that of a girl and woman.

With no one suspecting what her future diagnosis would be, the patient grew up without a clinical referral and, therefore, without a written psychohormonal record of her development in childhood prior to age 16:10, when the first psychohormonal referral was made. At that time, in a retrospective 'self-picture as a child', the patient gave the following items of information. 'Well, I was quite possessive; bashful too, but couldn't say I was really quiet. I had a lot of attention. I liked to do practically anything that had any interest at all in it. I was game for anything new, different.' She enjoyed sports and outdoor activities. Mainly, however, she and her sisters 'played house most of the time'. She would play the mother – 'We all were mothers. We had rocks for doll babies.' Her brother didn't join in, 'because he always liked to build things, and he'd be out cutting wood, to carve something, or build something. He didn't have a role in any of our games. Of course, you know, it was too sissified for him.'

She did not, herself, like to engage in boy's stuff: 'No, not really. I can't think of anything right off hand. Just drawing and painting; and then when I found out I wasn't any good at it, I dropped it and lost all interest in it. Reading. I've always liked reading.'

At school or play, if another child picked a fight with her, 'I held myself back too much', she said. 'I was very backward. I never began a fight. Perhaps that's why I have sometimes such a bad temper, now [at age 27]. I was so unaggressive.' She did not, however, consider herself too clinging and dependent, nor to be lacking in self-assertiveness.

She had always had an interest in clothes, make-up, and jewelry. As a child she liked to dress up, especially in frilly dresses, though she kept in style with slacks, shorts, blouses, and skirts, also. On the Draw-A-Person Test, her first drawing was of a female in a flouncy dress, posing as a fashion model. On good looks, she rated herself as average. She was by other people's standard very attractive in appearance and manner

One of the items on the agenda of the interview at age 16:10 was a thumbnail self-portrait, which elicited the following: 'Well, I would say, sometimes I laugh too much. Even over serious things, I might laugh until I cry. Plenty of people have told me that I have an awfully wild sense of humor. ... I would say that sometimes I'm selfish and mean. And if somebody says something about me, I might take it too hard. As for my outer appearance, I used to think I was a very skinny girl, and I always have had that – would you call it inferiority complex? ... Sometimes I don't put my whole heart into what I have to do, like with homework I'm pretty lazy. ... And if I've got a headache, sometimes I'll pretend it's killing me; and I shouldn't do that.'

The patient could justifiably have mitigated the foregoing self-confessed sins had she perceived them as being a sequel to carrying a full scholastic program in high school each day, and then working as a sales clerk every evening from 5 pm to 10 pm, and all day, Saturdays. Her father, whose own schooling had not been beyond fourth grade in a rural county, required that, after age 16 his children should work and support themselves if they wanted to keep going to school. The first year that she had to work, her school grades dropped, and she was obliged to go to summer school. Otherwise her school record was satisfactory. Her IQ, tested at age 18 on the Wechsler-Bellevue Intelligence Scale for Adolescents and Adults, Form I, was: Verbal IQ, 98; Performance IQ, 109; and Full IQ, 104.

In girlhood, she had envisioned herself being 'a painter, an artist, an airline stewardess, a nurse. I wanted to be many things – a mother most of all.' By the time she was 18, and working as a sales clerk, she was also training in business and secretarial work, but did not like it. Her ambition at that time was to be able to be employed in a nearby aircraft plant – 'engineering, and assembly line; things like that.' By way of explanation, she called herself a tomboy which meant: 'I've always liked to take things apart and put them together again. Like a music box that broke; I got it back together again. ... Some people call a girl like that a tomboy.'

There was nothing of the tomboy, however, in the immediacy of her response, 'Woman, definitely', when as a married woman at age 27, she was queried as to whether it's more fun to be a woman or a man. She had contemplated the male advantage of 'work all day, come home, sit down, and say, "Honey, have my supper on the table for me and fix me something cold to drink," read the paper, and go to bed. But I'd rather be what I am, being able to raise children, be close to them. I like to wash, iron, sew, house-clean, and cook. If I was a man, I wouldn't be able to do that. ... It really hurts me to think of my oldest sister, wanting children so badly, and can't have them' (because of her husband's infertility).

As a high school student, she and her fellow students had always gotten along 'perfectly', she said; 'I don't have any enemies'. She could count five or six girls as 'real good friends'. At age 16 she had 'what you'd call a first love'. Her boyfriend was in the navy, stationed nearby, and was new in her life. After a couple of years, he left her with a broken heart which has never completely healed. She stayed single until age 21. After marriage, her social life revolved around her role as housewife and mother of two adopted children.

In the early years of parenthood, there were no evident domestic problems by which the family might have been distinguished as different from their neighbors. In retrospect, it is possible that the mother's self-disciplined perfectionism was too stringent and demanding for the other members of the household, whereas her husband's insidious withdrawal and disengagement from domestic issues and confrontations was too laissez-faire, and too readily able to be misconstrued as victimization. The extent to which this reciprocal husband-wife dynamic adversely affected the lives of the two adopted children, the older a girl and the younger a boy, is unascertainable. The fact is that, from puberty onward, both became caught up in the counterculture of the 1970s. The daughter's future was in greater jeopardy than that of her brother. Both have survived, precariously, having realized less than the ideal that their parents strove for on their behalf. One of the ramifications of domestic disharmony with the children was that, in mid-life, the parents established separate living quarters for seven years, after which they reunited.

The patient was reared as a Protestant, but married into the Catholic faith. Allegiance to God has always been major in her approach to life, whereas church attendance and ritual have not.

Socioeconomically, the family is comfortably well off in the middle class. The husband has a bachelor's degree and is professionally qualified and employed as an administrator.

Lovemap biography

As in the case of the social biography, so also the juvenile era of the sexuoerotic (lovemap) biography was obtained retrospectively, beginning at age 16:10 with the first psychohormonal interview. Its noteworthy characteristic, throughout, is that it is not specific to the girl's syndrome so much as to the era in which she reached puberty. It was the postwar era of the early 1950s, prior to the so-called sexual revolution of the 1960s. Thus, as a teenager, the patient construed her encounter with sexual morality not in terms of the coital adequacy of her vagina, but in terms of whether or not she should remain coitally a virgin, until marriage.

In girlhood, her lovemap developed in the context of her parents' marriage which, according to what her mother disclosed, had been a true love match that had never faded. It was a match of opposites, insofar as her mother was socially outgoing, whereas her father was so socially reclusive that he was unable to be present even at the weddings of his own daughters. Only in a limited degree could he express his affection to his children, or talk to them intimately. Their mother, by contrast, had no such limitation. She overflowed with maternalism. Though she had no philosophy of sexual emancipation, when her daughters needed her to talk straight to them about personal sexual matters, she was not evasive. Thus this daughter did not have information about her diagnosis and prognosis hidden from her, once it was known.

Talking retrospectively at age 16:10, she identified her own sex education as having begun at age 9, when she overheard her mother telling her two older sisters about menstruation. Thenceforth, she was able to talk with all three of them. She knew that her mother 'would always tell us the truth'. When she was 10 her mother was pregnant with her youngest sister, from which she learned about how a baby is born. She learned about sexual intercourse by asking her mother, after 'me and my girlfriend were coming home from a Halloween party, and we walked in on her mother and father, and we saw what, you know, we shouldn't have seen. ... So then I knew.' At first she had been shocked, and wondered if that had been how she came into the world. She thought, 'I wasn't ever going get married if I had to go through that to have a child. But I didn't understand, then.'

She had, in fact, understood more at this early age than she was able to disclose during the initial interview at age 16:10 with an interviewer with whom she was meeting for the first time. At age 27, with the additional sophistication of another ten years, she summarized, as follows, her story of childhood sexual play: 'All children experiment. We would see and hear things, and of course we would try it to see what it was like – pretend that we were older and married. And we all took part in that.' She did not get into any pretend play with her sisters, but out in the woods she did do so with her brother, while they still lived on the farm. She recalled that his penis 'was big, and I wanted to know what it was like, and different things like that. But as for getting any feeling out of it for myself, no, I didn't.' Subsequently, she had a guilty conscience, 'But since then I have children of my own, and I've read books of psychology of bringing up children, and I've understood that all children have done the same thing.'

There was nothing particularly remarkable about her history of masturbation in early teenage, except that she tried it, knew from her older sisters that it was normal, and then tried to believe that she would grow mature enough to quit it. 'I did it,' she said, 'because I always believed that you shouldn't have sex before marriage. And I felt a need for it. So I experimented with masturbation to relieve an urge, because I felt it would be better to do that than to actually have sex relations.' She would have an accompanying fantasy of being married, and having intercourse with her husband, and would build up to 'a great feeling, and you are really satisfied.' The fantasy could be an actual movie. 'I can even go to a movie,' she said, 'and see a love story or romatic part, and pretend that I'm in the picture, in that woman's place, and I can get nearly, not quite, but nearly the same sensation.' Whereas imagery of sexual intercouse entered into daydreams and waking fantasies, there was no explicit imagery of genital activity in dreams while asleep, and no occurrence of orgasm while dreaming or asleep.

Homosexuality as a variant of erotosexual experience was broached in the first psychohormonal interview at age 16:10. The response was: 'I've met a few girls who were attracted to me ... and this one girl, I laughed it off, and told her to watch her hands, that my chest didn't belong to her.' Two years earlier, when she and her girl friend had been making up plays about 'our dream house, and being married,' she had felt a kindling of erotic feeling, and they had teased each other. 'I'd say, "You're no man. If you are, prove it!"' Subsequently, erotic attraction toward another woman would become nonexistent in either imagery or actuality.

At the age of 16, a few months before her first psychohormonal interview, the girl was confronted for the first time with what was for her the terrible anguish of virginity, namely, whether to save it until marriage, or to lose it in the wild, love-smitten passion of first limerence. For example, she and her boyfriend, Jeff, were at a drive-in movie, 'and I didn't have nothing to lean on, and we were kissing, and so he happened to be leaning on me, and we fell back in the car seat, and I had a hard time then, trying to hold myself. But, all of a sudden, I just became frightened.' There was another occasion, in the living room at home, when 'it would have went all the way. I felt that it was normal and not to be ashamed of, and I didn't care about anything at that moment, except that my brother knocked on the door. He came home early.'

'Before I met Jeff,' she said, 'I thought that every girl should be able to hold herself back, if she wanted to, bad enough; and that, if she had enough faith in God, and wanted to stay clean, then she could. But after I met Jeff, I learned that it's awfully hard, and that there are some people who are weaker than others. Maybe I'm just a little bit stronger. But I don't think I would still be the good girl that I am right now if we hadn't stopped going out together. If we were still going with each other, I probably wouldn't be the same person' (and still a virgin). 'He would get awfully excited, and then it would be my fault if something did happen, and I couldn't make him stop.' She knew at this time that she could not become pregnant, but intercourse, not pregnancy, was what she defined as the moral issue. When he proposed marriage, she demurred, as she had not finished high school. He went off with another girl, an acquaintance of her brother. Subsequently he returned, and they became engaged. Later he left for good and married someone else.

Looking back, years afterwards, she compared her first love with her love for her husband: 'It was a different kind, entirely,' she said. 'I'm satisfied with my life now, and I'm glad I have the man I have. But, of course, at the time I was younger, and I gave all. I gave my whole heart, my whole mind, and everything. But, then, when it was rejected, it hurt so much that I've never been able to give my whole heart to anybody. As far as my husband now, I didn't give all. But I seemed to understand him more; and he understood me more. And we enjoyed everything together, and it was more mature, more satisfying.'

Her husband was able to deal with the issue of infertility partly, it was said, because he had as a teenager seriously contemplated the priesthood and, therefore, had not expected to be a father. Not being able to have pregnancies was for his wife, by contrast, a malevolent blow of fate. When she found out about it, she said: 'That was really bad then. I thought that the world had come to an end. I always wanted children, and I loved them. But, I don't know, all of a sudden, it just came over me that there had to be some people on this earth to adopt children and give them homes, and I was one of them.' At age 20 she said, 'You have to take things as they are, and there's nothing you can do about it.' The poignancy was obvious, and it ran deep.

At age 19:4, in the interval between losing her first boyfriend and meeting her future husband, the patient had a woman-to-woman interview to discuss the surgical issues of gonadectomy and vaginoplasty. Whereas she was ready to consent to gonadectomy, she had doubts about vaginoplasty. The explanation of these doubts necessitated a disclosure which, for the patient, carried the full weight of a confession of guilt which she had not unburdened to anyone else. She had even told two other doctors, both men, that she was still a virgin, whereas, in fact, after Jeff left her, she had had intercourse with another boyfriend. She had 'no special feeling for this boy, and knew that I could never love him, even though he wanted to marry me.' Perhaps her very lack of commitment was essential to her being able to sacrifice her virginity in order to discover the competency of her vagina. She knew he spoke from the vantage point of having had prior experience when, after she explained something about her disability, he said he had found her vagina satisfactory. He had no apparent difficulty with intromission. She too had found their experiences satisfactory, and on several occasions had reached orgasm. Nonetheless, her guilt required her to tell him not to see her again, and finally he gave up trying to dissuade her.

Her guilt also forbade her utilization of what seemed a perfectly feasible alibi, even after it had been formulated by the interviewer, namely that she needed to find out if her vagina was good for intercourse, perhaps even without surgery. Her final decision was that, even though intercourse had been satisfactory without surgical lengthening of the vagina, she would never know if it could be even better. In that era, it was not known that dilation, with either a penis or a dildo, would probably have been a substitute for surgery.

The surgery was done a year later, at age 20:3. Six weeks postsurgically, there was another upsurge, or rather a veritable furor of guilt regarding sexual intercourse – not on this occasion because she had had it, but because her mother had the right to distrust her and not believe her explanation. She had, on an impulse, taken a day off from work, and gone on a long drive with one of the young men with whom she worked and who was on vacation. 'I just wanted to get away from everything, and try to feel young and carefree, just once,' she wrote.

On this occasion, the intensity of guilt-furor regarding sexual intercourse was not appropriate in a woman of 20 whose vagina had recently been lengthened and made fully functional as a coital organ. Guilt about loss of virginity and about being unchaste may have been only the surface fog of a deeper psychodynamic storm, namely, that no matter how good the vagina might be for intercourse and orgasm, it was useless as an organ for making a pregnancy and delivering a baby. It did not communicate with procreative organs, inside.

This hypothesis receives some support in that, in an interview at age 27:7, with a young, new interviewer, the patient had an apparent amnesia regarding her history of intercourse before marriage – the anguish that it had generated and, paradoxically, the greater anguish that it had covered up.

Still further support for the hypothesis may be adduced from the history of the patient's orientation toward oral sex. As a young married woman, she said she liked the foreplay of kissing and fondling, the neck and ears, for example, and less so the breasts. She definitely disliked genital fondling and kissing as either receiver or giver. Later in life, she was able to compromise to some extent on oral sex, but she was adamant that 'to reach orgasm just like that is obscene, dirty, and a sin.' In consequence, when her husband had trouble sustaining intromission long enough for her to reach orgasm, there was no acceptable alternative. She was left unsatisfied. Cumulatively, that wrought havoc not only with their sex lives, but also with the marriage. It contributed to their partial and temporary living apart at one stage of the marriage until, as grandparents, they reunited.

The forces dictating to the wife that her orgasm must be induced in the missionary position, and from penovaginal intromission only, were unyielding and self-sabotaging. Psychodynamically, they could be construed as signifying the imperative that the vagina must be used for something, namely orgasm, so as to compensate for its procreative unfruitfulness. If that imperative was thwarted, then the psychodynamic outcome was feuding with the sex organs. His role was being aggressively passive (impotent, ejaculatory premature, or nonparticipatory) as a counterpart to her being aggressively accusative. Happily, once they were able to recognize their feuding for what it was, the problem of sexual dysfunction went into remission. Their sex life changed from a mid-life famine to a banquet.

CASE #2: 46,XY AIS, REARED MALE

Diagnostic and clinical biography

It was in a referral letter from the pediatrics department of an out-of-state hospital for the armed forces that this boy's case first became known to members of the psychohormonal unit at Johns Hopkins. The letter requested a psychohormonal evaluation 'as the gender of this patient has been changed in the past and a change may possibly be considered in the future.' The family had been advised to defer making a decision for or against a change, pending a gender evaluation. The boy's age in a month would be 12 years. His main problem was breast enlargement. The diagnosis was 'probable testicular feminizing syndrome.'

The transcript of the clinical record that accompanied the referral omitted the years from birth to age 4:2. Thus there is no direct record of the appearance of the genitalia at the time of birth. If the clitoris had appeared somewhat enlarged, it was not sufficiently prominent to engender doubt that the sex of the genitalia was female. Accordingly, the baby was declared to be a daughter, and the sex of rearing began as that of a girl. The mother subsequently supplied the information that her pregnancy and delivery had been uncomplicated.

There is no information on record as to when masses or swellings may have first been noticed either in the labia majora or high in the inguinal canals, and there is only a brief retrospective mention of surgery of bilateral hernia repair, at age six months. Originally the repair appeared to have been successful, but there was soon a recurrence of the hernia on the left.

The next surgical intervention was an exploratory laparotomy at the age of 18 months. The circumstances that led up to this operation presumably included a suspicion regarding the correctness of the child's assigned sex, but the details are missing along with the missing clinical record. The information subsequently ascertained from the parents was that no female structures were identified internally. A biopsy was taken of the gonads, one in the abdomen and one in the canal, and each was reported as testicular in structure. The physician in charge of the case favored the option of changing the baby's sex of rearing from girl to boy. The parents 'leaned on his counseling', according to the father's retrospective recall, 'feeling that he knew much more than we did in this particular matter.' So they endorsed the change of assigned sex by giving the child a boy's name, using the male pronouns, and dressing him as a boy.

The first surgical admission for an attempt at masculine reconstruction of the feminine external genitalia was postponed until some time between the ages of 3 and 4. Although the operative record is among those missing, a subsequent note says that an attempt was made 'to release the penis from excessive tissue,' preparatory to trying to construct a penile urethra. That the outcome was only partially successful is an understatement, since the penis had the dimensions and shape of a clitoris.

For the next surgical admission, at the age of 4:1, the patient was transferred to a major military hospital, and it is from this time on that the clinical records are intact. The admission was for 'first stage reconstruction of the penis for hypospadias.' From now on, an organ that by age 11:11 would measure no more than 2 cm in length was invariably referred to as a penis, whereas its actual size was consistently omitted. At age 13:6 it still measured only 2 cm. In one report, at age 10:4, only one gonad, the left, could be palpated. Its dimensions were given (2×3 cm), but not the dimension of the 'small phallus with meatus at the end.' At age 5:6, a complete fiction entered into the clinical history with the statement: 'At birth the clitoris was quite large and prominent. At about $9^1/_2$ to 10 mo. of age the sex organ began to develop to the proportions of a penis,' to which was added, 'but without normal anatomy for a penis. Scrotum not present. Vagina appeared small.' The surgical task of creating a functional penis from an organ so grossly unmasculinized and small was formidable, and it ended in failure. As an adult, the patient could not stand to urinate, as the urine sprayed too widely.

From age 4:1 onward, the sequence of surgical procedures was as follows:

Age 4:1 First stage reconstruction of the penis for hypospadias.

Age 5:5 Second stage reconstruction: release of ventral aspect of penis with scrotal flap coverage; and V-Y procedure to reposition the two halves of the scrotum (which was shrouding the penis).

Age 5:11 Cystoscopy: revision of scar and left orchiopexy and herniorrhaphy, and biopsy left testicle. Cytoscopy repeated two weeks later (did not reveal a urinary cul-de-sac).

Age 6:1 Orchiopexy of right intraabdominal testis and incidental appendectomy. Postsurgical complication of pneumonia and collapse of the left lung.

Age 6:3 Third stage reconstruction: plastic repair of hypospadias. Postsurgical urinary dribbling.

Age 11:0 Bilateral testicular biopsy.

Age 12:2 Bilateral mastectomy.

Early in his hospital career, the patient received test doses of the hormone, chorionic gonadotropin, once at age 3:8 (dosage not on record) and again at age 4:11 (10,000 units). In each instance there was no response. That is to say, there was no enlargement of the clitoridean phallus, no coarsening of pubic fuzz, and no descent of the testes. With the knowledge of hindsight, this failure to respond should have been used to predict that the hormones from the testicles at puberty would fail to masculinize the body, in defiance of the change of rearing from girl to boy. The inability of the body' cells to utilize the testicular hormone, testosterone (i.e., their androgen-insensitivity), was tested too late insofar as the change of rearing was concerned, namely at age 5:4. At this time 'the patient received massive doses of male hormone to enlarge the penis in an attempt to provide sufficient tissue for further repair' – but to no avail (Fig. 2.3). The surgery was undertaken, despite the failure of this male hormone test. The predictive consequences of the failure, namely, failure to masculinize at puberty, were not at the time recognized.

The first sign of failure to masculinize at puberty made its appearance early, at age 10:0 in the form of nipple tenderness, followed soon by swelling of the breasts (Fig. 2.4). By age 10:3 there was bilateral breast enlargement with approximately 2 cm of glandular tissue beneath each areola. The endocrine consultant at the overseas military hospital where the boy was examined suspected that the diagnosis would be what was then called the testicular feminizing syndrome. In partial confirmation of the diagnosis the chromosome count proved to be 46,XY, as expected. Another testicular biopsy, the third in the patient's clinical career, was done at age 11:0 – possibly in the wake of a recently published warning that testes in the testicular feminizing syndrome carried an increased hazard of cancer. Microscopically, the biopsy specimens from each testicle were the same: 'Immature tubules containing some germ cells, and tending to occur in groups separated by thick, fibrocollagenous tissue septa. No interstitial cells were recognizable.'

At this juncture, the patient's father was transferred back to the U.S. mainland. The endocrine workup done at age 11:10 showed that urinary hormone levels were consistent with a diagnosis of androgen-insensitivity syndrome. A trial treatment with dihydrotestosterone, a potent form of testosterone then newly available, was considered, but postponed, contingent upon the final decision regarding the possibility of changing to live as a female, and then not undertaken, despite the decision to continue living as a male. There was no further laboratory record of hormonal measurements until age 23:7. As compared with male norms, the results then were: LH (luteinizing hormone) 44.4 mIU/ml and FSH (follicle stimulating hormone) 46.0 mIU/ml. Prolactin 11.8 mg/ml, within normal limits. Esterone 37 pg/ml and estradiol 66 pg/ml, both high. Androstenedione 300 ng/dl and testosterone 800 ng/dl, both beyond the upper limit of normal. DHEA (dehydroepiandrosterone sulphate) 350 mg/dl, within normal limits. The endocrinologist for whom these determinations were made interpreted them to the patient as signifying that, 'Hormone injections will do no good. As you can see from the hormonal results, you have above-average levels of male hormones. As mentioned in my previous letter, there is an enzymatic deficiency in the cells. Therefore the hormone cannot get into the cell and cannot do its biologic activity.'

New information concerning familial incidence was entered into the clinical record immediately prior to the psychoendocrine referral at age 11:11. The patient himself had two sisters, one older and one younger than himself, both anatomically normal, and no brothers. His mother had five sisters and no brothers. One sister, deceased, had married but had had no pregnancies. A second had never menstruated and had needed surgery to lengthen the vagina. A third sister had a child, then aged 17, with the same condition as the patient, his cousin.

The clinical history of the older cousin was that he, too, had been declared a girl at birth and subsequently redeclared and reared as a boy, after bilateral hernias in the groins had been discovered to be testes (Crawford et al., 1970). At age 15, he had undergone surgery (bilateral mastectomy) on account of feminine breast development. At age 18, his existence as a male was proving to be calamitous, so much so that eventually he would elect sex reassigment.

Throughout the years of childhood, the patient's clinical record is silent concerning both his personal sense of well-being as a boy and his knowledge of his condition, its treatment, and the expected outcome. No one, including his own parents, had been able to talk to him, according to the boy's own diffident responses during the course of his first psychohormonal interview at age 11:11.

His understanding of why he had been referred for the interview was: 'Well, all the talk about,' and then there was a prolonged pause, 'the latest – like, studies about me.' For him, every response was the equivalent of having a tooth pulled without anesthetic. His own words for what was wrong with him were, 'Um, well, that I need an operation.' Then he became blocked and could not specify what operation he would need, even when the interviewer said that he knew the two difficult things to talk about would be 'your chest, and down here, your sex organs.' He did manage to say, when queried, that one of his referring doctors had told him, regarding his chest, 'that it's a pretty difficult situation. I can't remenber the other thing.' Had they talked to him about an operation? 'I don't think so,' he said. Had he done much thinking about the swelling of his breasts? His response was reiterative, as if to be self-reassuring: 'Sometimes I think about it. Lots of times I don't. Most of the time I don't. I don't really worry about it. Most of the time I don't worry about it.' When he did think about it, the thought in his mind was: 'Am I normal?' And then: 'Will I have an operation or not? That's the only frequent thing I think about – the only important one.'

Estimating the possibility of an operation, he said: 'It's probably about a 50:50 chance that I might have one.' Then, with syntax gone awry, he based his 50:50 estimate on the expectation that the decision would depend on whether the doctors his father would talk to 'would know what they're doing ... or whether they're still studying, and everything like that. Sometimes they don't know what to do about it.'

Eventually, the boy did himself elect in favor of removal of the breasts. The surgery was done three months later, when he was aged 12:2. The postsurgical recovery was uneventful. Soon thereafter, the father was given another, distant military transfer. Subsequent psychohormonal follow-up was by mail and telephone.

Gender-coded social biography

The stigmatization contingent on having had one's sex reannounced at age 18 months, and the threat of future exposure in public, both were mitigated in this case insofar as the father's military career allowed the family to relocate. Potential community gossip was left behind. Within the family, there was a younger sister, too young at age six months to have explicit recall of the reannouncement of her 18 month old sister as a brother. The older sister, aged 3 at the time, was old enough not to forget the change. However, since it became an unspeakable topic in the household, whatever she recalled remained unascertained. Is was masked, perhaps, behind progressively severe depressive psychopathology which, in turn, masked the uncertainty and doom of having the same thing happen to her, as she watched the brother undergo a mysterious progression of unexplained surgical wounds.

In retrospect, the parents reproached themselves for having agreed to the sex change without having recognized that their baby's fate would be dismal, like that of his cousin, instead of cheerful like that of his mother's sister. It did not occur to them at the time that all three indiviuals had the same condition, and they had not been asked for a family pedigree. Their decision was posited on the principle that testis equals male. Naively, they did not pose until much later 'the big question in our life that he can never have a normal sex life ... even though my wife and I have reconciled ourselves to the fact that he can never have children.' These were the father's words. 'Hindsight, of course, doesn't pay off,' he added. 'I don't think I could foresee, like today. If I had foreseen the breast development, if I had even thought of breast development, maybe we would have gone the other way. ... I know I've thought we made a mistake when we decided to make him a boy.'

At the time the change was made, the decision was not easy. Here is another quote from the father's first psychohormonal interview: 'Of course we have some pictures when he was being reared as a girl, and he was a beautiful girl, he really was ... beautiful curly, wavy hair. ... There never were any qualms or worries about his – whether he was feminine, or male or female, then. ... Of course, after the biopsy, and the fact that he did have male testes, and it was decided to raise him as a male, um, you can't help but keep an eye on him and wonder, you know – how is he going to develop?'

'He was always very quiet, very shy, very much of a loner,' the father said. Otherwise the father had not considered him unmasculine in childhood, except that, 'in throwing a ball, he moved his hand as a female might do. ... To this day he doesn't like to throw ball. He doesn't want any part of that. As a participant in sports, he's not interested, and also not as a spectator. He'll sit down and work his stamp collection. If there's a football game on T.V., he couldn't care less.'

When asked about friendships, the boy identified himself as being socially isolated. 'I don't have anyone in particular,' he said. 'The only guys I have friendships with are at my bus stop, but I never go to their houses. I don't like to visit people. And I don't like to fool around with people. I only like work. Like, every day when I get home from school, I have at least three hours of homework. When I get finished, I'm tired, and all I want to do is watch T.V. ... I either watch T.V. or work. That's all I did last Sunday, was work. ... I do vacuuming. Sometimes I do gardening. Sometimes I clean rooms, or clean the bathroom, or wash floors.'

Apart from working, he had solo hobbies and interests. 'I collect stamps,' he said. 'I collect coins. I collect models and sometimes feathers. And I like animals. And I have an interest in one President of the United States, Lincoln. That's my one President that I really appreciate. I've been studying him for a couple of years. I've collected some pictures. I've just started on Washington, but I haven't really read his history.'

His interests did not run to dramatics and play-acting. He had dressed in costume for Halloween, but otherwise was not interested in wearing fancy dress, nor feminine clothing. He considered himself particular about the clothes he wore: 'I like long sleeves. I don't know why, but I just like them. And I like comfortable slacks, and I like good jackets. ... My favorite color is blue.' He did not classify himself as a tidy person 'because I always get my room messy; I always throw stuff on the floor – books, and I have a bookshelf, but I always throw my books on the floor.'

He began a self-description with: 'Well, I don't like fighting. I don't like to be mean to people, because then they get on to you.' He did not relate not being a fighter to having occasionally been called a sissy: 'People always call you a sissy if you look like, good, like nice, or if they think you're dumb.' His reaction was based on what he had adopted from his father's wisdom: 'He always said, if you think you're not, you're not.' He did not recall that anyone had called him queer or homosexual: 'If they did, I can't remember.'

About sports he said: 'Sometimes I like baseball, sometimes football. But I don't really care for them. ... The only days I go outside are to work in the garden, or to get some snow off the sidewalk. ... I don't like to move from one place to another every day. If I do, that's the only time I get pains in my neck, like I have now.' He was not a member of any organized social group. Nominally a Roman Catholic, he quit attending church services, but remained 'interested in the Bible. ... I read the Bible sometimes,' he said.

He did not explicitly connect his becoming reclusive with having breast enlargement, but his parents did. After his breasts became prominent, he had to drop out of gym class at school. He developed a stoop-shouldered stance. It partly disguised the enlargement of his breasts. So also did a loose-fitting shirt. He quit wearing a tight-fitting undershirt, because it accentuated his shape.

After his family relocated at another military post, 'He brought home one friend,' the father said, 'one boy friend, and that lasted only one day. ... He doesn't really have any friends. ... He spends a great deal of time in his room. Like I could never punish him by sending him to his room, because that's not punishment to him, you know. He lives in there.'

His mother, on the occasion of the boy's second psychohormonal visit, related his social isolation not only to gynecomastia but also to being self-conscious about having undeveloped genitalia: 'I think he feels that he won't be accepted in a lot of circumstances, and so to avoid embarrassment, he spends more time by himself than he should ... although he does enjoy sharing his hobbies and doing other things with his sisters. As a female he could have functioned in life. As a male, I don't think he can function fully, unless something is developed through research.'

Despite her misgivings, the mother had the impression that her son 'is very much a boy. All his hobbies and interests seem to lean this way. And I feel, bodywise, that he is very masculine as far as being heavy handed and footed. ... He would prefer to stay masculine.'

The father's first interview included a checklist of funny habits and odd behavior in his son's developmental history. It disclosed nothing noteworthy, other than being a recluse, as above indicated. The older of his two sisters, by contrast, was 'experiencing some emotional problems,' which has necessitated a professional referral. The patient did not get along well with this sister. He was closer to the younger one.

He himself was quite explicit about being a recluse. Its pervasiveness was self-expressed in the 'Draw Your Family' subtest of the Draw-A-Person Test. He explained the omission of himself: 'I don't like to draw myself. I feel foolish when I draw myself – like, why do you put yourself in a drawing if you're drawing the drawing?' On the preceding subtest, 'Draw Yourself and Friend,' he had given a quite explicit account of his negative self-image. 'I don't like to draw myself, so I drew myself like that – a back view. I don't get an image of myself. I don't appreciate drawing myself. It makes me mad. When I draw someone else, I try to make it exact, but when I draw myself, I get all frustrated, and I want it perfect.'

Academically, the father's statement was: 'We've had no problems with him. He's an average student. I think he could do better. But I don't think he's that interested in school.' The boy identified his grades at school as 'B's sometimes, C's sometimes. Sometimes A's and sometimes D's.'

The Wechsler Intelligence Scale for Children had been administered when he was aged 11:11. It yielded the following: Verbal IQ, 113; Performance IQ, 111; Full IQ, 112.

In young adulthood, he completed his education by graduating from college. He then took up further training and employment in medical technology. In his early twenties, his medical interests became self-directed. He became a serious and well-informed student of his own syndrome, and of his own clinical record, of which he obtained a transcript. He found a clinic specializing in sex counseling and psychotherapy, and engaged in personal sessions on a regular basis for a period of at least two years. The dilemma of being socially isolated was, however, persistent and unyielding. It intensified as his chronological age became increasingly discordant with his physique age and appearance, by reason of the fact that, being androgen-insensitive, he did not virilize. Consequently, his social age was stranded, in limbo, too advanced for the juvenility of physique age, and not advanced and full-rounded enough for the maturity of his chronological age. With the exception of those who had accurate and advanced knowledge, people responded to him as to a youth imposturing an adult. It was a no-win predicament.

Lovemap biography

The first installment of sexuoerotic information was obtained at the ages of 11:11 and 12:0 in the course of the two extended interviews of which the agenda was the feasibility of sex reassignment. Then followed a period of sporadic contact until the second installment was obtained between the ages of 23 and 25 in an exchange of correspondence initiated by the patient, together with a follow-up audiotaped interview by long-distance telephone.

In his first interview, at age 11, the boy was able to say only, 'yes', without further elaboration, to a query about having a picture in his mind of the appearance of the sex organs of a boy and a girl. What he thought about himself, in comparison with another boy or girl was, he said 'unimportant,' which he then explained: 'Well, that I just wasn't born right, or formed right. That's really all I think about.' He considered that he more closely resembled a boy than a girl, but he could not stand to urinate because his penis was too small, and there was a spray, not a stream of urine. Regarding surgical repair of his penis, he replied: 'I don't remember.' His two sisters knew about the appearance of his penis, but only once, while dressing after swimming, might a boy have seen it. He avoided exposure, and so had not encountered teasing.

His learning about reproduction had come variously from a program on T.V., from a human body book, and from his homeroom instructor at school. His vocabulary and concepts of reproduction both were inadequate. Thus, his conjecture of how birth takes place was: 'I think the doctor makes some kind of opening in the mother's body, and [the baby] slips out through, like a vent, over in the side.' His first conjecture about what makes the baby start to grow inside the mother was: 'The food that she eats.' The correction of this conjecture was: 'Um, like – it might sound stupid, but cells. ... All I can think of is cells, like, form into dozens. Then the baby starts forming.' He did not evidence any acquaintance with the concepts of egg, sperm, penile erection, intromission, or sexual intercourse in general. After he had listened to the elementary information, and its application to the irreversible handicap of a micropenis, his evaluation of its personal significance was: 'I don't think I'm going to get married.' Asked to elaborate, he digressed: 'If you get married, you have to work real hard on a job. If you don't get married, you just have to earn enough money for yourself – by just living in a hotel.' The strategy of giving a digressive answer, tangential to the question, appeared also in response to other queries that implied masculine inadequacy. For example, about living together with a girlfriend as a lover, he said: 'Well, I'm not interested in girls, that's all I can say. I'm not really interested.' About sharing an apartment with a boyfriend, the answer was: 'Well, what do you mean? Like if I'm in college? Well, maybe. The time hasn't come yet, but I don't really think so.'

Ultimately, he would make it explicity clear that he resented having to be confronted with information that he correctly perceived as exploring the possibility of male-to-female sex reassignment, so as to avoid the fate of living as a permanently unvirilized male. This possibility had been presented obliquely in the form of a narrative of an actual case that might or might not be self-applied. This is the Parable Technique (Money, 1986). The parable was told as follows:

I'm going to change the subject again and tell you the story of a boy who came to see me. Just today I got another letter from him. He's 19 now, and he's been coming to see me since he was a young boy with exactly the same kind of problem you have. When he was 10 years old, he was down here for a visit, talking. He came from New York, as a matter of fact, the same as you.

I had a feeling that there was something he wanted to tell me. So I told him a story about some of my other patients with his, and also your kind of problem. There was one boy in particular who had been telling me about some dreams. He dreamed that God had really intended for him to be a girl. Then he actually had one dream where, in the dream, he could imagine himself living as a girl.

The boy who was listening to this story was sitting there, in the same chair you're sitting in. He said: 'Yes, I sometimes think about that too.' Why? Because he had just a very tiny penis. That's why he wondered if he was really supposed to be a girl. Then we talked on, and he said: 'Well, I've read about some of the people who have changed their sex.' In the newspapers, there was a famous case. Have you ever read any of that? The famous case in those days was Christine Jorgensen, who sometimes comes on television. A couple of years ago, she wrote a book about her life. Her case was a very unusual one, at the time, of a person who started out living as a boy, and changed over. Now she's been living for 20 years as a woman.

The young boy who was sitting in your chair thought about living as a girl. Then he said: 'I wouldn't want to change unless you could guarantee that, if I would try to live as a girl and grow up as a woman, I could have my own children, my own pregnancy. If that's not possible, then I'm just going to have to stay the way I am.' Of course, it wasn't possible to guarantee pregnancy. So he decided it was better to stay the way he was. He's now a college boy. He's one of the boys who can use an artificial penis, a strap-on one, like I've already mentioned. Like you, he didn't want to have any extra operating done on his penis. He wanted to wait to see whether he would want to use it, or not.

Now what do you think about my story of his story? After a long pause, the reply came: 'Not much.' It's not a very usual story, is it? I asked. After another long pause, there was a mumbled response, and then a clarification: 'Well, it may be, and it may be not. Because I'm not sure.' He was sure, however, that the boy made the right decision, because: 'If he couldn't change, if he – if he couldn't be what he wanted to be, then he would just have to settle for what he was. So he did that.'

The narrative continued with a matching parable of a similar case, this one of a boy who had skin-graft surgery to make an artificial penis. In teenage, his body grew breasts, so he had another operation to flatten his chest. At age 19, he still didn't shave. His body couldn't grow hair. It could not make him look like a man, because it couldn't use male hormone. No one ever asked him if he had ever thought whether God had put him in the right sex, and someone should have. To this boy's story, the patient's response when asked for his feeling about it was: 'I don't have one.'

The next topic was operations for change of sex in people born with normal sex organs, for whom the patient had one word, 'crazy.' Agreeing that many ordinary people would share his idea, he said, 'I wouldn't want to do it either. ... But let them do anything they want.' About women who change sex, he was more lenient: 'Well, I might feel it's not too good an idea, and they might feel it is. So, that's where, that's why I can't really say anything, because they have a mind of their own. They have a different structure in which they want to change.'

He listened to explanations of why, in patients like himself, the voice failed to deepen, and hair did not grow on the face for shaving, or on the body. As applied to himself, he found each prediction inconsequential. Thus, equating deep with loud, he said 'I have a deep voice when I want it to be deep. My voice can be as deep as my father's is when I get loud and I want to talk about something.' Again quoting the example of his father, he regarded shaving as a burden, and said, 'I don't like to shave. ... Sometimes you have blemishes. ... Some people have a beard and something wrong with their face which makes it more terrible, and holes and bumps in it.' About body hair, his reaction was: 'I have some. I have some right now. Sometimes it grows, and sometimes it stops. And then it starts to grow all over again. And it starts to grow long.'

He explained his thinking about not wanting surgery for an artificial penis in the following words: 'Going through all that trouble – I just like it how it is. And my father will have to spend a lot of money, and I don't want that. It's too much for me. I don't want to go through that. ... I don't like people hanging over me, to hang on me. I don't hang on them. ... Sometimes I get mad and, like somebody tells me what to do, I get mad. And I get madder, I get more angry at the time they hang on, and they start telling me what to do. So I start just hating them for it.'

From among all of the things that doctors might be able to do for him, the only thing would be, he said, after a long pause of consideration: 'Well, change these.' He indicated his breasts. 'That's about it. That's the only thing I want.' He had already been told that an operation to flatten his chest would be feasible. Now it was time to offer a warning: 'The only thing about it is that we had better not make a mistake. You had better not make a mistake. When you're 16, or 18, or 20 years old, you do not want to be coming back, saying "I think maybe I made a mistake and should have kept my breasts."' There was no hesitancy in his reply: 'I won't. I won't make that mistake. Because I don't want to be like those other guys. ... I want to stay like I was, like um, born to be. I want to stay like – because if I'm, if like, if you're a girl, you don't feel confident, and you're always leaning on somebody. But if you're a man, like if you born as a boy, and you grow into a man, you work on your own; and you're not married, and you can join whatever you want. But, if you're a woman, like, some people won't give you jobs, and you can't do much except, like my mother, she's a housewife. ...'

As a man with a flattened chest, there would always be the problem of having to stand up to urinate. Preceded by a long pause, his response was: 'I'll be – it probably won't bother me, because I'm not bothered at all by it.'

Romantic attraction, should it happen later in teenage, would not, in his opinion, be contingent on penis size. His response to how he would deal with this eventually was: 'I'd try to avoid it. ... Avoiding has helped me a lot, and it's still helping. ... I'm not fond of nobody, and nobody's fond of me.' If the girl was in love with him and had let her feelings be known, so that he couldn't avoid her, then 'I would probably show her how really unpleasant I can be,' he said. 'She'd probably, like most girls do, she'd probably hate you.'

His answer to the possibility of being in love with another boy was: 'But how can you fall in love with another boy? I've never loved a boy, but I've liked him, that's all.' If a boy had a romantic and sexual interest in him, then not knowing how to deal with it, he said: 'I'd probably go to my father and explain it to him, and he would help me.' At school, he had not heard any stories of boys being homosexual, nor of girls being lesbian. 'I've heard that word, lesbian,' he said. 'My sister mentioned it, but probably didn't know what it meant.'

He did not have to consider relinquishing any feminine yearning to take care of infants and young children. 'I don't like baby sitting,' he said, and recalling his younger sister at the baby-sitting age, added: 'I just think they're a mess. They get into things. I'm against baby sitting.' When younger, he had not been interested in playing with dolls: 'I used to have this Indian doll,' he said, 'and I used to pull things off it. I used to pull it apart.'

Toward the end of the long two days of the initial evaluation, he summed up the best thing about it: 'Well, talking about what I think is right, and what is wrong. About what I don't want to be done, and what I do want to be done, so they won't like, get on my back. That's about it.' On the tape, he dictated a message for the referring surgeon that he wanted 'something done about these, his breasts.'

In accordance with routine policy, the visit ended with a joint interview in which the father, the only accompanying parent, and son were both present, together with participating psychohormonal staff members. The father was able to talk openly about the pros and cons of breast surgery to remain a boy, versus sex reassignment, quoting what his son had told him about his interview, when they both had talked together in the cafeteria and in the hotel the previous day and evening.

It was too soon for the boy to commit himself to a final and irrevocable timetable for surgery. He wanted to think about it for a while. A return visit, with the mother included, was scheduled for no later than four weeks ahead, and sooner if requested. That would allow sufficient time for a subsequent admission prior to his father's military transfer to the West Coast. It would also allow for the boy to escape from the limbo of indecision in which he had been trapped for over a year, ever since his breasts had begun to enlarge.

When he returned a month later, a few days after his 12th birthday, he gave the impression not of someone no longer entrapped, but rather of someone more firmly held in the grip of a damned if you do, and damned if you don't entrapment. His mind was not consonant with his body, and his body would not conform to his mind.

His mood had metamorphosed into sullenness and antagonism. His head hung low. There were long pauses when he had no words with which to communicate what he was experiencing.

Reviewing the upshot of the first meeting, the interviewer said: 'I told you that you were a very exceptional person so far as your life situation and your life choices were concerned, because you are one of the very rare people in the world who can actually choose the sex that you like to belong to. I think that's why you decided you weren't going to be in a big rush to get your chest operation that you had wanted pretty strongly.'

His reply was; 'No. I just didn't feel, like, up to it. I didn't feel that I wanted to get it quickly. ... I know I wouldn't – I told my father what kind of sex I would like to be, and that's male. And I still want to be it. ... Sometimes he mentioned it at home. He talked about, um, he talked about what, um – he said that if I would like female. I said no. And I disagreed with him. And so I dropped the subject, because I, I know what I want to be. So, and, nothing will change my mind.' Then followed 40 seconds of no response to the question of how he felt about the interviewer, now that he was bringing up the subject again. Finally, the reply was: 'I don't know. Well, um, as far as, and I'll say this, but, I think you're a rotten guy, because I told my father that, that you were trying to make me, make me do what he wants. And I think the same thing of you. So, I don't like people bringing up the same question, over and over again. ... You're trying to make me say what you want me to say. And I don't want to say that.'

In what ensued, the interviewer explained that his position was to support the patient's decision, now and in the future, irrespective of the outcome, but also not to withhold information on comparable patients' decisions and, with examples, their variable outcomes. Examples of the failure of hormonal virilization in cases like his own of micropenis and androgen insensitivity were reiterated, in view of the enormity of the sequelae of his decision for the rest of his life. His verdict was unswerving: 'This is the one thing that I want done. I want my chest flattened out, and that's about it.' He rejected the proposition that he was on the horns of a dilemma, saying: 'I told you what I want, and what you're doing is saying that I'm part of this, and part of that, but I'm not. ... You said we won't mention nothing about the other sex that I don't want to be. ... What you're saying is to imagine that I'm the other sex, that's what you're saying, and I don't like imagining that way.'

At this point, the boy had had enough not only of the interview, but also of temporizing. There was no neutral ground, for even postponement of surgery in order to pursue a year, say, of psychotherapy would, by reason of retaining breasts, have been an affirmation of feminity. There were only two alternatives: either to impose a decision by coercion, or to give the deciding vote to the patient himself. He had the vote, and it was to return home and undergo mastectomy.

Two months later, he had the operation, and three weeks after that made his final psychohormonal visit before leaving the area, as expected, for the West Coast. The scars from the operation were not completely healed, but he felt 'good', he said, as laconic as ever, to no longer have breasts. He had not had any dreams pertaining to surgery, nor to the appearance of his body.

There were still no dreams to be reported when, at age 13:5, the next comprehensive follow-up interview was done by long-distance telephone. Erotic imagery and ideation were absent not only in sleep dreams, but also in waking fantasy. There was one occasion only when he got what he would call a 'kind of excited' sexy feeling, and that was when he looked at a pin-up poster in his cousin's bedroom. There had been no similar arousal involving the other senses, no masturbation fantasies, and no incentive to masturbate. He said he had not yet had any evidence of either erection or escape of fluid other than urine from the penis. The evidence of orgasm, or even of pleasurable genital erotic feeling, was zero. The nipples were sensitive only to deep pressure. Romantic inertness remained unchanged. He objected to the idea of hormone pills or shots, even for a trial treatment, saying: 'If I feel the need, I'll try it.'

For the next three years, he maintained contact on an annual basis until, at age 16, he initiated clinical inquiry: 'Can my penis be enlarged or not, and can the scars be removed from my sides? I have discussed this with my father, and he told me to write to you.' In the follow-up letter, he was more detailed: 'Have been going through my medical records and find I am now getting more interested in girls and a marriage, a job, and the beard I once said I didn't like or want. I still find it hard to make home-friends, and my life is hardly active except for family activities.' Five questions were appended: 'Can the appearance of my penis be improved greatly by [plastic surgery] operations? How long would it take for the reconstruction? Will I have a beard without having to take medication? Is there any surgery that could ensure feel and erection of penis? I would like corrective surgery, and for my enlarged penis to look as close to the real thing as possible. How can I get both testes descended into the scrotum?'

The upshot of the ensuing correspondence and phone calls was that, since he expected to be living again in the East, he would make a return visit to Johns Hopkins for further discussion and an evaluation by a plastic surgeon. Then followed five years of silence. A letter requesting a progress report remained unanswered for another 18 months, until, at age 23:7, the response was: 'I must apologize for not responding sooner, but at the time I was not able to deal mentally with my syndrome. As is evident by the photocopied letter [the report of a complete endocrine workup] enclosed, I am actively looking into my syndrome, with the help of my counselor. I feel constantly plagued by my lack of secondary masculine sex characteristics. I am also writing to my cousin who has the same syndrome, and is currently living as a female, about her present feelings about her sex change. Despite my syndrome, I feel my life is successful. I'm presently a college student majoring in microbiology. Also I work part time at Mercy Hospital. In the future, I plan to pursue a career in industrial or research laboratory work.'

In another letter, eight months later, the patient wrote: 'After seeing a number of physicians, I understand my syndrome a lot better. ... I think when I was younger your conversation with me didn't sink in. You gave me a lot of new information, but I was both immature and overwhelmed by it, and didn't accept some of what you were telling me. A physician-biochemist at the University is interested in my androgen-insensitivity syndrome. He made it quite clear that, whatever research he is contemplating, involving myself, may lead to a dead end in helping me achieve my desire for secondary masculine characteristics. After much thought, I still will remain a male, even after almost a dozen letters from my cousin encouraging me to consider becoming a female, and the problems I must face when interacting with people.

'After reading many articles, especially the one by John D. Crawford (1970), I have gained much insight. For example, in Dr. Crawford's article, my counsin's diagnosis was incomplete form of testicular feminizing syndrome. ... Thus it's likely I have the incomplete form. ... In my case I make dihydrotestosterone. Thus it must be the cell receptor which can't recognize the compound.

'I thought I'd just keep in touch. Thank you for your last letter. Other people, including myself, agree with you that plastic penile surgery [phalloplasty] is less than satisfactory when the risks and after-surgery complications are considered.'

Enclosed with this letter were xerox copies of his own handwritten notes that recorded the information obtained from medical consultations. Concluding the note on plastic surgery, he wrote: 'I asked him about medical insurance coverage, and told him I was giving myself ten years to decide on a line of action. He stated that, in ten years, they may have better procedures.'

At the age of 24:4, the patient was referred to a local, West Coast psychiatrist by his female psychotherapist for an evaluation with a view to transferring to a male therapist. The intake interview was done by a specialist in sexology and gender. The patient kept his own minutes of the interviews. The following information is taken from these minutes: '... He was amazed at the volume of my military-hospital and other medical records and asked me to give him a brief summary. I talked of being born a female, changed to a male at two years old, followed by a series of operations taking approximately 5–6 years, with some gaps in between for healing; then of the operation to remove hairs from my urethra (they grew at puberty) because of crystalized urine stones causing urinary blockage. I said it all in a calm sort of monotone voice, without too much emotion, looking up at him occasionally.

'He asked me of my family, and I told him of my cousin Peggy [used to be Danny], and of Linda, my aunt [on the maternal side], both of whom have my diagnosis. Then I told of my parents and two sisters, Connie, the elder, and her mental condition [more likely severe depression than schizophrenia, though not formally diagnosed] and my younger sister, the perfect child of the family. ... I had a good relationship with my older sister when we both were younger. Then when my younger sister got bigger, the two girls became good friends and I was left out. My childhood friend was Jimmy until, after 5th grade, I more or less withdrew from the world because of my condition. I felt I couldn't shower with the other boys. As I became older in my teens, I withdrew more, and had only my younger sister to talk to once in a while.

'After my cousin Peggy changed sex, I was uncomfortable around her. I said I would feel uncomfortable joining a transexual group. I was more drawn to becoming a male. Secondary sexual characteristics were my goal. ... I'm not drawn to males. I like females better as people. I tend to associate males with dominance. ...

'I answered, yes, that I do masturbate, and that it's not caused by dreaming of sexual encounters. It comes on due to a bad day at work. I do it to relieve tension, or sometimes I just feel like doing it. ... I've never done any dating, and I plan to remain unmarried. I felt the tears swelling, but I knew I couldn't cry – tears resolve nothing. He said that it didn't sound as though I had planned a very happy future, but that he thought me an exceptional person with more to offer the world than my asylum from it. I half agreed with him. ...

'My impression of him was of a rather kind man. ... I really felt he asked some crucial questions, like dating, marriage, my future. He answered some too – like I shouldn't put myself in exile, because there would be someone who would love me. Also the transexual meetings held once a week (could I be interested?). At points I felt intimidated and uncomfortable. I crossed my arms, and I covered my mouth at times. He complimented me at various times (I was stunned) and at the end he told me what an exceptional, talented **man – not a man feeling like a freak** – I was, and I told him it was hard to accept what he had just said. ...'

The foregoing quotations are taken from a xerox copy that had been forward by the patient, along with other materials that he considered relevant to a follow-up interview conducted by long-distance telephone when he was aged 24:7. He had received a copy of the transcript of the tape of this phone interview which had been sent to him. The following excerpts are from the manuscript. Small capitals indicate that the speaker is the interviewer.

WHAT ABOUT HAVING HAD ANY ROMANTIC FEELINGS TOWARD SOMEONE? HAVE YOU EVER HAD THAT? Yeah, I think I have on one occasion. She was an older woman and, you know, she never knew about it. She had her own life and everything. I think she was kind of indifferent to me. So I just sort of canned that one. ...

IF YOU HAVE TELEPHONE PRIVACY, I'M INTERESTED TO KNOW WHETHER YOU DO GET A GOOD SUPPLY OF SEXUAL FEELINGS FOR YOURSELF, PERSONALLY, IN THE GENITAL ORGANS. Yes I do. I do. And I get erections – but lately, I don't know. It cooled off. I used to get more than I do now. I don't know whether it's work, or maybe just my feelings, whether I'm thinking about it, or whatever. I don't know. WHAT ABOUT LOSING FLUID IN THE WAY OF EJACULATION? I used to, maybe about half a year ago, but now I'm, I don't know, for some reason I can't do that. It's just not coming any more. AND THAT WOULD BE FROM PERSONAL STIMULATION, WITH THOUGHTS, OR WHAT? I think thoughts and personal stimulation. MOVIES OR TAPES? OR PICTURE BOOKS? No. I think it's personal, thought and then stimulation, masturbation. I think that's the only way I'd be able to. DID YOU CONSIDER THAT YOU'D REACHED A SEXUAL CLIMAX? ORGASM? I've often thought about that, but I don't think I have anything to compare to. So I really can't say.

TO GET A SEXUAL TURN-ON, SOME PEOPLE GET MORE FROM LOOKING, OR TOUCH AND FEELING OF THE SKIN, OR THE OTHER SENSES. HOW WOULD YOU RATE YOURSELF? I think – I guess I get my turn-ons by, just kind of personal imagery. I think I do that. I think when an attractive girl walks by, or something, I'm attracted to her, but I don't think – I think I get a positive feeling about her, but I don't think I, what do you call, ejaculate or whatever. NOT SOUNDS, MUSIC, SMELLS OR TASTES? No. I think it's mostly from my mind.

WHAT ABOUT HAVING SEXUAL DREAMS AND ROMANTIC DREAMS? I think the last couple of dreams I've had, they've always been, the sexual dreams have always been sort of violent or negative. ... None romantic. I've written down some of my dreams. I can send you one I wrote down five days ago, if you want me to.

(He did send it. It reads as follows.) September 6. Dreamed at midnight. Recorded at 12:30 a.m. These people climbed into an office and went in through the wrong place, a window possibly. If they were found, they would be asked to leave or hauled away. To stop being found out, they went inside another hallway. In particular this one girl went inside a room, to take a bath, possibly, or use the bathroom. She got inside the tub and this fat woman who was in the room, prior, raped her – oral sex. As the fat woman was leaving, now fully dressed, she passed my younger sister, who was or was thinking of becoming a psychiatrist. In real life she's getting her masters in social science. My sister asked her about the rape incident. How she knew about it I can't remember. Well, I think the woman came running out, the one who was raped, and told her. The smiling fat woman admitted the rape and said something like she had always been like that. She left. I think my sister was amused, possibly about having the fat woman in her psychotherapy. The only thing I can remember about my feelings is that I disagreed and was repulsed by the fat woman's rape and felt anger towards her. I couldn't believe my sister's reaction of being amused or fascinated by this person as a prospect for a future client.

... DO YOU KNOW HOW YOUR PARENTS FEEL WITHIN THEMSELVES ABOUT YOUR HISTORY? I've asked them. I've asked my mother and father separately. My father just said that they pretty much agonized over a change in sexes. I can't really remember what my mother said; but one of my aunts wrote her a letter, wanting to know about the syndrome because her daughter was going to have a baby, and her obstetrician wanted to know about it. So what I did was just compile a bunch of articles that I had and sent them to her. You know, I really haven't heard a word from her about it, whether she'd received them, or what.

Two weeks after the phone call, during which time the patient reviewed his copy of the transcript of it, he wrote a five-page response from which are taken the following items of information.

'From one of the reprints you sent me [Money and Onguro, 1974] I was surprised at the number of people with the partial androgen-insensitivity syndrome; also, that the majority of males were against using the prosthetic penile device. I find myself still against romantic involvement: dating, marriage. I'm quite positive my 'role-identity' is definitely male. I find myself not fond of domestic chores. I'll do them, but with the thought of finishing them as soon as possible. Here I possibily used my father as a role model. He's not fond of domestic chores, but dabbles into maintaining his cars, which is what I also do on my own car. I still consider myself a loner and avoid fighting.

'I find myself positive toward children and post-neonatal infant care, but still put child adoption out of my mind.

'I've discussed with my present psychiatrist about my having a very negative attitude toward my youthful appearance. I feel others are attacking or insulting me when they question my looking young, and ask me which sex I am, or refer to me in the female gender (dear, Miss, or sweetheart). This frustrates me, since I've had a short haircut for the past year. My voice, having such a high pitch, adds I think to their perception. I have tried lowering my voice. It does help at least in making me feel more positive towards myself.

'I agree with you that "withholding of information can be extremely traumatic." I knew from age 18 months (sex exploration operation) something about me was different when a little boy, in the same hospital room in a bed opposite to mine, stood up in the nude and smiled and said something to me. At that time I thought I was male, or possibly just different. I haven't decided whether, as I look back on the incident, my brain refers to me as 'he' simply because after that time I was reannounced as a boy. Because I was different, I didn't dare stand. I had a sheet over me.

'My parents always avoided the discussion of my sex, and it was only when I saw you (1971) that sexual discussion began. The only time my father discussed anything was to say that to make me female (this was before mastectomy) all they'd have to do is 'cut it off' (my penis). I was horrified and angry at even his suggestion. Since then, I've always been the one to talk openly about anything concerning my syndrome, whether it be the cause, or new research. My parents were interested, especially about the day you called two weeks ago, wanting to know what was discussed. I told them some of the things we talked about, but avoided subjects like masturbation, ejaculation. They'd turn into stone if I hadn't. ...

'My cousin, Peggy, just sent me a letter. She speaks even more against medical opinions: "I've been disgusted with doctors for many years. They do not listen to me. ..." I don't want to alienate her from me, so I can't write and say that doctors are to be used for what advice they have to offer, but that it's up to you to make the final decision.'

A year and eight months later he did make his own final decision. Another two years went by. Then there arrived a letter, from his cousin. It was addressed, Dear Dr. Money, and read as follows:

'I became acquainted with your work over a decade ago as a college student. I was in a very difficult situation at the time and, to a large extent, it was your writings which gave me the knowledge and motivation to get myself out of it. It was only later that I found out that my cousin, Godfrey Kline, had seen you. You see – I was born with the same condition as him, and was also raised as a boy, but as an adult I changed to being a girl. Godfrey once told me that he had thought about such a change for himself, but didn't find the idea attractive. I guess it was foreign to the way he thought about himself.

'I know just what sort of problem Godfrey had to go through, so it was only a small surprise to hear of his suicide in May 1985. I think that his death was a waste; he could have had a happy and successful life if he had not been reassigned as a boy. That the same thing had happened to me I don't consider as tragic; the doctors didn't know very much back then, but had to find out what to do the hard way. In Godfrey's case, seven years later, the knowledge was there, but the doctors didn't bother to avail themselves of it.

'Godfrey said in a letter to me, "Dr. Money really has been a kind foundation of support from my early teens till now." More importantly, I think, you offered him a chance to avoid mastectomies and a painful future of trying to be a male. Such a chance was never offered to me and, as a result, my life has been more difficult than it needed to be. I'm still happier now than I ever expected to be, but most of my problems should have been avoided.'

So also should her cousin's problems have been avoided. Then they would not have plagued him until, too gross, they were self-terminated.

EXPOSITION

The androgen-insensitivity syndrome (AIS) was earlier known as the testicular feminizing syndrome, and still earlier only as one of the forms of male pseudohermaphroditism. Male pseudohermaphroditism and male hermaphroditism (the preferred term) are synonymous. Both terms refer to that type of birth defect in which the organs of procreation are intersexed or ambiguous, except for the gonads, which are testicles – not both ovaries, not mixed as ovotestes, and not combined as an ovary on one side and a testis on the other. In the nomenclature of the 19th century, still used, only the mixture or the combination of ovarian and testicular tissue in the same individual met the nosological criterion of true hermaphroditism. If there were only two ovaries, or two testicles, then the prefix, pseudo-, was added, and the diagnosis was female pseudohermaphroditism or male pseudohermaphroditism, respectively. This now outmoded nomenclature had no term for cases of a fourth category of hermaphroditic ambiguity of the sex organs, namely that which is associated with vestigial gonadal streaks that are neither ovarian nor testicular.

A special feature of the androgen-insensitivity syndrome of male hermaphroditism is that, when insensitivity to androgen is complete, the external sex organs do not look to be deformed, for the pudendal anatomy is wholly female. The syndrome qualifies as being a form of hermaphroditism by reason of the incongruity between the feminine external genitalia and the nonfeminine or masculine internal genitalia and the masculine gonads. It requires the close inspection of a genital examination to show that, beyond the orifice of the vagina, the vaginal canal is short, and ends blindly. There is no cervix, no uterus, and no fallopian tube on either side. In the place of ovaries, there are gonads that are, under the microscope, testicular in structure. In some AIS cases, these gonads try to descend into the scrotum from the abdominal cavity. Since there is no scrotum, they have the appearance of lumps (or hernias) in the groins, either high in the inguinal canals or, if lower, in the labia majora. In other AIS cases, the gonads remain in the ovarian position within the abdominal cavity, and their secretive existence may be first suspected only in teenage, secondary to failure of the onset of the menses, which in turn is secondary to the absence of a uterus from which to menstruate.

Absence of the internal female procreative organs in the AIS dates back to the period, early in embryonic life, when the testes secrete antimullerian hormone which instructs the mullerian ducts to vestigiate, instead of to differentiate and grow into a uterus and fallopian tubes. This mission accomplished, the testes continue in prenatal life to secrete androgenic, masculinizing hormone which, in the AIS, has its masculinizing mission totally defeated. This defeat is secondary to a genetic defect, an X-linked recessive, transmitted in the maternal line. Present in every cell of the body, this genetic defect destroys the ability of androgen-hungry cells to do their masculinizing work. The cells either fail to take up molecules of the masculinizing hormone, testosterone, or its derivatives, or else, having taken them up, fail to respond to them within the cell nuclei. In consequence, instead of creating Adam, they create Eve, insofar as nature's primary principle is Eve first, Adam second. Adam requires the addition of the masculinizing factor.

The masculinization of Adam, or his demasculinization, takes place not only between the loins, but also between the ears, in the sexual centers and pathways of the brain. Experimental animal evidence completely confirms this assertion. It is on the basis of this animal evidence that one extrapolates to the uncontrived experiments of nature in human beings, and arrives at the conclusion that AIS babies are born with a sexual brain that has been neither masculinized nor defeminized. Thus, the brain and the feminine external genitals both very well predispose the baby to be assigned and reared as a girl.

Since sex is assigned neonatally on the basis of visual inspection of the sex organs, it is exceptionally rare to have a 46,XY baby with the complete AIS reared as a boy. For this reason, it is the man of the matched pair here presented who has special heuristic importance. His case is virtually without replication, whereas there are not only dozens but hundreds of 46,XY androgen-insensitive babies who are reared as girls and live their entire lives with their femininity unquestioned. One of these was selected to be the other half of the matched pair of cases here presented. Her case was selected on the basis of its having been known to the clinic since age 6 with follow-up until, at the present, the age of grandmotherhood.

In the history of their lives, and in the quotations from their interviews, the two patients differed in the polarity of their sense of sexual membership, one as a member of the female sex, and the other as a member of the male sex. They differed also in the security of membership accorded by other people. Membership credentials were not questioned by other people in the case of the woman, whereas they were in the case of the man. His greatest obstacle was the phenomenon that I have called the tyranny of the eyes: what the eyes see dictates that the beholder simultaneously construe what it is that he/she is beholding, and then act accordingly.

The tyranny of the eyes is manifested in, for example, the case of dwarfism, when the beholder's immediate impulse is to respond to the physique and height age, which is juvenile relative to the birthday age and to the social and mental age. Consequently, people of dwarfed stature are talked down to, and subjected to juvenilization. The converse of juvenilization occurs when the eye beholds a child of, say, 5 or 6 years of age with precocious onset of hormonal puberty. Again the beholder responds to the physique and height age, not the social, mental, or chronological age, and misconstrues the child as a mentally retarded adolescent.

Female impersonation is another example of the effect on the beholder of the tyranny of the eyes. It is possible for a gynemimetic, that is a male who lives the role of a lady with a penis, to give such a virtuoso performance as an impersonator of a prostitute that male sexual partners do not discover the impersonation.

In Case #2, it is possible that the patient could have become proficient in the cosmetic techniques of theatrical makeup used by professional impersonators, applying them to his own advantage. When this strategy was brought to his attention, it proved to be impossible to implement, insofar as, like others similar to himself, he lacked the flamboyant confidence and theatricality that would be required to put it into practice. Consequently, he was condemned to be a victim of the tyranny of the eyes of those many people who, until they were correctly informed, misconstrued him as being a youth in the early stages of puberty, or in some instances as a woman, perhaps a lesbian, trying to pass as a male.

To deal with this predicament, there was only one strategy that he could implement, and that was to become a recluse. Opting out as a recluse is to be a victim rather than a victor, and runs counter to the cultural stereotype of masculinity as being, if not aggressively dominant, then at least unsubmissively self-assertive.

According to the cultural stereotypes of masculinity and femininity, sexuoeroticism is more dominantly assertive in males than females, and more compliantly submissive in females than males. On this criterion, the sexuoerotic component of masculinity in Case #2 did not measure up to the cultural expectancies of masculinity. As a male, the patient gave the impression of being sexuoerotically too inert behaviorally, and too apathetic subjectively. There are some who would say that inertia and apathy are synonymous with lack of sexual desire. However, just as there is no way of measuring sexual drive other than by its behavioral manifestations, so also there is no way of measuring sexual desire other than by its manifest expressions.

To the extent that sexual drive and desire appeared deficient in Case #2, the poverty of their manifest expression may well have signified not that they were missing, but thwarted. This interpretation is supported by the data of Case #1, in which there was ample expression of sexual drive and desire. It became manifest in teenage by the yearnings of sexuoerotic attraction and the attachment of falling in love (an experience missing in Case #2), and by the counterbalancing moral yearnings of remaining a virgin until marriage. In marriage sexuoerotic satisfaction, including orgasm, was taken for granted as a birthright, even in the face of situational impediments to the husband's copulatory competence for which the wife insisted on sexological couple therapy. The successful outcome of therapy was based on the hypothesis that the inadequacy of coitus for fertility, not for pleasure, was a key to their copulatory dysfunction.

In Case #2, it was not copulatory dysfunction that was an issue, but the total copulatory disability of the defective genitalia. The genital defect was so stigmatizing that it alone may have been held accountable for inhibiting any form of sexual expression with a partner, either heterosexually or homosexually. However, there is an alternative hypothesis, namely that the overall demasculinization of androgen insensitivity in the androgen-insensitivity syndrome includes demasculinization of the sexual brain.

There is some support for this hypothesis in the data from other types of penile insufficiency, for example, other syndromes of hermaphroditism characterized by penile insufficiency; syndromes of congenital micropenis; traumatic ablation of the penis; and the syndrome of female-to-male transexualism. Among affected individuals with these various syndromes, those who manifestly express sexuoerotic drive and desire are more prevalent than those who do not. For a higher prevalence of manifest deficiency of expressed sexuoerotic drive and desire, one turns to other syndromes in which, prior to hormonal substitution treatment, there is a sex-hormonal deficiency. Examples are: in women, Turner's syndrome; in both men and women, hypopituitarism (with or without growth hormone deficiency and statural dwarfism); and in men, a testosterone-resistant subset of Klinefelter's (47,XXY) syndrome.

Considering the evidence of these various syndromes, one may entertain the hypothesis that in fetal and neonatal life, the sexual brain in the androgen-insensitivity syndrome is demasculinized. Subsequently, in adolescence and maturity, the cells of the sexual brain, though insensitive to androgen, are sensitive to estrogen, which is secreted by the testes, if they have not been removed, and to exogenously administered estrogen, if they have been removed. With its sexual pathways hormonally activated, the brain will readily allow the expression of demasculinized sexuoeroticism if there is congruence between its own prenatal demasculinization, the prenatally demasculinized (i.e., feminized) anatomy of the copulatory organs, and the postnatal feminization of the cerebral cortex secondary to assigment and rearing as a girl.

By contrast, if there is incompatibility between prenatal demasculinization of the brain and genitalia, on the one hand, and postnatal masculinization by rearing, on the other, then sexuoerotic expression in maturity may be thwarted: prenatal demasculinization and postnatal masculinization may cancel one another. Such cancellation is not inevitable, witness in Case #2 the patient's own cousin who had the AIS condition and who was able to undergo, in young adulthood, a male-to-female sex reassignment, thus reverting to the sex in which she had originally been declared as an infant. By contrast, there are some cases of androgen insensitivity (unpublished; see also Ch. 5), with more neonatal visibility of genital ambiguity than in Case #2; and with male assignment and rearing, the individual is able to sustain a compromised sex life as a man.

The complexity of the brain in relationship to sexuoeroticism experienced as drive or desire is demonstrated in patients with spinal cord injury, paraplegia, which results in complete severance of round-trip communication between the brain and the genitalia. Whereas paraplegic men and women do not forget about their prior sexuoeroticism, and do not experience it again in the same way, they are not harassed by it in the same way as are patients whose genitalia have been traumatically mutilated, but not disconnected from the brain. On the basis of this information, it would be wrong to assume that, in Case #2, there was an absence or even a diminution of sexuoeroticism.

The manifest evidence of sexuoeroticism in Case #1 poses a challenge to psychohormonal theory. It is known from experimental animal studies (Clarke, 1977) that estrogen will activate not only feminine mating behavior in a normal female, but also masculine mating behavior in a prenatally masculinized female. It is also known from clinical evidence in human beings that androgen is the sexuoerotic or libido hormone in both sexes (Money, 1961). Conversely, estrogen acts as an antiandrogen in men, and was formerly used as an adjunct in the treatment of sex offenders. Because it induces breast growth, it has now been replaced by synthetic progestin which also is antiandrogenic, but not feminizing (Money, 1987). The unsolved challenge of Case #1, and of all other similar cases, is to be able to explain their evident sexuoeroticism despite the inability of the cells of the body to respond to androgen.

In their divergent ways, both of the cases in this chapter point to the need for a specialty of sexology in medicine. In Case #2, the professional who enforced the change of sex at age 18 months was relying on folk wisdom derived from barnyard castration in the era before written history, namely that testicles cause masculine behavior. With the knowledge of hindsight, and with the availability of a consultant fully trained in the specialty of medical sexology, he would have managed the case differently. Still today, there is no specialty training in medical sexology. The prospects for training in pediatric and adolescent (ephebiatric) sexology are getting worse, not better. Thus, mistakes like those of Case #2 have still not been eradicated.

One of the functions of a sexological specialty in medicine would be public sexological enlightenment. The shame and secrecy that today obfuscate the management and rearing of children with sexual problems, including birth defects of the sex organs, might then be lessened. Children would be emancipated to reveal themselves more accurately and honestly, with fewer fears of reprisal. Entrapment in no-win situations would be diminished, and the desperation of suicide would not be the only alternative.

The principle of suicide exemplified in Case #2 is the principle of 'if you can't beat the enemy, join them.' The enemy are those whom you experience as designing your life according to their own specifications. The life you live is not your own, but theirs. They are annihilating you, and you can find no way of escape. The person whom you should be, according to the dictates of your own criteria, will not be able to exist. So you join the enemy and thwart the successful completion of their annihilation of you by jumping the gun. Without warning and without consultation, you annihilate yourself.

BIBLIOGRAPHY

Amrhein, J.A., Meyer III, W.J., Jones, H.W. Jr. and Migeon, C.J. (1976) Androgen insensitivity in man: Evidence for genetic heterogeneity. *Proc. Natl. Acad. Sci. U.S.A.*, 73: 891–894.

Brown, T.R., Maes, M., Rothwell, S.W. and Migeon, C.J. (1982) Human complete androgen insensitivity with normal dihydrotestosterone receptor binding capacity in cultured genital skin fibroblasts: Evidence for a qualitative abnormality of the receptor. *J. Clin. Endocrinol. Metab.*, 55:61–69.

Clarke, I.J. (1977) The sexual behavior of prenatally androgenized ewes observed in the field. *J. Reprod. Fertil.*, 49: 311–315.

Crawford, J.D. (1970) Syndromes of testicular feminization. *Clin. Pediatr.*, 9:165–170.

Fordney, D. (1978) Dyspareunia and vaginismus. *Clin. Obstet. Gynecol.*, 21:205–221.

Frank, R.T. (1938) The formation of an artificial vagina without operation. *Am. J. Obstet. Gynecol.*, 35:1053–1055.

Manuel, M., Katayama, K.P. and Jones, H.W. Jr. (1976) The age of occurrence of gonadal tumors in intersex patients with a Y chromosome. *Am. J. Obstet. Gynecol.*, 124:293–300.

Money, J. (1961) Components of eroticism in man. I: The hormones in relation to sexual morphology and sexual desire. *J. Nerv. Ment. Dis.*, 132:239–248.

Money, J. (1986) Longitudinal studies in clinical psychoendocrinology: Methodology. *J. Dev. Behav. Pediatr.*, 7:31–34.

Money, J. (1987) Treatment guidelines: Antiandrogen and counseling of paraphilic sex offenders. *J. Sex Marital Ther.*, 13:219–233.

Money, J. and Onguro, C. (1974) Behavioral sexology: Ten cases of genetic male intersexuality with impaired prenatal and pubertal androgenization. *Arch. Sex. Behav.*, 3: 181–205.

Morris, J.McL. and Mahesh, V.B. (1963) Further observations on the syndrome 'testicular feminization'. *Am. J. Obstet. Gynecol.*, 87: 731–745.

Wilkins, L. (1965) *The Diagnosis and Treatment of Endocrine Disorders in Childhood and Adolescence,* 3rd Edn. Springfield, Charles C Thomas, p. 321.

CHAPTER 3

Matched pair concordant for vaginal atresia, assigned sex and feminine lovemap status, but discordant for chromosomal and gonadal sex: AIS and MRK syndromes compared

SYNOPSIS

The androgen-insensitivity syndrome, also known as the testicular feminizing syndrome of male hermaphroditism, is featured again in this chapter as it was in the preceding chapter. The matching here, however, is not on the basis of concordance of diagnosis and discordance of sex of assignment and rearing, but the other way around. The two cases are matched concordantly for assignment and rearing as girls, but discordantly for diagnosis. The diagnosis in Case #1 is androgen-insensitivity syndrome (AIS), and in Case #2 Meyer-Rokitansky-Kuester syndrome (MRKS).

Concordance of matching for assignment and rearing in both cases is based on the fact that each was born with female external genital anatomy. Although it was not recognized until teenage, the two were concordant also for the defect of vaginal atresia, that is to say a vagina that ended as a blind pouch and did not connect with a cervix. They were concordant also for having a malformed uterus. In Case #1, the uterus was a cordlike structure with no interior cavity, as is typical in AIS. Though the same cordlike structure is typical in MRK syndrome also, in Case #2 it proved to be atypical and to have enough of an interior cavity to permit an unexpected, late onset of menstruation, with accumulation of menstrual blood that had no outlet.

The variables on which the two cases were discordant originated in chromosomal genetics. The definitive chromosomal sex for AIS is 46,XY, and for MRKS it is 46,XX. Thus the two cases were sex-chromosomally discordant from the moment of conception onward. Only the Y chromosome carries the gene for the testicular determining factor (TDF) which signifies that the gonads will differentiate as testes. Thus, whereas the gonads in Case #2 were ovaries, in Case #1 they were testes. Very early in prenatal life, these testes secreted their mullerian inhibiting hormone (MIH) which prevented the development of internal female organs, as it does in all cases of AIS.

The factor responsible for incorrect formation of the internal female organs in MRKS cases remains still unknown.

As well as secreting MIH, the testes in Case #1 secreted their other hormone, testosterone, the androgenic or masculinizing hormone. In the AIS, by definition, the fetal cells that should form the male reproductive organs are insensitive to androgen and fail to do so. Thus, in Case #1, the homologous organs of the female, the clitoris and labia, formed in the place of the male penis and scrotum. At birth, Case #1 and Case #2 were concordant for genital appearance, and so both were declared to be girls and were reared as girls.

At puberty both of the girls were concordant for hormonal feminization. In Case #1, the body rejected the masculinizing androgen secreted by the testes and accepted only the feminizing estrogen that, though normally secreted by the testes, is overpowered by androgen in normal males. In Case #2, there was no androgenizing hormone secreted by the ovaries. Consequently, the body feminized at puberty under the influence of ovarian estrogen, in the way it ordinarily does in girls.

Heuristically, these two cases demonstrate that the developmental differentiation of femininity of the gender-identity/role, overall, and of the lovemap, sexuoerotically, is not contingent on the chromosomal sex, nor on the gonadal sex, per se. The influence of the gonads is by way of the use that the body makes of the hormones they secrete, initially in prenatal life, and again at puberty.

CASE #1: 46,XY WOMAN, AIS SYNDROME

Diagnostic and clinical biography

When, at the age of 21, this patient had her first appointment in the Johns Hopkins Psychohormonal Research Unit (PRU), she did not know that information concerning her condition had been explicitly withheld from her by her family, so that she could remain, ostensibly, in the protective custody of ignorance. She also did not know of the chain of family events that led up to her appointment.

The precipitating event was that the patient's mother's younger sister had, two years earlier, delivered her third baby who, like the first, was born with inguinal hernias. The baby's chromosomal count was 46,XY. Reporting retrospectively, the mother said: 'The pediatrician, when he confronted me with the diagnosis, insisted that the baby be operated on immediately to have the gonads removed ... as they would grow as in any normal male. I told him that was absolutely untrue, because I have other family members, and I had them as an example, and I had done a little research on it. I more or less told him that he didn't know what he was talking about.'

One of the other family members was the mother's niece, the teenaged younger daughter (and sister of the patient herein reported) of one of her two sisters. This girl had recently been diagnosed as having the androgen-insensitivity syndrome. She was the first of the affected members of the family to have been so diagnosed. Her parents kept the diagnostic information secret until the mother, panicky on behalf of her sister and the new baby, disclosed her secret to her sister, whom she advised also to maintain utmost secrecy regarding her baby.

The family's prognostic wisdom was based on the fact that the two mothers with AIS children had a sister who had adopted two children, who had never menstruated, and who had been told that she had a malformed uterus. They had also two maternal great-aunts with the same history, one in her sixties, and one in her seventies.

The mother of the new baby was the first member of the family to contact the PRU. 'I'm going to college,' she said, 'and I was in a class of the sociology of the sex roles. We were discussing chromosomes. The professor had a copy of your book [Money and Ehrhardt, 1972]. It helped me a great deal to understand the problem a lot better.' It also prompted her to seek an appointment for herself and her husband. Her baby was 20 months old at the time.

One item on her agenda was to know how to help her older sister who had the two grown daughters with AIS. 'She's acting on an emotional level. She's not being rational about it, at all. When the doctors first suggested that everybody move, she was going to move. I mean, she was ready to pack up and make her husband quit his job, sell their house, everything, just move. ... I told her she had to just calm herself. She's absolutely dead set against telling her daughters anything. She just does not want to discuss it. She's afraid that doctors will exploit us – as we have been in the past. She's very distrustful of doctors, and she just cannot listen to reason.'

The older sister became able to listen to reason when the younger one returned home and disclosed the benefits to her and her husband of their discussions in the PRU at Johns Hopkins. The sequel was that the older sister made an appointment for herself and her husband, who is referred to herein as the father of the two AIS daughters, though he is legally the step-father. Both parents were married for the second time when the two girls were aged 4 and 5, respectively, long before their diagnosis was suspected. Subsequently, they had one more child, a normal boy.

The couple came alone. Their purpose was to ascertain whether their two daughters might safely learn something about themselves, diagnostically and prognostically, without being traumatized as the parents themselves had been two years earlier, when the younger of their two AIS daughters had a hernia operation.

The father succinctly summarized his own traumatization in their vernacular, saying that he 'shit a brick' when the surgeon bluntly told him and his wife that their younger daughter had testicles. 'The mention of testicles in a woman is almost hideous,' he said, and he reflected especially on how damaging it might be for the husbands of his wife's AIS sister and two great-aunts to know that their wives had testicles. His own wife had 'walked around in a daze for days,' after the surgeon who had operated on their younger daughter's hernia told the parents 'that he had done a biopsy, and they were both testicles. ... She'd get this pressure in her mind that one of the kids would find out; and she holds herself responsible for it, you know.

She wanted to move. But you could go to Alaska, but if one of the kids married and decided to go to her own gynecologist – there's no way you can control people's lives or destinies; only up to a point.' The effort at control had already been costly, as he had forfeited medical insurance benefits rather than risk a leakage of the diagnosis from the office where the claim would be processed at his place of work.

'I hate to use the expression,' the mother said, 'but a freak of nature is what I would call it. That's not good terminology ... but I don't know what I would tell them if they would persist too much, on the subject.' The explanations about XY women that she herself had received in the course of her PRU visit had brought her to the realization 'mainly, that I don't have a choice, and they really do have to know the truth, so that they can handle it in their future lives when I'm not around, with other doctors and hospital situations ... but emotionally I'm still very much against it. I'm fighting it, but it may be the only solution. ... I'm sure my husband is going to talk me into coming back with the two girls.'

Entrapped in their own anxieties, the parents had not put themselves proverbially in the shoes of either daughter so as to imagine what they might either know or conjecture about themselves, and to what extent they may have recognized and resented their parent's lack of candor in not speaking about the unspeakable. The older of the two daughters, the one who is the subject of this report, lived with the knowledge of a hernia repair on one side at age 12, and the other at age 16. On neither occasion had the surgeon mentioned male testicles. All that the father recalled was 'that he had replaced one of her ovaries that had descended – or something,' but he could not specify which operation. He did not know what the girl herself might have been told at either time. He did know that she had developed breasts – 'I mean really developed,' he noted. He believed that she had been told she could not conceive children, because she had the same condition as her younger sister, with undeveloped ovaries. She knew, of course, that she did not menstruate. 'As smart as she is,' her father said, 'she's even written term papers and stuff on chromosomes. Maybe she just simply never related it to herself.' At college she studied and did tutoring in marine biology, so that she had library access to reference sources on her chromosomal and diagnostic status. The mother, like the father, discounted the idea that their daughter might have read about her own syndrome.

It was a month later that the daughter, a 21 year old woman, had the opportunity to speak for herself. 'I came down,' she said, 'because my parents were concerned about our condition, and to get a better explanation of what it is all about. ... I knew that I hadn't gotten my periods, so I probably couldn't have children; and I knew it was in my family. I had heard my aunt talk about her condition, and I assumed it was the same as mine. Hers was that her uterus wasn't large enough.'

Initially she had been reluctant to accept the Johns Hopkins appointment because: 'I thought I was just going to a gynecologist. I thought it would be kind of futile to come all this way for that, because my sister had gone to one – they didn't really know much about it. If they did, they didn't explain themselves well. I've gotten a lot out of this visit, because I understand my condition now. It wasn't just like going to the doctor.' She summarized what for her were the main points of her new understanding: 'A vestigial uterus, and because of that I would be unable to become pregnant. It was caused by hormones; by an ovary not being, not forming eggs, which is caused genetically.' Then she added that she had learned how to explain her condition to various people – 'How to handle it with doctors, the guy I'm going to marry, or guys I might go out with. Just everyone, I guess.'

The information she subsumed under 'how to handle' was recapitulated at the end of the second day in the concluding conference interview in which this woman, her younger sister, and her parents were present, together with the PRU staff. What the four family members had heard individually was now spelled out for all four to hear and share mutually: The name of the condition is androgen-insensitivity syndrome (AIS). Genetically, it is chromosomally XY (illustrated in a picture of a chromosomal karyotype). Chromosomally there are a minority of XY girls and women as well as a majority who are XX. There are also XX boys and men, as well as many other chromosomal variations in both sexes, such as XXY and XYY men; XXX and XO women; XO/XY men and women, and many other mosaic types, not to mention broken and transposed chromosomal segments. For self-definition, an XY woman may characterize the Y chromosome as an X with its arms broken off, for it is incapacitated and unable to do Y work. Consequently it does all its work to make Eve, not Adam.

The same applies to the gonads, the sex glands, both in prenatal life and in adulthood. Under the microscope, they have the histological architecture of testes, except that they are incapable of making either sperms or eggs. In prenatal life they are able to secrete the hormone that tells the uterus not to develop. Though they secrete also the hormones of masculinization, these hormones are unable to do masculinizing work, for all the cells of the body, being more feminizing than the cells in normal females, refuse to be masculinized. Thus it is actually not scientifically precise to call these sex glands testicles, because they cannot do the testicular work of either fertility or masculinization. However, in a patient's medical record, they almost always are still referred to as testicles, so that a patient who reads or hears this term must make an automatic correction, and call them gonads, or sex glands, which are neither perfect ovaries nor perfect testicles.

Because they are imperfect, and even though they carry the full work load of feminizing the body at puberty, the gonads have an increased risk of developing cancer in adult life. Therefore, it is widely recommended that all patients consider the wisdom of having the gonads surgically removed. The experts are not all in agreement, however, as to when, or even if the gonads should be removed. The operation should never be enforced, and never done without the patient's intellectually informed consent.

In the femininely perfect example of the AIS, the external sex organs are entirely female. The vaginal cavity, however, may be too short, in which case it can be lengthened by self-dilation, progressively by gentle copulation, or, if necessary, by plastic surgery.

Girls with AIS grow up to undergo normal romantic and erotic development in teenage, and to have a normal sex life, independently of infertility. Their motherhood is by adoption or step-parenting.*

After evaluation in the PRU, a patient may maintain contact by mail or telephone, as well as by appointment in person. No comment or query is too trivial. A patient may also request a double appointment, one for the self, and the other for a lover or marital partner.

In the course of the two-day psychohormonal evaluation, the patient was seen also for a gynecological evaluation. The report said that the breasts were well developed. There was no axillary or pubic hair (which is typical in AIS), and no facial or body hair. In each groin there was a scar from a prior hernia repair. The right gonad, but not the left, was palpable by rectal examination. The external genitalia were normally female. The vaginal cavity was 7 cm in depth, and ended blindly without a cervix. A blood sample was drawn for cytogenetic study by fluorescent banding which gave a chromosome count of 46,XY, thus confirming the diagnosis of AIS. The physique and appearance were of average proportions for an attractive-looking female. The voice was feminine.

The patient elected to postpone a decision regarding gonadectomy, despite the urgings of the gynecologist to the contrary; and she continued to do so at the time of follow-up at age 24. Her decision was influenced in part by the continued good health into their seventies of her maternal great-grandmother's two sisters with the androgen-insensitivity syndrome.

She summed up her two-day, Johns Hopkins experience in its entirety: 'I'm glad I came. I can understand it better now. I came across things like this before, and I knew that I fit some place within the terminology. I was glad to get a fuller explanation from people who work with it. And I think it was good for my mother, especially.'

Gender-coded social biography

The patient was described by her father as always having been outgoing, conspicuously more so than her sister who was more the quiet type. There was no history of tomboyism. In high school, 'she was a cheerleader, always boyfriends and stuff,' the father said.

She did not encounter any difficulties with agemates in connection with not getting

*Today it is known that it is physiologically and technologically feasible to have an ectopic pregnancy, in a mother who has no uterus, carried to term, and viably delivered by laparotomy (Jackson et al., 1980; Teresi and McAuliffe, 1985; Barsky, 1986; Norén and Lindblom, 1986).

her periods. Her friends knew she did not menstruate. She didn't feel she was missing anything, but rather that she had it good. She and her younger sister both knew that they didn't menstruate: 'When we were young, we were close and would play together. We both kind of accepted it, so we never bothered to mention it.'

She summarized her dating history: 'I had about three or four first dates in high school; and then I dated one guy particularly, and we used to go out with a bunch of people. You know, it was a large date. And then I dated more different people when I was a freshman in college. Now, since the end of my sophomore year, I'm dating only one guy' – the man whom, in four more years, she would marry.

Her college interests included acting, tennis, and skiing. She rated her physical energy level as medium, with the qualification that, when awake, she was always doing something. Academically she did very well.

She projected for herself a long-range professional career, along with being married. At the time of her last follow-up at age 24, she had been married for two months. She and her husband had begun dating six years earlier and for several years had lived conjointly. They had married because 'it was time to decide to make a commitment,' and because it was easier to be accepted socially as they advanced their respective academic careers. She had graduated with a master's degree in social work and planned on pursuing a doctorate. He had a junior academic appointment.

Domestically, she expressed a strong interest, stronger than that of her husband, she believed, in parenting: 'That's something I would want to happen,' she said, 'to complete my life.' Her opinion about domesticity was: 'I like a clean house, but I hate to clean it. I'd like to know more about how to cook.'

To a query on clothing style, she replied: 'Jeans, they're my first preference; and then slacks, pantsuits, things like that.' She liked dresses, but found them impractical in general. She was interested in clothes. She used makeup on a daily basis, and wore jewelry and perfume. Her appearance was stylishly fashionable, and femininely good-looking. She rated her own looks as 'average college student-type appearance,' with which she was not dissatisfied.

She considered herself independent, though she cared about what other people think, and could be influenced by them. She specified the advantages of being a woman as: 'Feelings. I look at males and they seem to have been deprived.' The disadvantages: 'Females are not brought up to be assertive enough. I am in the process of trying to work on this.'

Correspondingly, she specified the advantages of being male: 'Money. They can get more of it, easier. And they think more spatially.' The disadvantages: 'Males feel too much responsibility to meet people's expectations.'

Lovemap biography

At the time of her first PRU interview, and while still at college, this girl had begun her sex life. Her first and only sexual partner was her permanent boyfriend, with whom she would live together for a time and then marry after they had known each other for six years. She had told him about not being able to have her own pregnancies, and about the familial incidence of her condition – information which she had also confided to her closest roommates at college.

Sexual intercourse was self-reported as having been 'a little painful at first, but nothing to worry about.' She was affirmative about having orgasms, which she described thus: 'I guess I would say it's like warm little beads going through your insides, and up inside your body, starting from the toes.' After intercourse she 'felt relaxed and satisfied with good feelings' about herself and her partner.

Regarding premarital sex in youth, the father summed up his position: 'Well, it's not right. But then again, what's – you know, life is short. [He refers to when the First Lady, Betty Ford, was confronted with a diagnosis of breast cancer.] If someone wants to experiment with sex, or marijuana, or something, you know, why not? You know, life is short.' Expanding further, he said: 'If I walked in and caught them in my bed, or on my couch, I'd – what the hell? That's going too far. But, I mean, she's an adult. She's over eighteen now. She's a woman. She's an adult. Who am I to tell her what to do?' He believed that he and his wife thought the same way, and that she was relieved to find out that her daughter could have intercourse – 'Naturally she asked her. I mean, hell, you know.'

Nothing either positive or negative was retrieved regarding childhood sexual play, sexual learning, or sex education. Sex education at home had been conventionally parsimonious. The history of sexual self-stimulation by hand, or any other way, was self-reported to be as infrequent as once a year, and never to orgasm. There was no history of sexual attraction towards, nor of sexual experience with other females.

Responding to a PRU checklist at age 21, the young woman checked yes to daydreams and fantasies in the following categories: romantic, erotic, wedding, pregnancy, motherhood, infant care, and career, with nothing added in the category of other. In sleep dreams, the positive check marks were for romantic and mothering dreams, and nightmares: and the negative check marks were for erotic, wedding, and pregnancy dreams, and others.

Referring to a checklist of the sensory channels of sexual arousal, she summarized as follows: 'Touch, yes. Smells, no. Taste, no. Reading material, no. Pictures, no. Music, selectively. Seeing the one you love, yes; unclothed more than clothed.'

The erotically sensitive parts of her body she listed as legs, back, abdomen, neck, ears, breasts, clitoris, and vagina.

She rated herself as an affectionate type of person, and as being above average in sex drive – 'Oh, five or six times a week; with a lot of foreplay; and some variation to forestall boredom.' She considered foreplay essential.

Giving an update on erotic responsiveness when she returned three years later, at age 24, she said she was turned on by 'either oral or manual stimulation before intercourse – anything, at the moment, that I like. I like to feel my husband's skin, the smell of his skin, and things like that.'

Everything considered, sexually and otherwise, she had a good sense of well-being, maritally, and in general.

CASE #2: 46,XX WOMAN, MRK SYNDROME

Diagnostic and clinical biography

The first contact that this patient had with Johns Hopkins was a phone call made to the Psychohormonal Research Unit by her father on the recommendation of his surgeon. The surgeon had recognized not only the condition of atresia of the vagina, but also that the girl would benefit from a referral where psychosexological evaluation could be combined with the establishment of a differential diagnosis, an endocrine evaluation, and a program of surgical treatment.

The history antecedent to this referral was that, since the age of 11 and probably at a much younger age, the girl herself had known that the depth of her vagina was deficient. She had not said anything about it until, at the age of 16:6, she told her mother that she would need to see a gynecologist. She knew that late onset of menstruation was not viewed with alarm in the family, insofar as her father's mother had not menstruated until age 17. Even though her three sisters, then aged 20, 14 and 13, had all begun to menstruate at age 13, her own primary concern was not menstruation, but the fact that, eight months earlier, intromission had proven absolutely impossible when she had resolutely tried to have sexual intercourse.

An appointment with the family physician yielded the advice that there was nothing wrong and nothing to worry about. Dissatisfied, the mother took her daughter to a gynecologist whose examination demonstrated the depth of the vagina to be only two inches. She advised an endocrine evaluation which was by-passed in favor of a surgical evaluation. Because of the possibility of a misplaced bladder in the midline, the surgeon recommended an exploratory laparotomy. The father sought a second surgical opinion, the upshot of which was the referral to Johns Hopkins. At this time the patient's age was 17:2.

After the gynecological examination at Johns Hopkins, the diagnosis was given as a 'classic example of the Rokitansky syndrome' of vaginal atresia, amenorrhea, and agenesis of the uterus in association with normal ovaries, normal female hormonal function, and a normal 46,XX chromosomal karyotype. The developmental history included onset of puberty at age 11, normal-sized breasts slightly larger than those of her two sisters, and no menstrual cramps as well as no menses. Pubic and axillary hair were normal. The height at age 17:2 was 155 cm (5 ft. 1 in.) and weight 50 kg (110 lb). Although no uterus had been palpated rectally, at the age of 17:5, in the course of surgical vaginoplasty 'a soft, cystic enlargement, probably one of the ovaries, about 5 cm in diameter, was felt in the midline.' It was elected not to take action other than to keep track of this soft enlargement, in the expectation that it would, if a functional cyst of the ovary, be self-reducing.

The vaginoplasty was done according to the standard McIndoe procedure, with a split-thickness skin graft, taken from the right buttock, used to line the extension of the cavity of the vagina. When the form that held the grafted skin in place was removed after eight days, the graft was seen to have taken quite well.

Nineteen days later there was an unexpected postsurgical complication in the form of a fluid-filled cavity bulging into the anterior vaginal wall. The fluid proved to be blood and mucus. It was released, and a surgical drain was put in place so that any further discharge could leak out. A foul-smelling discharge failed to clear up until, nine months later, an exploratory laparotomy revealed the source of the problem to be the soft mass which was formerly thought to be a cystic ovary, but which was actually a small uterus. It had no vaginal outlet. By a coincidence of timing, the first menstrual period had commenced shortly after the operation for vaginoplasty. The necessary surgical correction was made, so that the menstrual outflow could escape through the surgically constructed vagina. After healing was complete, there was no further recurrence of symptoms. The vagina was, according to a gynecological report, of very adequate dimensions, with a length of 8 cm. In the ensuing years, menstruation was 'pretty normal' but at irregular intervals, sometimes two to three months apart.

Gender-coded social biography

On a written questionnaire at age 19, the patient rated her relationship with her parents in childhood as good, in teenage as bad with many fights, and in young adulthood as much better. She experienced early teenage as a period in which there was a discrepancy between her value of independence and her parent's value of protectiveness, but she did not consider that her parents were intentionally unfair. Of her father she said: 'I know he likes me a lot. As a matter of fact, I've been, like, his favorite. And so we fight – more than any of my sisters fight with him.' On the Sacks Sentence Completion Test she wrote: 'I feel that my father seldom understands my way of thinking. My mother and I have different point of view.' Counterbalancing these criticisms, she wrote, 'Compared with most families, mine is bearable.' The latter appeared to be a realistic appraisal, for her family of three sisters and two parents did seem, by and large, to accommodate to one another without acrimonious excess.

If left to her own devices, she was the go-it-alone kind of teenager who could meet adversity by dismissing it with a shrug of 'so what?' and a blend of resilience, daring, and maybe a bit of defiance in making the best of what can't be changed. 'When the odds are against me,' she wrote on the Sacks, 'I never give up.' And, 'When luck turns against me, I don't let it get to me.' These were not idle boasts. They accurately represented the way in which she coped with finding a solution to her sexual anomaly. The sexual mores of her time required that as a sexually active teenager, she did not disclose to her parents the intimate history of how she had discovered that she had an imperfect vagina which would not function in sexual intercourse. 'She is quiet about her private life,' her father said, 'and makes her own decisions.' She did not establish the logic of a connection between the teenage strife that existed between her and her parents and the self-imposed isolation that the very nature of her inability to copulate as a teenager imposed on her. 'I was basically a loner,' she wrote, 'but socially outgoing, also.' There was only one friend, a girl, whom she rated as a best friend. They had known each other since age 3, 'We understand each other pretty well,' she said, 'and are open with each other.' At high school, her friendships were mostly casual, and with boys. She did not get seriously involved with a male friend until age 18. With him, she was 'very close and had a very good sex life.'

When there is a discontinuity of philosophy and understanding between the members of two generations (or, alternatively, of two cultures), one of the contingencies of the formulation of a new code of values and a new meaning of existence may be the visionary experience of the pharmacologically altered state of consciousness. This concept may well have applied to the high-school generation of the affluent middle-class culture of which the present patient was a member. In her case, the discontinuity of values between her and her parents pertained specifically to sexuality and eroticism. She did, in fact, define new perspectives for herself, while high on one or another of the drugs that were in fashion among the high school crowd of the era. She did not have money to buy her own. When friends offered them, she tried them, but 'only when I'm in a good partying mood, and only at night.' She did not become drug dependent. Her educational achievement did not deteriorate.

Her academic level and her IQ were consistent. On the WAIS (Wechsler Adult Intelligence Scale) administered at age 17:2, the scores were : Verbal IQ, 124; Performance IQ, 108; and Full IQ, 118. The Verbal-Performance discrepancy was attributable to specific superiority of verbal reasoning (Specific Factor IQ equivalent, 141), not to specific inferiority of either praxic or numerical reasoning, both of which were average.

As a senior in high school, she had an ambition to teach karate and to pursue her early childhood affinity for animals by becoming an exercise girl for horses. She took karate lessons, and enrolled in a vocational school class on race horses. Her recreations included swimming, motorcycling, camping, boating, and waterskiing. Though she wore blue jeans predominantly, and always while working with horses, she had a more varied wardrobe to select from, as the occasion demanded.

'I definitely get along with guys a lot better than I get along with girls,' she said in one interview, 'but I always thought that sometimes a girl and a guy can be more alike than two girls, so I was really glad when I found in my reading that someone else thought the same. When I – I mean there's no way I want to be a guy, and I definitely know I'm a girl. But I definitely get along better with guys.' She found other girls' psychosexual and psychosocial level of development to be younger than her own, especially when she was a college freshman.

After graduating from high school at age 18:5, she enrolled at the local campus of the state university, taking courses in chemistry, physics, history, and contemporary morality. Predictably, on the criterion of the IQ differential, she did poorly on chemistry and physics, and so changed her major from science to humanities. The most prevalent grade was B.

During the academic week, she lived in the college women's dormitory, and went home on weekends. She did volunteer work at the student gynecology clinic, counseling women on the use of birth control. She had never before realized 'so many people are up-tight about sex.' On weekends she went home where, on Saturdays, she was employed in the local pet clinic, in charge of the exotic animal room.

Her future plans at age 17:5 included marriage, but not at the expense of forfeiting a professional career. About housekeeping, her comment was, 'I'm pretty lazy, but clean,' and about motherhood and childcare, 'not too swift.' She characterized marriage as 'an extremely close relationship where you don't have to be constantly with each other. You don't have to push each other, and you do separate things, but you have a good sex life together. And there may be things like major goals and everything, the same, so that you could strive together for them. And also you can't just be lovers. You also have to be good friends – be able to talk together, be best friends.'

Pondering the issue of the possibility of future infertility, at the age of 18:3, she said: 'A lot of people, little girls, are trained that they're going to be a mother and have children. They play house all the time. And that's what their life is all about. I just never thought that way. I never played too much of those games and anything like that. I was usually off to myself, anyway, so I guess I never received that type of training.'

There may be here represented the coping strategy of finding the redeeming feature. There had been another, though different manifestation of the same strategy a year earlier in which the black cloud, as then declared, was infertility contingent on absence of a uterus, and an imperfect vagina that was not healing well, postsurgically. The silver lining was the feminine perfection of the breasts and body contours.

'Do you remember,' she wrote, in a letter asking for advice, 'when I told you that I felt really strong about females not being able to go topless? Well, I started going topless on the beach – not too close to other people, and not acting lewd or anything. At first it was just me, then my girl friend, then a few other friends. We met a few other guys without suits and got into some really good conversations. ... Further down there's a gay beach where there are more people with my ideas. ... Some people gawked. Some thought it was disgusting. Most thought it was cool. Some even came up and complimented us on our guts. I felt really good and comfortable. But yesterday the cops got us, me and Pam. 'It's just a $5 or $10 summons; just go and pay it, and your parents won't know about it,' said the sweet pigs. I'm so fuckin' pissed. I feel like I've been arrested for picking my nose. ... I feel like telling them I am genetically XY and I got too much female hormone in puberty, and so I developed breasts. I have to go to court. The easiest thing to do would be just to pay them. But I would feel like I was giving money to the "Be ashamed of our bodies and go back to Victorian times league". ... I think if someone took it to the Supreme Court, they might make it legal.'

Eventually when she told her parents: 'My father kept saying, "You're not as smart as I thought you were." He always says that. My mother kept saying things like, "Why should other people have to be subjected to your body? Is it that perfect? Why should they have to look at you?" They both convinced me I am an exhibitionist! ... They kept relating everything to sex which really has little to do with it. I can't tell you how frustrated I got because they would not listen.'

She and Pam succeeded in retaining an ACLU (American Civil Liberties Union) attorney to make a test case of their summons. Afraid of publicity, her friend dropped out, leaving her to take time off from school only to find that the court date had been put forward several times. In the end, the judge dismissed the charge, and she let it go at that.

After the postsurgical vaginal complications were cleared up, there was a period of 10 years of no further psychohormonal contact until a long-distance audiotaped interview by telephone. By then she had been married for several years to a fellow college student of her own age. They had lived together prior to marriage. He was employed as a metropolitan police officer. She was in her third year of radiology school, maintaining a straight-A average. She financed her education and simultaneously sustained her lifelong close relationship with pets by working as a veterinary technician and animal groomer, and by breeding exotic pets to sell.

At the time of marriage, she had wanted no pregnancies, and he had agreed, even though 'he is a lot more family-minded than I am. He really loves kids. I'm not really a kid lover.' She had subsequently required emergency treatment for what proved to be an ectopic, unplanned pregnancy.* That left her with only one fallopian tube which itself was apparently blocked. When she and her husband had changed their minds in favor of pregnancy, it had proved to be necessary to induce ovulation. There were repeated attempts at in vitro fertilization, but no pregnancy. After graduation from radiology school, she planned another attempt at in vitro fertilization. If it failed, she did not favor adoption. Rather she would pursue an advanced degree.

Lovemap biography

As a young adult of 17:6, this patient reconstructed the kind of sex life she had as a child between the ages of 5 and 11. 'When I was really young, we'd have the typical "I'll look at yours, and you look at mine" type of thing. Me and my friend would get into playing veterinarian a lot, and we'd always be having kittens. ... We even got to the point where we'd actually mount each other like an animal would – not

*See footnote on p. 62.

undressed or anything. We'd try to do it as close as we could as animals would be doing it. We never tried it as people.' She estimated her age as 8 or 9 when she first learned that human beings have sexual intercourse. At that age the idea of 'having intercourse scared me, but I was really curious and everything.'

As a 17 year old with a serious interest in the science of ethology, she had a history of always having had a deep attachment to animals. 'As a child I was always pretending to be an animal,' she said. 'Even now, with wild animals ... I'd make them feel that I'm part of them, instead of part of the human world, and if other human beings were to come by, I'd think of them as being intruders to us.' She attributed her sexual philosophy to her attitude towards animals. 'I relate everything to animals, and it seems to me that just about anything that's open and free and natural with animals should be the same with people.' She considered that more of her sex education had derived from animal behavior than from printed or spoken information. However, with the advent of puberty at around age 11, there had been some quasi-furtive reading of erotic stories. She recalled that she and her closest girl friend would 'each have a book to read and be really getting all hot over the book, and somebody would come downstairs. We'd throw it behind the couch and whip out National Geographics.' Her parents' policy had been to answer questions, but not to volunteer sexual information. When she was 13, her mother provided her with a book about menstruation, but did not discuss it.

At around this same age she went through what she called 'seventh grade bullshit which I never believed in in the first place. ... You go into the woods with anyone you can, and they feel you up and they feel you down. You don't really enjoy it, but you do it just to find out what it is. ... And he was so glad to go and tell his friend. 'Wow, I did it,' you know. I guess the guys got something out of it, but I never did.' As she got older, she hated 'anything that's going to cause a rush job – definitely not. If it comes to back seats of cars and junk like that, I hate it. Absolutely not.'

While other 7th graders were perfecting the dimensions of their newly adolescent sexuality, this girl already knew, no matter how inchoately, that something about the depth of her vagina was imperfect. Five years later, when she was a senior in high school, it continued to be a virtually unspeakable monstrosity in her life, even when she gave an otherwise remarkably open-minded sex history in the course of her initial psychohormonal evaluation. The monstrosity was declared not in an interview, but in a letter written after she had returned home from the second surgical admission, the one necessitated by postsurgical complications following the first admission.

'Before I had this operation,' she wrote, 'I did a lot of suffering. I can easily remember how I felt. I knew there was something wrong, but I hoped there wasn't. I was always trying to find my damn hole, first when I was eleven I tried so often to insert a tampax, then trying to get my fingers in, and then in final desperation (when I had first met Dan and knew I'd try balling him soon) I spent some horrible times clawing at myself until I was bleeding, hoping I was ripping a tough hymen, and clawing even more when I could still not get in. It was really painful, but I was so very badly determined. And there was no one to talk to about it. Later on I used to dream and fantasize about balling, and it would be so great, and I'd have nothing to worry about, because I had a nice big juicy hole!'

Her letter went on to express her distress and to give a graphic description of the unremitting, foul-smelling, greenish discharge leaking from where a surgical drain was in place inside the vagina. 'But in spite of it,' she continued, 'I have been even happier so far than I knew I would be since I have acquired a hole. I thought I was happy with my life and living it good before, but now it is twice as intense mentally and physically. I get delight out of such little things now, and my senses seem to be working so intensely, I seem to be so much more aware of them, especially smell and touch. Even just to pet a cat, or to come out of the ocean soaking wet and then lay in the sun, I can feel it warm me and I love the goose bumps when a soft, warm breeze passes over my still wet body, hitting the body hairs, making them rise. It is all so intense I can't possibly describe it all. I can't wait to make love, and I hope nothing holds me back.'

The postsurgical discharge did hold her back for another eight months, at which time she was readmitted for a third surgical procedure, the exploratory operation that revealed that her internal reproductive organs did indeed include a small uterus without a vaginal outlet. Then, in the recovery room she had a scary, déjà vu experience in which what actually happened seemed to be a fantasy come to life: 'I had doctors holding my legs and hands. Then I grabbed one of the doctor's hands and tried to pull him away. Meanwhile, they were pulling some kind of packing out and putting stitches in, and putting a clamp in, and I could feel that, it was really super painful. ... When I think about it, it seems like a true fantasy.' Subsequently she asked a doctor what had happened and he told her: 'Oh, you were bleeding a little bit, and we had to put a clamp in and put some stitches in.'

This recovery-room episode evoked a very early memory: 'I remember when I was about five years old, my friend told me that her sister had a hernia or something like that, an operation, and they had to cut her vagina with a knife, or something like that. She didn't really tell me the details. But I used to sit there and have her tell it to me over and over again. I'd get really horny over it,' she laughed, 'just for her to tell me about that. I guess she did too, because she kept telling me, you know. ... I think I base my fantasies of being held down on that story.'

The first disclosure of the fantasy of restraint and submission had been audiotaped nine months earlier. 'It was not necessarily sexual, but just like when they were trying to help me ... or perhaps to get me pregnant, and breed me for some kind of experiment. It was like you had found an animal and tried to hold it down to help it, you know. It wouldn't realize you were helping it. It would bite you. And that's exactly the thing I was relating to.' At a later date she added, 'I was not usually a human being in the fantasy, I was more an animal. It wasn't specific what type, it just wasn't a human type thing. And I didn't want it to happen, but I didn't have a choice, and a lot of times I was passive because I was drugged or something. So I didn't fight it that much. Usually I was pretty scared. They'd be doing all different kind of things, just you know basically doing different tests, just handling everything down there, you know touching everything, probing into everything.

'Just thinking about that would get me pretty horny, you know. I never have a fantasy to the point of orgasm unless I'm sleeping, but it's enough to get me really horny. And when I'm actually having sex – I have never really had a fantasy while I'm having sex. ... When I think of having sex, I think of it as something like I can really let myself go and do whatever I want. So usually it's sort of a crude, animalistic type thing where I just feel I'm having raw sex like an animal is having – unless it would be with somebody that I really care for.' At this time, at age 18:3, actually having sex was subject to the constraints of the adverse postsurgical complications.

In the fantasy of being forcibly constrained while struggling, she always saw herself not as an animal but a human. Whereas the fantasy by itself alone did not precipitate orgasm, the two coincided for the first time early in adolescence. 'I guess I was about eleven or so at the time,' she said. 'I was getting in bed. I don't, you know, let my dog sleep on my bed. But she jumped up on the bed, and I said, 'What the hell, the dog can stay.' Okay, so just as a natural instinct, I guess, she just started nosing around, and I was going to kick her away – you know, tell her no. But it felt pretty good, so I just let her go at it, you know. I can't remember if I had an orgasm from it the first time I let her. It was either the first time or the second. ... I wouldn't actually think of the dog as a person. I'd be thinking basically of being held down for my own good, and the more I struggled, the better it would be. ...

'When I did get an orgasm from it, it was good, but I had definitely felt it before, I don't know where, which is what really surprised me. So after a while, you know, I just kept it up every night. ... If she didn't know where to lick at the right time and the right spot, I'd put butter there and that did the trick. ... I was really nervous you know – like my mother coming in or something like that. But she'd just come around and sit by the bed and start begging, and I'd say okay, and she'd hop up. After a while, if anyone started walking by the door, the dog would jump down. ... After a while, since I knew that I couldn't have sexual intercourse, it seemed a lot easier with the dog, because there was no explaining to a dog.

'Then I just started masturbating (either in bed or with a jet of water under the shower) by myself, without her, because I'd begun to feel guilty about it. ...'

The guilt was kept concealed and private until, at the age of 16, she and Pam, her closest girl friend, while tripping one evening, began sharing sexual secrets. 'She wanted to tell me something she had wanted to get off her mind for a long time, which was that when she was younger she had, you know, first intercourse with her brother. It was kind of a bad experience because she felt so guilty about it. I told her I didn't think it was that bad, that weird, because in animals I know it's like a natural behavior. Animals do that, and people are just kept from doing it because it's against society. Anyway, you know, because of that I told her about the dog.

I didn't know what kind of opinion she'd have. It flipped me out, because she just stood there and started laughing. I can't believe it, you know, but she said, "I've done the same thing." And that made me really happy, you know.'

At around this same time, she found from her reading that her fantasy of being bound and tied down sexually was not too grossly atypical. She used Alex Comfort's book, The Joy of Sex, as a lead-in to talk about bondage fantasies with a friend: 'When he came to that part,' she recalled, 'he said, 'Have you ever done that?' and I said no, but I've thought about it, and he said well, he had, and it was really good and everything.'

It was also at around this time that she experienced in actuality not the bondage part of her fantasy, but the struggling and fighting part. The partner was a friend older than herself. 'I was in the apartment,' she said. 'I had waited for him. After he had kissed me, he went into the kitchen and came back, and all of a sudden he started beating me up, you know. And I went along with him. We got into fighting each other, and chasing each other around for, like, fifteen minutes. After that we started getting into, er, regular sex – oral sex, not intercourse, because I couldn't then. But we did have anal intercourse.' Her way of explaining her vaginal incapacity was to say that she had a constricted womb.

The next item on her sexual agenda was to see if intromission could possibly be achieved. She half knew that her partner in this enterprise would be a friend, 10 years older than herself, whose younger brother was the boyfriend of her closest friend, Pam. The four of them were together at his apartment. 'He asked me if I wanted to come with him to feed the doves. I said alright. I knew what was coming on. We both knew it was coming. ... I didn't know what to tell him, or anything. I figured I might as well take a crack at it. Maybe I just had a tough hymen. So the next thing you know, we didn't have too much on and we were laying in bed. And I said to him. "Dan, I'm a virgin," and he said, "I don't care." I knew it was going to be a tough time. After a lot of foreplay, I was pretty calm, but I was still nervous about finally getting it in, you know. And then we got into oral sex. It was really good, and I didn't expect it. Then we tried actual intercourse. And we tried and we tried. It hurt like hell. We kept trying. We stopped for a while and just relaxed, and drank some soda, and talked. I felt really bad, you know. Then after a while we tried it again and it still didn't work, but he came anyway. I didn't have an orgasm because I'd been really worried about it. As long as he came, and got satisfied, I was happy over that. I could have enjoyed it a lot more. I was upset, but I wasn't crying or anything like that. I rarely cry. It takes a real lot to make me cry. ... Dan helped me feel better. He told me how nice I was, and how pretty, and don't worry about it.'

After she returned home from the first vaginal operation, she read 'over 20 books and magazines' dealing with sex, and was exuberantly considering ideas of how she could get practice in using her new vagina. For instance, she and her girl friend, Pam, thought about the idea of making a group with Dan and his younger brother, Bobby, Pam's boyfriend, but didn't follow through. 'I prefer having emotional things there,' she said. 'I like the loving part that goes with it, and it's not just sex. ... I would like to find me a guy that I can love, maybe not forever, but it would be a real love relationship.'

Now that she was able to envision herself as being Pam's equal in vaginal intercourse, despite the disappointment of postsurgical infection, she was able for the first time to tell her everything regarding the operation. 'She was really thrilled that you doctors could do something like that,' the patient said in her next taped interview. 'So we got to talking about all the things that we used to do when we were little, and everything. ... And then she told me about the night when she slept over, about two years ago, and made it with the dog, my dog, while I was sleeping. Finally, like, I really wanted to make it with her, and she wanted to make it with me, but neither of us were really going to come out and say it. I was in a kind of messy state (with discharge after surgery), and she said, "Well, where's the dog?"' With butter as an enticement, the dog licked Pam. 'I was just really turned on. The dog was still licking her, and after a while it was just me. It was, you know, pretty good and all, but it's not that I – I really don't feel like a lesbian or anything like that. ... I just thought it was something different, you know, and something that I'd probably do again, but not something that I'm really into.' Pam was not able to go along with it until she had an orgasm, so neither of them had one.

Telling everything about her surgery to Pam broke the ice for her to tell it also to Dan. Even though she knew that she would not 'be going out with him forever, because he was a lot older,' and already had another girlfriend, she invited him over. 'It was really strange,' she wrote in her sexual diary. 'We blew a few joints, and I told him everything. We were completely open with each other. I found out more about him and related to him better than the whole time I was going out with him two years ago. I told him about how I was not really at ease when I had had sex with him, and he told me he had really enjoyed it, hole or no hole. I wish I'd known that then. Anyway, he told me he'd had a hard-on all night thinking about me coming over, and a lot of times he got a hard-on just when I came over. I wasn't surprised, but I thought it was pretty funny. ... Anyways, we started to get it on, and it was almost really good. We did erotic touching and kissing and oral sex. I had explained to him about the form and the discharge, and that I could not have intercourse yet, and this did not bother him at all. He just licked my thighs, lips, and clitoris. It was really good, because we brought each other to orgasm. But I still did not have intercourse. But we were open and responsive and relaxed. And we also laughed, which I love to do.'

The next rendezvous with Dan was delayed because the travel schedule of his regular girlfriend was changed. In the meantime, she was alone at home and Pam slept over. 'After everybody left my house at around 1:30,' she wrote, continuing her sexual diary, 'I said, "Let's get into some bondage." Pam said, "Yeah, far out, what do you want to do?" So I showed her. (Actually this is really easy to tell you when I'm writing it, you know?) I got some string and tied her hands to the head of the bed. She was lying on her back, her legs free. I got some body lotion and my old lady's mink stole (she'd flip if she knew). And then I started talking up a fantasy: that I had her drugged and she had to do what I told her. I treated her as a helpless earthling that I didn't want to hurt, but I had to perform tests on her. I was from another planet, and I had caught her for experimentation. I tried to calm her, while she moaned and wriggled. I went over her with my hands, my tongue, the mink, and the lotion. I also went down on her, because I felt I knew just what a chick would like, because I'm a chick myself. We both really got into the whole thing. It was so good. Then we switched places and she did the same to me. I loved it. By the time we were done, it was 4 a.m. Amazing. I have enjoyed this more than anything, not because it was with a chick, but because we felt so free and open about acting out a really good fantasy.'

Although there would be another seven and a half months before the final vaginal operation and the ultimate cessation of unhygienic vaginal discharge, the two months following the first vaginoplasty procedure had been sufficiently successful that she was able to review them in her sexual diary as follows: 'What a fine Labor Day weekend! I'm so sorry the summer must end. I've gone boating, waterskiing and swimming a lot. I got a good job, and I've changed my head quite a bit. There's so much to learn. I also got a hole. And I also used it. I've really missed a lot, you know. Balling is great, but I'm sure I don't have to tell you that. I feel I can get the most feeling in the doggie position. Female on top is really good too. I got Dan into rabbit pelts (furs) and he really digs it. I dig it with guys better than with chicks, probably because they have cocks (I talk like a real lady, don't I?). But I really did dig it a lot with Pam because, as Pam puts it, "We've done everything together, so why not sex?' So I am very close to her. In other words, if I can find a guy and have a love relationship, then sex would be all the better. But it's okay in the meantime.' Her age when she made that diary entry was 17:7 years. Her sexual odyssey in the preceding six months had covered a long distance.

The end of the year marked the second and final installment of her sexual diary. One entry read thus: 'You gotta hear this! Okay, are you ready? I was talking with my friend Cliff (whom I have balled before), and he was telling me that all cunts are located differently, and on different angles, which I already knew. Then he told me that they all feel really different, that some were really nice and comfortable, and some were just alright, and some were uncomfortable and scratchy, which I didn't know. And then (now get this!) he said that I shouldn't worry, that mine was the best he ever felt and, I quote, "God was really good to you, he gave you some gift and you really know how to use it." Man, I almost shit in my pants (excuse my French). I mean I was really laughing to myself. That really blew my head off. Now, what do you think of that?'

This final entry in her sexual diary was written also as a letter to the young psychohormonal physician to whom her case had been assigned. She addressed him in the final paragraph. 'I can't complain about anything, really. I'm secure. I have good friends, a good outlook on life, and I feel I'm a very stable person. So I'm just riding the bumps a bit, and taking things as they come. Writing to you every so often helps me smooth the bumps a bit, like shock absorbers. Have a really fine holiday. Tell everyone the same.'

After her last gynecology admission and follow-up at age 19:6, she had no further need of any clinical services, and her sexual diary was discontinued. She had already met the man who would become her husband. In a sex history questionnaire filled out at this time, she indicated that the most sensitive genital areas were the clitoris and the lips; that arousal induced vaginal lubrication in about one minute (later in life she sometimes needed to use K–Y jelly); that the best word to decribe orgasm was intense; that she climaxed most of the time; that she had experienced multiple orgasms; and that the interval before orgasm could be five minutes or much longer. She rated her sex drive as stronger than that of most women, and her feelings after sexual intercourse as very satisfied and conducive to looking for more sex. The body parts that were most sexually arousing were listed alphabetically as abdomen, back, buttocks, clitoral region, vagina within, and vaginal opening. On a five-point scale, the maximum enjoyment rating of 5 was assigned to foreplay, oral sex, and satisfying the partner; 4 to intercourse; and 3 to masturbation. To induce a romantic and sexy mood, with 9 as the maximum, her ratings were: seeing the person you love, 9; touch, 9; reading material, 9; smells, 8; seeing some part of the person you love, 7; sounds (talking in whispers and exciting conversation), 7; erotic movies, 7; tastes, 5; and still pictures, 3. She could be aroused by an erotic movie only if she found the male involved attractive. Though she might take a look at a man on the street and say he'd be really nice, she would not get really horny. If she responded to a female in a pin-up poster, it would be to compare herself with the other female.

After the initial exuberance of having a vagina and learning how to appreciate it, her sex life became conventional and, with her husband, monogamous. Their coital frequency varied according to work schedules. Typically it was four or five times a week. Sometimes she would have a sexy dream almost to the point of orgasm and then wake up her husband. 'He's a lot more responsive than I am. He'll never say no. Anytime I want sex, he's in the mood, whereas I'm not always.' The dream content 'seems more like feelings than it does visual,' she said. 'I can recall the feelings, but I really can't recall any content to them. ... I'm definitely more oriented to touch than anything else.' By contrast, she could get turned on by sexy stories, of which the content would be 'bondage. I could be either dominant or submissive, not real heavy, just light. I usually don't care for typical, mushy love things. Just reading straight stuff like that doesn't interest me.'

A bondage story would be 'more exciting if it's explicit, although if it's built up well, it could be implied; but it's more exciting if it's explicit. And if I'm being told the story out loud, I like it explicit. ... I've trained my husband to do it. It's almost like foreplay. In fact, he can probably turn me on fastest if he does that. It took me a while to teach him that. ... Whenever we make up a bondage story, it's real short, because I start hearing it and get horny so fast that I just want to get down to it.

'Let me see. I'm alone in the house and somebody, an intruder, comes in. I'm afraid of him, or whatever, and he just sort of makes me submit – either ties me down, or tells me that I'm going to have to stay still, or he's going to do something, you know, actually more telling me what he's going to do to me, if I don't do what he wants me to do. ... You know, it's funny because it has nothing to do with what's really – I would hate that, if it happened, real. It would drive me crazy. I would be totally the opposite. But sexually I like it. I mean, if my husband were ever to try to push me around in real life, I would just go bananas. But sexually, you know, it's different.' Even sexually, the narration of bondage was far more arousing than enacting it, and being tied even 'just lightly. I don't like being tied. It's almost like going through the motions of doing it, but not really doing it. Then when I actually have sex, I'm free.'

In the technical terminology of the paraphilias, hers was narratophilia. The theme was bondage and submission, but it was not paraphilically enacted as masochism. As is typical of paraphilia, hers had its origin early in childhood, possibly as early as age 5, and represented the solution to the problem of having no hole. It is probable that the success of vaginoplasty aborted the development of a full-blown masochism which would have dictated clawing, ripping, and bleeding of the vagina as an erotic practice. Likewise, it may well have aborted the evolution of oral sex with either the dog or a woman partner as an alternative to failure to be able to hold a penis.

The end result, narratophilia, with a theme of submission and bondage, is a typically feminine paraphilic outcome to the challenge of retrieving genital lust from the threat of extinction. In this case it was a playful, not a noxious outcome, and it was compatible with the reconciliation of lust and love in an affectionate and lasting relationship with a partner whose own erotic arousal was not inhibited by the game of being the narrator.

EXPOSITION

The lovemap biographies of these two people represent two quite different lifestyles in teenage. Nonetheless, each quite clearly announces itself as a woman's biography. Their shared femininity does not have a shared origin in either chromosomal sex or gonadal sex. The one that is more femininely orthodox belongs to a woman who is, by conventional criteria, chromosomally and gonadally male, and the one that is less femininely orthodox to a woman who is chromosomally and gonadally female.

What the two women share in common, despite their chromosomal and gonadal difference, is a similar prenatal history of not having been exposed to the influence of hormonal masculinization – the one because the cells of all the organs of her body, including the brain, were insensitive to the male hormone that her testes secreted; and the other because she had no testes to secrete male hormone, but ovaries instead.

In both instances one may assume, on the basis of strong evidence from experimental animal studies (summarized in Money, 1988), that the sexual nuclei and pathways of their brains remained unmasculinized and, therefore, feminized. Thus, one is entitled to make the further assumption that each was born with a brain predisposed to accommodate to the postnatal social inputs of being reared as a girl – identifying with females and complementing with males. Neither brain would have accommodated so readily to the influences of rearing as a boy.

The feminine orientation that became differentiated in childhood was reinforced by the hormonal feminization of the body at puberty. In adolescence each girl's femininity survived the somatic insult of a vagina that was too shallow and that did not menstruate. In addition, it was not dependent on ovarian hormonal cyclicity which was present in the girl with MRKS, but not in the AIS girl whose testicular secretion of estrogen was continuous, not cyclic (Van Look et al., 1977; Aono et al., 1978).

Looking ahead to the age of menopause, the evidence from other cases, though sporadic, indicates that, in both instances, femininity will mature with age. Whereas it can be anticipated that the MRKS woman will undergo a normal menopause, there is presently a lack of clinical reporting from which to anticipate that the AIS woman will do so, if she retains her gonads. However, it is known that if she elects to have them removed in young adulthood, the symptoms of postsurgical menopause will be rapid in onset, and will need to be controlled by estrogen replacement therapy.

In sexological science there is, as yet, no adequate theory which would explain why only one of the two girls, the one with MRKS, developed a lovemap that in street language would be considered mildly kinky, and in biomedical terms would be classified as mildly paraphilic. It is evident from the girl's own narrative, however, that its dual imagery of zoophilia and bondage are explicitly derived from her body image and her personal odyssey as a girl reaching adolescence with an impenetrable vagina. The paraphilic fantasy, sometimes converted into sexuoerotic performance, salvaged lust from being completely forfeited during the years when vaginal penetration was impossible. Paraphilic heightening of sexuoerotic arousal with imagery in fantasy of submission, though it occurs in both sexes, is characteristically more prevalent in the fantasies of women. Unlike most paraphilias, hers did not become an all-consuming addiction. Instead, its strength diminished after the ordeal of getting a functional vagina was concluded.

The concordance of these two representatives of the AIS and the MRKS on the criterion of femininity is not a random chance occurrence in one matched pair. That they are like others with the same diagnosis has been shown in a group comparison of nine cases each of AIS and MRKS (Lewis and Money, 1983, 1986; Money et al., 1984).

BIBLIOGRAPHY

Aono, T., Miyake, A., Kinugasa, T., Kurachi, K. and Matsumo, K. (1978) Absence of positive feedback effect of oestrogen on LH release in patients with testicular feminizing syndrome. *Acta Endocrinol.*, 87:259–267.

Barsky, L. (1986) Holy hormones ... male pregnancy? *Chatelaine*, 59(8):62ff.

Jackson, P., Barrowclough, I.W., France, J.T. and Phillips, L.I. (1980) A successful pregnancy following total hysterectomy. *Br. J. Obstet. Gynaecol.*, 87:353–355.

Lewis, V.G. and Money, J. (1983) Gender-identity/role: G-I/R. Part A: XY (androgen-insensitivity) syndrome and XX (Rokitansky) syndrome of vaginal atresia compared. In: *Handbook of Psychosomatic Obstetrics and Gynaecology* (L. Dennerstein and G. Burrows, Eds.). Amsterdam-New York-Oxford, Elsevier Biomedical Press.

Lewis, V.G. and Money, J. (1986) Sexological theory, H-Y antigen, chromosomes, gonads, and cyclicity: Two syndromes compared. *Arch. Sex. Behav.*, 15:467-474.

Money, J. (1988) *Gay, Straight, and In-Between: The Sexology of Erotic Orientation.* New York, Oxford University Press.

Money, J. and Ehrhardt, A.A. (1972) *Man and Woman, Boy and Girl: The Differentiation and Dimorphism of Gender Identity from Conception to Maturity.* Baltimore, Johns Hopkins University Press.

Money, J., Schwartz, M. and Lewis, V.G. (1984) Adult erotosexual status and fetal hormonal masculinization and demasculinization: 46,XX congenital virilizing adrenal hyperplasia and 46,XY androgen-insensitivity syndrome compared. *Psychoneuroendocrinology*, 9:405–414.

Norén, H. and Lindblom, B. (1986) A unique case of abdominal pregnancy: What are the minimal requirements for placental contact with a matenal vascular bed? *Am. J. Obstet. Gynecol.*, 155:394–396.

Teresi, D. and McAuliffe, K. (1985) Male pregnancy. *Omni*, 8:50–56ff.

Van Look, P.F.A., Hunter, W.M., Corker, C.S. and Baird, D.T. (1977) Failure of positive feedback in normal men and subjects with testicular feminization. *Clin. Endocrinol.*, 7:353–366.

CHAPTER 4

Feminine lovemap in two 46,XY cases concordant for female external genital differentiation, female rearing, and eunuchizing puberty

SYNOPSIS

In this chapter, as in Chapters 2 and 3, the two cases are matched on the criterion of being concordant for female external genital appearance. They are also concordantly matched insofar as each was declared female at birth and reared as a girl, with no suspicion of a prenatal error of sexual differentiation until, in teenage, each girl failed to develop breasts and to menstruate. This failure of pubertal feminization was secondary to a lack of the feminizing hormone, estrogen, and to the absence of ovaries from which estrogen is secreted.

Lacking not only estrogen, but also androgen, which is responsible for pubertal masculinization, the body developed in each case by growing tall and by developing pubic and axillary hair (under the influence of adrenocortical androgen), but no secondary sexual characteristics that were either female specific or male specific. In other words, if hormonal substitution treatment had not been given, the adult appearance of the body would have remained similar to that of a juvenile, and similar to that expected in an untreated eunuch whose ovaries or testes had been missing since childhood.

The deficiency in female development led to the discovery, some years later, that ovaries had failed to differentiate in embryonic life insofar as the two cases were concordant for chromosomal sex which proved to be 46,XY. Traditionally, the 46,XY chromosomal pattern has been labeled male, but now the terminology, 46,XY female, is acceptable.

Whereas the two cases were concordant for 46,XY chromosomal sex, they were discordant for gonadal sex, and for the actual genetic error responsible for their different gonads, namely streak gonads in Case #1, and testes in Case #2.

In Case #1, that of Swyer's syndrome (Espiner et al., 1970; Swyer, 1955), the gonads had failed to differentiate as either ovaries or testicles. In their place were inactive streaks of tissue. This form of gonadal dysgenesis is attributed to a missing gene, the TDF gene on the short arm of the Y chromosome, named TDF for testicular determining factor (Page et al., 1987). Dysgenetic gonadal streaks lack the power to secrete masculinizing hormone not only at puberty, but also prenatally, in fetal life. In consequence, prenatal differentiation of the sexual anatomy follows the template of Eve, not Adam, and is female both internally and externally, except for the absence of ovaries. The uterus is able to menstruate if, at maturity, cyclic hormonal replacement therapy is given. There is, as yet, no recorded instance of pregnancy by in vitro fertilization and implantation of a donated ovum, though such a pregnancy is feasible.

In Case #2, the syndrome, being the first of its kind on record (Park et al., 1976), does not yet have a definite name. It is presumed to be the developmental result of an abnormal and biologically inert variant of luteinizing hormone (LH) which fails in prenatal life and again at puberty to stimulate the gonads which differentiate as testes. Under the microscope they appear to be normally formed and potentially capable of secreting testosterone if stimulated with a bioactive form of LH. This syndrome of bioinert LH exemplifies yet another instance of the ascendancy of the Eve Principle over the Adam Principle in embryonic and fetal life. The 46,XY chromosomal sex programs the embryonic gonads to differentiate as testes, insofar as the TDF gene is not missing. The testes are able, prenatally, to secrete their mullerian-inhibiting hormone (MIH) which vestigiates the embryonic mullerian ducts and thus prevents their differentiation into a uterus and bilateral fallopian tubes (hence menstruation is forever impossible). By contrast the testes are not able to perform their second prenatal assignment which is to secrete testosterone when stimulated by LH, insofar as the LH that reaches them from the pituitary gland is defective and inert. Without testosterone, genital masculinization is thwarted, both internally and externally. Since feminization takes place independently of the gonads of the fetus, the remainder of sexual differentiation is feminizing. The outcome is that the baby is born looking like a girl and, therefore, declared a girl, and assigned to be reared as a girl. When the age of puberty is reached, it is marked by the appearance of axillary and pubic hair, the growth of which is stimulated by adrenocortical androgen. The testes again fail to make testosterone, as they had done in prenatal life. Lacking testosterone, the body fails to masculinize. Since it also lacks a source of female hormone essential to pubertal feminization, it also fails to feminize.

In teenage, despite the lack of hormonal feminization, each girl of this matched pair maintained the G-I/R (gender-identity/role) that had differentiated concordantly with her female genital appearance and female sex of rearing. Athletically, and on the criterion of dominance assertiveness when challenged, each characterized herself as a tomboy. Both were delayed in reaching the dating and romantic stage of development, as if waiting for a mature feminine attractiveness to match their feminine attraction towards males. Eventually, even before their pubertal hormonal deficiency had been corrected, they could be romantically delayed no longer, and each began dating. Soon thereafter, they sought the treatment that upgraded their feminine appearance and erotosexual competence as women.

The treatment intervention in the two cases induced two different hormonal histories. In Case #1, there was a straightforward transition from not having to having female hormones. Sexuoerotically, the effect was to facilitate feminine sexuoeroticism that was already functioning. In Case #2, the unanticipated first effect was to release not female, but male hormone. This male hormone had a disastrously rapid masculinizing effect on the voice and on the growth of facial and body hair. It did not alter the patient's already feminized sexuoeroticism, but rather intensified it, and strengthened her apprehension that gonadectomy would somehow impair her sexuoerotic responsivity as a female. To avoid further masculinization from her own gonads, she consented to their removal. It then took a long time for her to become reconciled to the effects of the hormonal disaster, and to become sexuoerotically adapted to female hormonal replacement therapy.

CASE #1: SWYER'S SYNDROME

Diagnostic and clinical biography

At the time of her first psychohormonal interview, this patient was 25:6 years of age. Recalling the teenaged years, she said that there had been times 'when my sisters would tease me and say: "Girl, when are you going to get some breasts?" They would do it in a joking manner; and of course they would use the words, "Yes, she's going to be very slow in developing."' This was the beginning of the patient's progressive recognition of failure to become pubertally feminized, for there had been no precursor signs during childhood. Apart from her sisters' joking, there was no other family recognition of her dilemma. 'Neither my mother nor my father ever came to me or mentioned anything,' she said. 'But they knew. ... My mother saw me physically without clothes on from time to time. She would just look at me. ... I'm certain it crossed her mind, you know: "She should be developing, but she's not." Then she took on the idea of me perhaps being a slow developer, and with time she would see if I did develop. ... You know, my mother and father, I think, were too protective of me to even talk to me about it. They wanted so much to protect me and my feelings that they felt they could do damage talking to me about it.'

At age 16, she applied for a part-time government job that required a physical examination, so she went to the clinic where her sister worked. She recalled the doctor telling her: 'I know someone who could help you. It appears there is a hormone deficiency.' She was referred to an endocrinologist and a gynecologist at the university hospital. 'I went to the ob/gyn, and the pelvic exam. It took two doctors to hold me down. And then, you know, we're dealing with interns here. I don't have anything against them, but they brought I don't know how many people in there to see me. It made me feel like a freak. I felt very uncomfortable, but I didn't say anything. My mother questioned, "Why do all these doctors have to go in there with my daughter?"'

For chromosome testing, she had to return four of five times to give more blood, but no explanations were given, and the findings were not disclosed.

'The only thing they could come up with,' she recalled, 'was to admit me and to go inside and find out what the problem was. ... They said I had ovaries, but they were not delivering enough hormones to my body. Their final conclusion was: "We're going in. We're going to open up the entire belly." They wanted to see what was going on, and they would have to remove everything, the reproductive, you know, whatever was there. They would have to remove everything and give me hormones. I remember this so well, because I was going through that. But I listened very carefully.'

She was terrified. 'Then I did not go in for surgery. I just was not ready. I told my mother. I even took off and left home one day, the day before I was due to go into the hospital, and stayed out all night.' She spent the night with a girlfriend whose mother was away on vacation. The next day, she went home crying. 'I said, "Mommy, I can't do it. It's too much." And she said, "They're only trying to help you." I said, "I can't do it."'

Eventually, 'because of my mother, they got me to the hospital,' two days prior to surgery. After having been medicated and taken to the preoperative area, she panicked. 'I said, "No, I can't do it. I don't want to do it." They took me back to my room. ... I got them to let me go home for the weekend. My mother came and got me, and I never went back. They kept calling and talking to my mother, telling her that I really needed the surgery and things could get worse if I didn't get it. I mean they were really trying to bring me in up there. I just left it at that. And my mother never forced me again. I went through high school, college, and everything.'

As an athlete in college, she had regular medical check-ups. 'The physician there never said anything,' she recalled. 'It sounds strange, now that you mentioned it, but I was very conscious of that while I was being examined. I wondered if he would say something, but he never did.'

In her senior year at college, at age 24, she was required to have a physical examination prior to taking up student teaching. 'I went to a medical center where I saw a doctor who confronted me about my not having breast development. He asked me if I had a menstrual cycle and had I been to see about it.' It was this doctor who referred her to The Johns Hopkins Hospital where she saw a gynecologist. He was the first person to explain her condition to her. He inspired her faith and trust. 'Like I say, he told me everything. He just gave me the good with the bad and I really respect him for that.' Her understanding of her condition from what he told her was 'that from my chromosome tests, I assume that for women, it should be two X's, but there is an XY. And the Y tells him that the ovaries must be removed because they are not producing enough hormones throughout my body. And if they are not removed, I run a risk of catching cancer, ovarian cancer. After the surgery, he told me they would start me on hormonal therapy which would cause breast development, and which would cause a menstrual cycle, but I would not be able to get pregnant. I have a clear understanding of what's going on.'

At the time of the gynecological examination at age 25:2, the patient's height was 6 ft. $2^1/_2$ in. (189.2 cm), and weight 222 lb (100.7 kg). The bone age was delayed. The chromosome count was 46,XY. The breasts were prepubertal with no masses. Pubic and axillary hair were sparse. The labia and cervix were infantile. The uterus was not palpable. There was no menstrual history. The clitoris was at the upper limits of normal size. the vagina was rugated, nonvirginal, and without lesions. The hormone levels were listed as follows:

- Luteinizing hormone (LH), 427 ng/ml (greatly elevated)
- Follicle stimulating hormone (FSH), 1935 ng/ml (greatly elevated)
- Plasma testosterone, 56 per 1000 ml (upper range of normal for a female)
- Plasma 17-β estradiol, 4.6 ng/dl (at the low extreme of the range for the normal female at the low, follicular phase of the menstrual cycle)

The diagnosis was established as Swyer's syndrome (Espiner et at., 1970; Swyer, 1955), also known as pure gonadal dysgenesis. Because the dysgenetic gonadal streaks in this syndrome are at risk for eventual malignancy (gonadal blastoma), their surgical removal was recommended.

The diagnostic findings and the rationale for surgery and for hormonal replacement therapy were explained to the patient and her husband, together. An admission for surgery was scheduled. Following admission, the husband stayed with his wife until late in the evening when he went home, so as to be ready for work early in the morning. After he left, his wife had an overwhelming panic attack, secondary to an uncontrollable phobia of not regaining consciousness following general anesthesia. Had she not been discharged, she would have walked out of the hospital.

Following this episode, she was referred for psychohormonal counseling. In the first session, she revealed that she was phobic not only for general anesthesia, but also for heights and elevators. Her fear of anesthesia abated when, with the cooperation of the anesthesiologist, it was decided that she could have surgery with a spinal rather than general anesthesia. At age 25:8, she was readmitted and was able to tolerate going through with the operation, to her own great satisfaction. According to the surgical report, there was a hypoplastic uterus, hypoplastic fallopian tubes and streak gonads bilaterally, 'with no evidence of gonadal blastoma or XXX dysgerminoma.' The pathology report on the gonads was confirmatory.

Subsequent to surgery, the patient was placed on ethinyl estradiol and Provera. Two weeks later, at age 25:9, physical examination revealed bilateral breast bud development, diameter 5 cm $\times$ 5 cm, at Tanner stage 2 (onset of puberty). Pubic hair growth was rated at Tanner stage 3.

Three months later, on a replacement dose of ethinyl estradiol 0.2 mg (days 1–25) and Provera (medroxyprogesterone acetate), a synthetic progestin, 10 mg (days 16–25), the breasts had doubled in size to 9 cm × 10 cm bilaterally. The uterus was palpable in the midline. External genitalia were unremarkable. LH and FSH continued to be elevated.

At age 26:5, nine months after the beginning of replacement hormonal treatment, the breasts measured 12 cm × 12 cm, bilaterally, and their Tanner rating was at stage 4. The external genitalia were normal for an adult female. The clitoris was considered to be a little enlarged. There was increased vaginal lubrication.

Fifteen months after beginning replacement hormones, at age 26:11, cyclic withdrawal bleeding had begun, accompanied by menstrual cramping. Pubic hair had increased to Tanner stage 4. Breast development was at Tanner stage 4, with glandular tissue readily palpable.

Subsequent physical examinations have been essentially the same. There have been minor adjustments of replacement hormones as the patient experienced some dysfunctional bleeding, which may have been associated with admitted inconsistency in taking the medication.

Gender-coded social biography

Throughout the development years of infancy and childhood, the patient grew up as the youngest in a family of 12 children, four girls and eight boys. She was different from her sisters only in that she was a tomboy. 'Well, you know, they say when you climb trees and you're into sports, you're a tomboy. I was a tomboy. I played all sports. I ran. I would fight the boys more than I would the girls.' She described her three sisters as 'just dainty girls.'

She recalled that she underwent a statural growth spurt 'between maybe thirteen and fourteen years old when I started sprouting up. The other years, I always saw myself as the same size as the other kids that I played with. Then all of a sudden, I just started taking height. And it was like, wow, that girl's tall.' Her height, at 6 ft. $2^1/_2$ in., also distinguished her from her sisters, who are between 5 ft. and 5 ft. 2 in. tall. Of the eight brothers, the two tallest are 6 ft. and 5 ft. 9 in.

In high school, 'I felt very awkward,' she said, 'until I began to realize, hey, you've got something going for you. The basketball coach would take me before he would take anyone else on the team. ... When I started seeing other girls of my height, that made me feel better. Then I started seeing tall guys, and I said, that's even better. But I felt very awkward around most of my friends, because you could always see me out of the bunch.'

At college, 'when I started playing college basketball, I started coming up against girls my height or taller. I began to accept me for who I was, what I am. If you don't like it, that's going to be it. I'm the one who's got to live with me. And I love me. I'm satisfied with me and what God has made me to be. So if you don't like it, then you have to deal with it.'

With regard to absence of breast development, during college years her recollection was: 'I have had friends to notice. They never made any wisecracks, but they noticed.' Those who would ask about her lack of breast development 'would just make me feel so belittled. I wished there was a shell I could crawl into. I never lost a friend because of it, but I hated the idea of someone confronting me. I don't like wearing tight tops.'

Her husband had told her, 'Breasts don't really make the woman. You'll always be more woman than any woman I've ever known.' She added, 'What he doesn't know is that when I was younger, I did feel that I was different from the next, from the next female.'

Her reaction to not having menstrual periods was ambivalent. 'No, I never told anyone. Well, my friends, my girlfriends would talk about a menstrual period. In a sense I would feel, I felt like left out; but then, I've seen my sisters, and they have gone through, you know, those severe cramps, and in a sense I was glad not to have one. But I couldn't tell my friends that. I felt that they would look at me in a very different way.'

'When I compare myself with the next woman,' she said, 'I see her as having nothing on me except for the fact that she has a menstrual cycle, and she has breasts. When it comes to wearing a leotard or something of that nature, something very tight, I will not wear it. I like blouses. I like suits. I like slacks. I like corduroy pants. I like to wear jeans. I never wear a padded bra. I don't want to be known as a girl wearing falsies, you know.'

After her breasts grew, she said that she felt freer to 'wear cooler tops and I don't hunch in trying to hold my chest in. I walk straight up.'

The one thing that had pushed her into seeking medical help as an adult was being 'tired of people noticing at times that I had no breasts. And they'd say, "Oh, girl, where your breasts at?" And that sort of bothered me, and made me say I need to go see about this. ... The biggest hangup,' she said, as she prepared herself for surgery, 'is the fact of being put to sleep, anesthesia. I have always asked the question: "Do you go into a normal sleep?" The doctor said, "No, you're like unconscious." The surgery itself, there's not that much fear. ... I've tried and tried and tried to really prepare myself to come in and get the surgery completed. ... Since you were able to arrange a spinal, in which I would be conscious, I felt much better. I had a more positive attitude.'

The hospital experience of having little choice but to ride an elevator as a patient also aided in abating the phobia of elevators. Although she still was not ready to ride an elevator alone, she said they did not hold the terror for her that they had previously.

She became a college student after having been selected for a basketball scholarship. She not only excelled in athletics and physical education, but also did well academically, and qualified with a certificate to teach. After graduating, she ran the sports program at a summer camp. She considered that estrogen replacement therapy might have had a secondary effect on athletics. As she put it: 'I've become a little dainty, too, I think. Some things just aren't for me, whereas before it was like, I would play anything. But it's like no, that's not for me.'

As a teacher, she takes graduate courses in special education. On the side, she has qualified for a license as a real estate agent — and has found a suitable home to purchase. 'I say, well, it's getting near that time when we're considering adopting and a lot of times they want to know that you can accommodate kids. I said we'll have the house.'

She met her husband in college, where he too had a sports scholarship. His was in football. They married while still in college and, after graduating, he transferred his energies to a full-time business career. He accompanied his wife to the clinic on one occasion and agreed to a brief joint interview. He was a gentle and considerate man, not in the stereotypic mold of a football jock, and readily adaptable to talking about the romantic and erotic attraction he and his wife shared. He was pleased that his wife had overcome the hurdle of surgery and that her breasts had developed. He had not seen her formerly flat chest as a problem for, as he put it, he was a 'buttocks and leg man as far as my wife's concerned'. He and his wife appeared to be well attuned to one another. They were at ease together.

Lovemap biography

According to the evidence of her own recall, the patient's sexuoerotic development in childhood had been benign. Regarding sexual rehearsal play, she said: 'Well, there was a group of us, maybe seven or eight years old. We were in my basement, and we said, "We're going to play mothers and fathers." The boy would say, "Yes, Sherrie is my wife." Everybody had husbands and wives. It was like they couldn't wait until it was time to go to bed. They were saying, "Okay, wife, let's go to bed." They wanted to hug and everything. And I'd say, "No, no." They were talking about "We're going to have children." I said, "The children are asleep over there," where we had a doll baby. I really didn't like playing mothers and fathers. It meant that you have to go to bed all the time. But I guess that's what kids see. As children, we saw our parents, the husband and wife, sleep together. But I didn't like it, and that's where mothers and fathers would end for me, right there.'

She did not get into playing sexual games of any other type, but she did learn from her sisters: 'Well, of course, you know, you sit down with your sisters, and they tell you of their experiences. And then I did have a sex education class in junior high school. But I was immature to the point where I couldn't even grasp what I heard. I didn't even want to think about it. And you know, you hear people talk about, "Yes, my first time it hurt." And to me I didn't see sex as a pleasure. I saw it as something that would hurt you, because all you heard was that the first time, it's going to hurt. Other than that, my sex education was limited to the fact that, you know, when you get older, you have sexual desires. This is what happens to a little girl. This is what happens to a little boy. If the opportunity presents itself, they come together. And then you go on from there. It was just a very basic thing.'

The initial basis of this woman's attraction to her husband when they first met in the college athletic program was, she said, his physique and his moustache. 'When I first saw his moustache, I could have just sucked his lips in,' she said. He was attractive also because 'unlike other athletes, he was someone I could hold an in-depth conversation with. He was not loud. And he was well-dressed.' His response to a query about what initially attracted him to her was: 'Her personality. She was fun to be with.'

'Before I met my husband,' she said, 'I never kept a boyfriend too long, because when it came to having sex, that's something I could not do, because of me not having breast development.' With her husband it was different: 'It was something about him that just told me, when we first met up in school, that he was very nice. He was good to me when I first met him. I felt obligated in a sense that, if I wanted to keep him, I was going to have to do everything that it takes to keep a man. I wasn't going to be able to lollygag and play around, and jive with him, you know, and lose him. It was so — it was different. It's like it just happened all at once. But I waited three months before I did anything with him.' When, at the end of that three months, they had intercourse, it was her first time, and he has remained her one and only sexual partner.

Thinking back to when they first dated in college, her husband said that he never saw his partner's lack of breast development as a problem for himself. Nonetheless, he recognized that for her it was a problem, and that he 'would have to take it slow, so that she would realize he was interested in her for herself, and not for her physical attributes.'

Before they actually embarked on a sex life together, she had 'never tried masturbating,' she said, 'but I used to think about sex all the time. I also got an urge, a feeling, that I wanted to have sex.' Intercourse was 'painful at times. ... It's like every time was similar to the first time. ... I guess the doctor knows that I don't lubricate that much.' Hormonal replacement therapy proved succesful, as expected, in correcting this deficit.

The effect of estrogen on lubrication became noticeable within two and a half weeks. 'One day I was sitting around the house and I was just thinking about us, and how enjoyable sexual intercourse is. ... I had to go to the bathroom, and I urinated and wiped myself. I saw all this white stuff on it. I saw that it was lubrication. ... This past week, me and my husband, we became sexually active again,' for the first time since surgery. 'There was a lot of lubrication that I never felt before, coming from myself. I enjoyed it more because there wasn't much, there wasn't pain accompanying it. I was just able to really let myself go, which I felt was very good.' Her husband had noticed the change: 'Yes, of course he has,' she said. 'He's told me so.' For him it was a positive gain, as was also the development of her breasts. He did not experience the loss of their juvenile appearance as a loss of something that had been a source of attraction.

After hormonal replacement treatment, there was a change in the frequency of sexual intercourse. Formerly, it had been 'about twice a week. Sometimes maybe once. I'd like to have it on an average of perhaps three or four times a week. But my husband is a very busy man with his work.' Six months posttreatment, sexual frequency had increased to 'almost every day.'

The wording of statements about orgasm was more sedate in the before-hormone era than afterwards: 'When I reach orgasm, I can feel it coming on. It just feels good. And my reaction is a little more vigorous. ... I would say it starts at the opening of the vagina. ... It's centered here, but then I feel good all over, in a sense.' Six months later, the statement was: 'I know when I'm reaching the orgasm because I feel good, and I feel this feeling coming on, like this good feeling. At times I try to hold back, but it's not easy. You have to let go. It feels good! ... The feelings that build up to a climax are stronger. I'm more turned on than I was before. ... I would say that it took longer for me to reach it before. I had to work harder to reach it. It was like once, and that was it. ... Now I can have multiple peaks that come at intervals in a sense. Maybe, one time, and then, after a while, again. It's like it has to build up again. The most I would say is three in maybe half an hour or 45 minutes.'

After four and a half years on hormonal therapy, the experience of orgasm continued as before. 'We like to reach orgasm at the same time,' she said. 'And regardless of whether I'm in the mood or not, he has a way of making me get there. ... He knows when I'm there. A lot of times he can be ready, and if I'm not there yet, he can maintain until I'm ready, because that's what we work toward.'

Her favored coital position was with the man on top, but at times she 'would get on him.' The most exotic thing they do is oral sex. 'I would say he's crazy about it, but I'm lukewarm. I'll do it and enjoy it, but I'm not real high on it.' His own statement was that he enjoyed fellatio as part of foreplay, and that he thought cunnilingus a real turn-on for his wife.

In the period before hormonal replacement therapy, the absence of ovarian hormones had not precluded responsivity to erotic arousal. 'My husband is attractive to me,' his wife said. 'He turns me on. I mean, looking at him makes me want him. I guess that's why I married him.' Though each could be turned on by the sight of the other's nudity, in the wife's experience, 'touch, the feeling, is much stronger,' she said. 'There have been times when I've gotten desire for sex with my husband just playing around with my chest, and, you know, rubbing me on my legs, and running his fingers through my hair, and what have you. The neck, yeah. And sticking his tongue in my ears. ... The vaginal area is very sensitive. And the breast area. You know how a man will get to sucking on you and playing with you.'

Apart from the sight of her husband, vision, as in seeing erotic pictures or movies, was not an important channel of erotic turn-on. Smell also was insignificant, except for 'this cologne he wears that's so masculine, a smell like he's a thorough man.' Taste had no significance. Hearing, like smell, was by association with the husband: 'My husband's into jazz. A lot of that sets the mood, really. Like a quiet evening at home, and it's just the two of us. Then all of a sudden the thought is there. This is the time to make love.'

CASE #2: 46,XY SYNDROME OF BIOLOGICALLY INERT LH (LUTEINIZING HORMONE)

Diagnostic and clinical biography

This patient's first Johns Hopkins contact was by means of a letter which she wrote to the author of *Sex Errors of the Body* (Money, 1968), after she had found the book in her university library and read it. She wrote:

'In the course of my research on my particular problem, I have been very favorably impressed by your book, *Sex Errors of the Body*. Although I had planned to complete my doctoral studies before seeking a final medical solution, my condition has become psychologically unbearable, and I can procrastinate no longer. At the risk of oversimplification, my problem appears to be uterine and vaginal atresia. I have sought the counsel of several gynecologists and have received various, sometimes contradictory, diagnoses and prognoses. I believe Johns Hopkins is the logical place for me to receive the most thorough examination. Could you please advise me on the appropriate means of arranging such a consultation? Thank you *very* much — for your factual, compassionate book and any aid you can give me.'

The upshot of this letter was that the patient decided on an evaluation at Johns Hopkins which, regrettably, took place in two stages, 11 months apart, because of problems in synchronizing the schedules of all concerned. Her own schedule as a doctoral student was very exacting.

The first evaluation was on an outpatient basis, and was exclusively gynecological. The second evaluation was on an inpatient basis, the admission being on the clinical research service. It was during this admission that the patient was first seen for a psychohormonal evaluation, a year after her letter had been received. She was 26:9 years old. She related what she knew of her past clinical history.

During her juvenile years, she had had no reason to suspect that anything was different about herself, because she did not recognize herself as visibly different than other girls of her age. It was only at the expected time of puberty, at age 13, when other girls were beginning to menstruate, and she was not, that she began 'intuitively to have this feeling that something was wrong, though I would say that hard and fast knowledge did not come until around age eighteen.' Insofar as she could sort out the stages of her pubertal development, she believed that she might have had the beginnings of breast swelling at around age 11, though breast growth had been negligible. By age 13, it had ceased, and the chest was flat. Beginning at age 13, pubic and axillary hair developed normally in amount and distribution for a female. The covering of hair on the head was abundant.

By age 16, because she had not menstruated, she and her parents decided that she should see a gynecologist. She recalled that he told her she had an infantile uterus, and that he gave her a prescription for Enovid, a combined estrogen and progestin replacement, to be taken on a cyclic basis for six months. She lapsed from treatment, so as to escape the attributed side effects of nausea, vomiting, and weight gain, which meant also the loss of breast enlargement. Neither she nor her parents pursued the possibility of an alternative estrogenic prescription.

No further action was taken until, at age 20, she consulted a gynecologist again. He reported to her that she had no vagina and that, upon rectal examiniation, he had been unable to palpate a uterus or ovaries. He advised surgical exploration of the abdominal cavity (laparotomy), which she did not have, as her family left the country on account of her father's work. She was 22 when she decided on further action and saw another gynecologist, overseas. He confirmed that she had no vagina and no uterus, but not that the ovaries were missing. He recommended surgical construction of a vagina (vaginoplasty), to which her reaction had been: 'What he told me was so farfetched that it absolutely horrified me and my parents. He was talking about inserting a glass tube which would remain for six weeks and then would be replaced by this kind of mesh arrangement. At the time that sounded very negative.' Vaginal reconstruction, as advised by yet another doctor, she summed up as follows: 'And the other time I was told about vaginal reconstruction, it was made into a two-operation affair where they would take a piece of my small intestine in the first operation, and culture it, and then graft that. That was a rather serious sounding operation, and not one that I was too sanguine about.' On both occasions, she opted not to undergo vaginoplasty. At this stage of her life she had an incomplete image of her own genital anatomy as compared with the norm. Ten years later she had become free enough to use a mirror for self-inspection.

Although she had rejected the idea of vaginoplasty at the age of 22, she did undergo augmentation mammoplasty at that time. The implant was cosmetically successful for four years, after which the silastic material had to be removed rather urgently, because of repeated infections. This procedure was done in her 26th year, just months after she had returned home from her first Johns Hopkins gynecological examination.

The findings from first Johns Hopkins examination were as follows: The length of the clitoris, 1.5 cm, was at the upper limit of normal, though it had the illusion of being larger, because of a protuberant, fleshy fold of skin, bilaterally, that in a normally formed vulva would have been the labia minora (Fig. 4.1). Below the glans of the clitoris, there was a funnel-shaped pouch, or urogenital sinus, which ended blindly, and within which was located the urethral opening in more or less the normal female position. The pouch had a depth of 3–4 cm. Between the pouch and the anus there was a smooth flat area, fused in the midline, uninterrupted by a vaginal opening. Though the labia majora were poorly formed, they had not become wrinkled, and the midline fusion did not create the appearance of a labioscrotum. Externally,

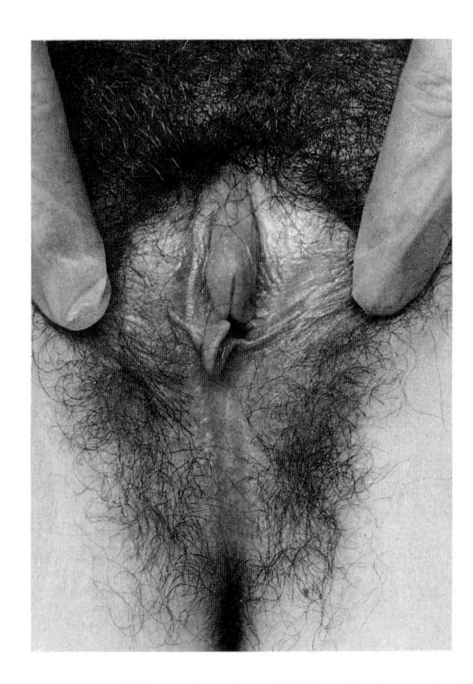

Fig. 4.1. Case #2. Female appearance of the vulva at age 26:8, prior to surgical lengthening of the vagina.

the visual appearance of the genitalia, overall, was not sexually ambiguous, and not masculine, but feminine, though with something amiss.

No uterus or gonads could be detected on rectal examination. Nor were gonads detected in the inguinal canals. Pubic hair was normal with a female escutcheon. There was no facial hair or body hair, and the voice was feminine. The nipples and areolae were infantile, with bilateral mammoplasty scars on the lower edge of each breast. Height was 170 cm (66.9 in.) and weight 57.7 kg (127 lb). Buccal smear was negative and the chromosomal karyotype was 46,XY.

Based on the foregoing findings, the provisional diagnosis was androgen-insensitivity syndrome. This syndrome, as said before, was known to the patient through her prior reading. Thus it was no surprise for her to know that she was chromosomally an XY woman, and that she could explain the Y chromosome to other people as the equivalent of an X with its arms broken off. 'My chromosomes aren't me,' she said. 'They simply aren't. They are a biological fact, but as far as my psychosexual identity is concerned, they aren't me. All that it means is that I can't pass the chromosome test to enter the Olympics.'

It was also no surprise to her to learn the status of her gonads as being histologically testicular in structure and, though azoospermic and sterile, androgen secreting. She had a strong conjecture that their hormonal function was a prerequisite of her erotic feeling and libido. Their removal, as routinely recommended as a precaution against cancer, meant to her 'a violation of my body.'

The gondads were still intact, when 11 months later she returned as a volunteer subject in a clinical investigation aimed at determining, in the adrogen-insensitivity syndrome, the body's nitrogen balance in response to exogenously administered androgen and gonadotropin. The unexpected response of the patient's body to the hormones administered was to masculinize. The patient, along with everyone else, now knew that she was not androgen insensitive. The diagnosis needed to be revised.

The research protocol for the nitrogen balance study was comprised of a control period of 10 days of constant diet after which a 100 mg intramuscular injection of testosterone propionate was given daily for five days. Following a second control period of five days, a 50 mg injection of dihydrotestosterone was given daily for five days. The third control period of five days was followed by 5000 IU of human chorionic gonadotropin (HCG) given daily for three days (Park et al., 1976). There was an unexpected marked increase of plasma testosterone to a normal male level after the administration of HCG.

About one month after discharge from the hospital, the patient telephoned to say that something had to be done about the changes she had experienced after leaving the hospital. The changes enumerated were very extreme deepening of the voice, rapid growth of facial and body hair, and unsightly facial acne – in other words, the delayed onset of masculinizing puberty.

She was seen one week after the phone call and the symptoms of virilization were documented. For the first time on physical examination, the gonads were found and were palpable in the lower inguinal area. At this time the patient said that she believed the rate of increase of symptoms had plateaued and regressed somewhat. Plasma testosterone levels were obtained and found to be 'low for a male subject and normal for an adult female.' On the basis of the elevated level of luteinizing hormone (LH) found by radioimmunoassay, it was postulated that the patient produced a biologically inactive LH, and that the component of active LH in the injected chorionic gonadotropin had triggered the testicles to secrete their own androgen. Accordingly, the diagnosis of androgen-insensitivity syndrome was revised to that of a hitherto unknown and unnamed syndrome of fetal and pubertal demasculinization secondary to inert LH secreted from the pituitary. Consequent to the rapid physiological masculinization she had begun to undergo, the patient consented to be surgically gonadectomized, despite great reluctance through fear of losing her libido. The operation was done two months later. Postsurgically the patient was put on estrogen replacement therapy with Premarin, 2.5 mg daily, by mouth.

Things did not go well for the patient during the next three years. She was beset by constant, debilitating despondency, fatigue, and weakness. In response to a letter of inquiry and concern written from the Psychohormonal Research Unit after a five month period of not hearing from her and 10 months after the gonadectomy, she wrote, 'With regard to my physical goal, complete feminization, success is more elusive. I'm afraid, literally, to risk losing this dream — it's too fundamental, and its realization too much out of my own hands. This (irrational?) fear that the operation will not work out well is a result of my previous experiences. I wanted a bosom. Okay, since I'd been told there was no other way, I tried plastic surgery. Not only were the immediate results less than satisfactory, but subsequent massive infection developed and corrective surgery was required. Meanwhile, I have discovered Johns Hopkins. A definitive diagnosis is finally obtained. This I can handle. I agree to participate in a study and what happens? Nine injections later, I'm into an operation I was deeply opposed to, and which has evidenced none of the results I hoped for. Every aspect of my life except my stubbornly healthy body suffered badly. You've already heard that list. So, here I am, victimized by my own half-realized fears and unwilling to risk another disappointment. Here's the rub, as I see it. The very goals that sustain me perhaps in subtle ways may ultimately destroy me.' She ended the letter by saying, 'Looking over the text of this letter, I am smitten by the urge to destroy it. I am disgusted by my ambivalence, but stuck with it. Believe me, if I could change my reinforcement schedule, I would, so we'd all be well done with this mess.'

In reply to her letter, she received supportive letters from the chiefs of both psychoendocrinology and endocrinology. Her reply was delayed for one and a half years. Then she wrote requesting an appointment: 'In particular, I am interested in discussing my hormonal treatment with regard to the possibilities for inducing breast development and some improvement in moodiness. Naturally, I would like to discuss further the details of vaginal reconstruction. I do find that the longer I delay, the more I dread the risks involved. The alternatives, however, have become no more attractive than they ever were.'

She was offered appointments in gynecology, endocrinology, and psychoendocrinology. She had come to the conclusion that her basic fear was, 'Probably, that I would have no longer any excuse with regard to participating in sexual adventures." She realized 'that it's not going to get any better — indeed, it's probably going to get worse.' Therefore, she set an appointment for a follow-up evaluation, and for a surgical admission for vaginoplasty as soon as it would fit into her busy work schedule, which was five months later. She underwent the procedure well, with little discomfort. This was three years after gonadectomy. Four months presurgically, hormonal replacement with estrogen had been augmented by the addition of a synthetic progestin, Provera (medroxyprogesterone acetate), 2.5 mg daily, so as to more closely approximate the hormones of normal women. According to the patient's subjective appraisal, the combination did effect an improvement by uplifting her mood and overall sense of well-being.

Six months after the admission for vaginoplasty, she elected to be readmitted for augmentation mammoplasty, her second such operation.

Gender-coded social biography

This patient was the youngest of four children. The oldest, her only sister, was many years older than she. The two sisters were not close, whereas the patient did have a close bond with her two brothers who were closer to her in age. 'They treated me like a treasure.' Then, chuckling, she added: 'Or something they wanted to annihilate! ... My family is very demonstrative. That means demonstration just as much of angry emotions as it does of love emotions. ... I had fights with my brothers right through teenage years. We were given to physical demonstrations of both affection and anger.'

On a sentence completion test, she wrote the following statements: 'My family is far more cohesive and nuclear than most. They spoiled me excessively. They treated me as a princess, slightly mad! My mother is lovingness personified. Her whole life was lived along somewhat libertarian lines, in terms of her generation. She and I have a deep and abiding faith in one another. My father is a fantastic human being, whom I trust. I wish he were capable of greater happiness in life as it is.' She wrote also that she would 'give anything to forget the time that I precipitated a rift between my two brothers,' regarding the older brother's wife. It could not be healed, as the older brother and his wife were together killed in an automobile accident two years later, when the patient was 20. This tragedy may have been a factor in the timing of further medical intervention on the patient's own behalf.

As an adult, this patient took responsibility for her own health care, unaccompanied. Thus there was no opportunity to interview family members about her childhood. She gave information herself, in retrospect. She identified herself as having been a tomboy, 'because, more than anything else, I really wanted to be like my two older brothers. At that point I don't think I had any idea about sex roles. I just wanted very much to be a boy. So I played baseball. I was very good at baseball — all kinds of sports and activities. Physically, I indulged in boy kinds of things. I even had a paper route. ... I even made the newspapers about being the little girl on the Little League. I was a shortstop – nicknamed Peewee, after Peewee Reese, the Brooklyn Dodger, because I was a peewee, kind of small.'

In addition to being competitive in sports, the patient was also a fighter as a juvenile. 'When I was little, I was kind of a bully,' she admitted, 'because I had my brothers to back me up. I would initiate a fight if someone did something I didn't like. I knew I wasn't going to get hurt too badly. If someone tried to fight me, I'd fight them back. ... I would be dishonest to say that I wasn't very pleased if I drew blood. ... I think I had my last serious fight at about age nine.'

Only 20 percent of her time, she estimated, was given over to traditional girls' activities: 'I still played dolls with my girlfriends. I still had little girl visitors. But I really wasn't interested in that. ... I knew I was supposed to be. I liked to play dress-up, as a very busty lady. ... Yes, I wanted to have a dress on. Don't ask me why. I think that's probably inconsistent with wanting to be a boy. But I wanted that too — a busty boy.' Her dressing up did not extend to games of wedding or pregnancy, nor of playing mothers and fathers: 'No, we didn't play that. We were a little bit more exotic, and more likely to be somebody's mistress, or in a harem – ideas supplied by my reading. ... Maybe I was corrupting those other naughty little girls who liked to play dolls. Oh, and sometimes there were wicked stepmothers, too!'

Between the ages of 5 and 12, boys 'accepted me. They didn't have many choices about that. Then I began to get negative vibrations that I was a drag. My brothers wouldn't take me with them anymore, and the boys, when I'd come up, they'd just stop playing. ... I was furious, just absolutely furious. Oh! Shunned and rejected. I hit them and kicked them. I'd throw things. I threw a beautiful shot and caught my brother right on the side of the head with a coke bottle.'

For about a year after the boys excluded her from their activities, she retreated into social isolation. 'I read a lot,' she said, 'I've always been a reader – $2^1/_2$ to 3 hours a day, still. ... I thought deep thoughts – they were deep thoughts then. I'd think about what would happen if I died, and how sorry they'd all be. Some very morbid things you think about at that age. ... I wrote a great deal, too. Poetry. Garbage. But some of it was good. Some of it was published. ... And then, of course, I really focused in a lot of school work, and began to get turned on about it.'

School had always been 'very easy,' she said, attributing it 'not so much to a great intellect as a lot of precocity, and growing up in the kind of home I did. Also it was a small school, a small town, and not that much competition.' She herself, by contrast, was fiercely competitive: 'Competition was more a way of life in our family, because of games and everything. But the realization of my deficiency, I think, gave me a sort of complex. ... As a woman unable to fulfill a stereotype, I was likely to work harder in other areas — getting my Phi Beta Kappa through graduating magna, setting records on tests, writing the best paper, and being the only girl to get her paper selected. That's introspection, but it's probably pretty valid. ... I'm also very competitive with other women, about the opposite sex.'

Though she downplayed her intellect, it did in fact prove to be superior, not only in terms of academic achievement, graduating with a doctorate, and being appointed to a faculty position, but also on the criterion of IQ. The Wechsler Adult Intelligence Scale was administered when she was still a graduate student, at age 26: Verbal IQ, 132; Performance IQ, 111; and Full IQ, 122. When the special factor quotients were calculated, the IQ equivalents were: Verbal Reasoning, 141; Numerical Reasoning, 102; and Praxic Reasoning, 101. The discrepancies are indicative not of nonverbal impairment, but of specific verbal ultrasuperiority relative to nonverbal normalcy.

During the year of social isolation at age 12, there was a transition towards more feminine interests: 'This began with an interest primarily in clothes. You know, up until then, it had been pretty much an old pair of blue jeans and somebody's shirt, preferably a grubby shirt. But then I began to get interested in dresses. And dancing — I loved dancing. ... As I left baseball, football, basketball, I moved into tennis, swimming, dancing.'

During the high school years, there was not much dating: 'I was different, you know. I felt that and, consequently, everybody else did too. Still, I was president of my senior class. ... The girls decided we were going to take over from the boys, who had been dominating the elections. ... I was into all kinds of activities in high school, primarily group activities. There was just not that much dating.'

After high school, she did begin to date, and began a sequence of four close romantic relationships from each of which she abdicated when there was no backing away from commitment to genital intimacy. She was not yet ready either to disclose the secret of having vaginal atresia, or to address the issue of surgical vaginoplasty.

About marriage, her response was: 'Oh, I would like to get married. I think it's highly probable that I will. ... I'd say the chance is 75%.'

Lovemap biography

The tomboyism of childhood was extreme, according to this woman's retrospective account of herself, but it did not include those gender-related components of development that are romantic, sexual, and erotic. On those occasions when she played with other girls and with dolls, she did not play the role of a boy or man. She played a feminine role, not of mother or wife, but of mistress or lady of the harem. When she played dress-up, she endowed herself with a bust, even though she recognized the inconsistency with wanting to be a boy in sports. Because she grew up in a household in which her parents 'were always very open about sex,' the reproductive dimorphism of the sexes was no mystery to her. Moreover, it had its own discrete purpose. The genitalia were, for her, not the final criterion of excellence in sports, fighting, and academics, in competition with boys. When the boys with whom she played reached puberty, however, their genitalia became a prime criterion of romantic and erotic bonding with girls. In her role as one of the boys, she became persona non grata. The rejection of her erstwhile playmates precipitated her into a limbo of social isolation and personal suffering. She was a misfit with either boys or girls of her age whose romantic and erotic attractiveness to one another was awakening.

It is possible that her own romantic and erotic awakening from this position of isolated misfit might have manifested itself as a lesbian attraction to another female. That would have circumvented the heterosexual problems consequent to not menstruating and being infertile, and possibly even the problem of having no breasts and no vagina. But such a solution was not to be. Far from being confronted with a homosexual yearning, the patient said at age 26, in response to a query about encounters with female homosexuality: 'I don't think so. Maybe without realizing it. But I haven't ever come up against the situation.' Guessing the percentage chance that she might be bisexual, she said: 'Oh, I don't think so.' But if it were the only way of saving her life, she laughed: 'Oh, well, that would be no hassle, assuming that I was in the mood of wanting to live. I just don't see that scene as mine.'

Within the family, there had been during the years of childhood no equivocation regarding her status as a daughter who would grow up to be heterosexual. It was later that she learned the family saga of her precipitous birth at home with no assistance except that of her aunt who was quoted as having been confused and screaming: 'It's a boy, it's a girl — it's a girl, it's a boy.' The patient's comment on this quotation was: 'It's a bitter kind of black humor, but I often think to myself, huh, gender problems right from the word go! Then there was my older sister who had two brothers and had said she would not look at the new baby if it was a boy. My uncle teased her and said: 'Come in and see your new little brother.' She was walking away, so they had to tell her it was a little girl. I can laugh at these things ... but they were painful to me at one time.'

From teenage onward, it was the patient's fate to have to cope with both heterosexual attraction and being heterosexually attractive, despite the immediate handicap of not being capable of copulation, and the long-term handicap of irreversible infertility. 'I was,' she said, 'an absolutely unsightly teenager, and, unfortunately, my beginning interest in boys preceded their interest in me. O yes! My word, I had dreadful crushes. People had them on me, but those I wanted didn't. ... I'm very competitive with other women. If there's someone they're dating whom I would like to date, or someone I'm dating whom someone else likes to date, I have to admit I feel competitive.'

Her first major affair with a boyfriend reached its zenith in engagement at age 17. 'I think my parents displayed a great deal of wisdom in allowing this to run its course,' she said. 'It was smarter not to force me into eloping, which I probably would have done. ... I often tease them that it was because they knew I was born with a chastity belt, so they could afford to play it loose. They knew I wasn't going to get pregnant.'

The joking reference to the chastity belt had its deeply serious side too: 'Invariably it was a factor in making me unwilling to go into a relationship with an eye towards marriage.' With increasing maturity, it became progressively difficult to sustain excuses for chastity. This was particularly so in the case of the third and fourth affairs with a long-term boyfriend. In each case the partner was a fellow professional, was seriously love-smitten, and was intent on marrying. Here are the patient's own words concerning the breakup of her fourth major affair, at age 27: 'We went through graduate school together. We've been close for years. It was one of those long and enduring kinds of relationship that was composed of scar tissue as well as any other kind. I felt I had to finish it in fairness to him, because there was no place — I was not going any place that he would be able to go with me, and vice versa. My physical problems made certain limitations on the relationship. In addition, I thought I was very bad for him because I made him feel very inadequate sexually. I've never given him an adequate explanation. It wasn't him, it was me. Quite frankly, the guilt was appalling, and my reaction to that guilt would be to take it out on him, which made me feel more guilty. You see a lovely vicious cycle going there. ... I was totally miserable and absolutely devastated. ... I miss him a lot, still.'

Together, they had talked about pregnancy: 'He was very big on children. At one point I had told him I was sterile. He really loved me, so he said we'll adopt some — which, of course, made me feel all the shittier' (about the way she was treating him). She herself had always been positive about adoption, as her mother had been an adopted child, and she had two adopted cousins.

Whereas the fact of sterility was speakable, the fact of having no vagina was unspeakable. Like all forbidden topics, its unspeakability had no rational defense: it was simply, by definition, unspeakable. By extension, it became off limits to visual or manual intrusion, whether in the course of erotic activity, or for surgical vaginoplasty. The patient dealt with erotic activity by invariably proscribing removal of her panties, restricting genital petting by hand, and not participating in oral sex, either cunnilingus or fellatio. 'I would like to be able to reconcile the discrepancies between what I believe in, what I consciously or unconsciously suggest in an interpersonal relationship, and what I am able to do,' she said in an early interview on the timing of vaginoplasty. 'I don't really believe there is a [moral] distinction to be made between petting, which I have indulged in, and intercourse. And that's why I say I've been pretty dishonest [in making excuses].'

The patient's own reply to whether she had ever been in love or not was: 'Probably not. I don't think it's possible to be really in love with someone, and to have always a nagging feeling of insecurity, and not being honest — all kind of guilt things pertaining to the relationship. You can think you're in love, but perhaps in retrospect you weren't — I have loved people, but I'm not sure I've ever been in love.'

In establishing as well as discontinuing long-term boyfriend affairs, the patient was definitely not a passive recipient, but an active initiator. She initiated other, casual dating relationships, also, but she was not sexuoerotically driven either to find a partner, or to masturbate. She localized the strongest erotic feeling genitally in the clitoris, though her whole body had always been tactually responsive, especially the neck and below the jaw.

About orgasm in real life and in a dream, her answer to both was: 'Yes. I think so. If you read *Cosmopolitan* [the monthly magazine], I'm quite sure I've never had an orgasm. From their description, it should be something between seven on the Richter scale and a tidal wave. But, from girl talk, and talking with men, women, people, you know, I think I have.' She could not be exactly sure about how orgasm became incorporated into a dream. 'Sometimes there is the feeling of knowing — a detachment from the dream. ... I remember one time I had a dream about horseback riding, something that suggests manipulation down there. ... In another dream I was in a disco with a round little open window in the middle and a man stepped through it. The next day I thought that was pretty crass sexual imagery.' She dreamed of herself sexually as 'myself as I would wish to be, rather than my real self.' There was no clear visualization of having a vagina, but there were 'dreams in which there was physical sensation of weight, of hair, of pressure, and of motion.' Dreams of sexual intercourse did not, however, figure prominently. As compared with dreaming, 'the orgasm of petting,' she said, 'is more intense, because it's emotional, and it's another human being. It's touching and feeling, and so on. That, I would say, probably, subjectively, has a much more intense kind of feeling.'

Asked to find words to describe the feeling of her orgasm, she said: 'Well, first of all there's the feeling of heat in the genital area. And tightness. And, uhm, just, uhm, heightened sensitivity, most of the time in almost any part of my body, but particularly in the more or less erogenous zones. Then when having an orgasm, it's a tremendous feeling of relief.'

Whatever the nature of this feeling, there was no doubt as to its authenticity, according to a subsequent report, four years later, of two orgasmic experiences in the course of postsurgical care following the operation for vaginoplasty: 'Basically, it was simply a matter of experiencing orgasm twice. Once when he was swabbing the area. It was just lovely. And I twitched. He said, Did I hurt you? And I said, No, quite the contrary. ... The second time was an even stronger experience. I was sort of under the influence of Demerol at the time. When the form that had been put in during surgery was taken out, and he began to irrigate the vaginal canal, I just had the most lovely, pulsating orgasm you can imagine. Without thinking, I just kind of groaned pleasantly and asked him if he was a happily married man! The next thing I knew, I was in the recovery room.'

After recovering from surgery, the patient was in no hurry to use her new vagina for fear of injuring it before it was fully healed. Subsequently, she was very pleased with its functionality. There was some erotic feeling around its orifice, but the chief source of genital erotic feeling continued to be in the clitoris.

It was not possible for the patient to give an estimate of the frequency of her orgasms. Their occurrence was variable and situationally related. One period, when she was 20, of eight months away from home she remembered 'very well, because I don't think I had an orgasm the whole time I lived there.' She was not averse to masturbating, but not very much given to it either.

Considering the sensory channels of sexuoerotic turn-on, her first answer was 'tactile. I'm very touchy-feely oriented. ... Visually, naked men don't really turn me on. Nor do naked women, for that matter. But, uhm, pornographic literature or X-rated movies, you know, can be psychologically very stimulating in that respect. A well written description can be very exciting.' Concerning her response to the image she might be reading or seeing, she said: 'I suppose it's probably a wish to participate, or a wish that I could do that — sometimes the response is almost hostile [because I can't].' She confirmed that men's and women's responses to explicitly sexy movies are 'very definitely' different: 'When you're with someone else, you know the two of you react differently. I've been virtually attacked in a movie seat. Some people get very excited, and suddenly you have an eight-handed date. And it just never occurred to me. I found it exciting in a, you know, detached kind of way. But evidently, he was there, all the way.'

Inquiry regarding the imagery and ideation of the patient's erotic arousal did not reveal any evidence of a paraphilic nature, that is of so-called kinky or bizarre sex.

During the patient's voluntary admission for participation in the nitrogen balance study, the following entry was made in her psychohormonal history:

'I dictate this note because her request for information on personal medical matters sheds light on the relationship of hormones and eroticism. She was pleased that the doctor in charge of her research protocol had talked with her in honest frankness. She was particularly interested in knowing that the injections she had been receiving were androgen, because of the general good feelings she had been receiving from them. She described this feeling first as one of being high. It was quite obvious to me that she was using lady-like expressions and euphemizing. I asked her if she meant simply a feeling of euphoria, or whether she meant also sexy feelings as well. She meant very definitely both kinds of feelings. She mentioned in particular that there was one doctor over there with big brown eyes. She had been flirting with him in particular. I made further inquiry as to whether the flirting referred to something in a romantic sense or whether perhaps it also referred to sexual feelings in the sense of in the pants, so to speak. It was definitely the latter, the patient said. As she discussed these matters I got the impression from her general manner and bearing that she was talking about something that had been pleasant for her. ... She wondered what changes she might expect to have to undergo if she should submit to a gonadectomy. She is very definitely against this operation. If I may infer from the conversation she produced, then I would say that she is against it partly because she views it as a tampering with the body and mutilation. She also is against it because she would not like to lose any of her present sexual feelings, and feels that castration may bring about such a loss. I mentioned the possibility of substitution treatment with not only estrogen but with supplemental androgen also. I did not really make much headway with this suggestion, however, as the patient gave expression to her own misgivings.'

Additional evidence of the patient's sensitivity to androgen, initially manifested subjectively as a positive erotic effect, was soon manifested objectively as a negative effect, namely bodily virilization, which was extremely rapid as to create an endocrinological crisis. Although the level of androgen in the bloodstream fell to its pretest value after the termination of the test, gonadectomy was nonetheless advised as imperative, in order to minimize further virilization. Despite her distrust, the patient agreed to be castrated.

The night before surgery, the patient recalled thinking: 'This is the last night I'm going to be the same me. ... I was really thinking that something dramatic was going to happen. Well, nothing did. I was afraid I was not going to have any libido, but I know I had an orgasm in my sleep a few nights after the operation. So I presume that's all functioning, except that, on the fifth night after surgery, I woke up having a hot flash, a diffuse feeling of heat all over, and perspiring, dripping all over. I had another incident exactly like it the next day.'

Six months later, in a written progress report, optimism had turned to despair:

'The most postive changes have occurred in relation to hair growth. Almost immediately following surgery in January, the hair on my head stopped falling out. By April, positive changes in thickness, texture, and general condition were discernable. This trend was accompanied by the return of a normal (rapid) growth rate of head hair, and has continued to the present time. Concomitantly, as you predicted, slight reduction in facial and body hair has been observed. Areas still evidencing considerably heavy growth include the upper lip, thighs (particularly upper inside areas), forearms, and hands. Following removal, however, the hair in these areas has required progressively longer intervals for regrowth. Since May, it has been necessary to tweeze my upper lip only once a week. Curiously enough, pubic hair growth appears retarded. In March, the distinction between the area shaved for surgery and the unshaved area was still quite distinct. Gradually, the hair growth extended to cover the total original area, but is generally more sparse than prior to surgery and falls out during baths.

'In the immediate post-operative period, facial skin became dramatically drier. Eruptions disappeared and genuine crinkles appeared around my eyes. Ugh. Application of various moisturizing agents alleviated feelings of tautness. By the end of March, however, eruptions began to reappear and such an alternating, irregular pattern has continued to the present time. Generally speaking, body skin is significantly drier than ever before.

'As anticipated, the deepened voice remains unchanged. Vocal quality now rivals the luxuriance of the hair on my head for casual comment among friends and strangers alike. This vocal change presents no special problems except that my reasonably fine singing voice has been reduced to an uneven growl not unlike that of a bull alligator in mating season. Distressing pitch breaks are almost totally absent, and with the passage of time, vocal control and flexibility in my new range are steadily increasing.

'Proceeding to erogenous zones, no ill effects seem to have accrued as a result of surgery. That is to say that my incisions have healed nicely and are relatively unobtrusive and insensitive to stresses. In the immediate post-operative period, I seemed remarkably robust: climbing stairs, standing for long periods, and general physical activity presented no problems. Changes in the breast area are less heartening. Actual tissue increase, if any, is slight and the lower edge of both nipples has turned very dark. Occasionally, heightened sensitivity is experienced, followed by general tenderness or soreness in the area. Since my bosom (or lack thereof) is an issue of high priority, I intend to ask for a frank assessment of the possibilities here. Anything you might know would, as always, be appreciated.

'Concerning erotic functioning, the data are unclear. Referring to my notes, I find frequent mention of the absence of at least two components of my sexual response. First, a sharp decrease in erotic dreaming has occurred. Prior to surgical intervention, I frequently (average four times weekly) had erotic (fondling, intercourse, etc.) dreams. These dreams regularly involved concrete sensory impressions, particularly vivid thermal, tactile, and visual responses. Following surgery, such dreams have become exceedingly rare (less than once a month). With one exception, these infrequent dreams seem subjectively less intense than those previously experienced. Frequency of general dreaming seems unchanged, only the content.

'The second missing element is less amenable to description. I refer to it as my self-starting mechanism, the capacity to respond (in terms of perceptible changes — increased heartrate, stiffening of the clitoris, moistness in eyes and skin, etc.) to visual or auditory stimuli. Previously, it was possible for me to induce these responses by concentrating on something ripe with sexual connotations: an attractive male, vividly written narrative, even a disembodied voice. One of the greatest assets of this mechanism was its effortlessness. As often as not, it simply occurred without conscious exertion of will. Since January, though, this ability has been sharply diminished. Even if I consciously strive to be responsive it is sometimes impossible. This I do not like. I have progressed from a very easily turned-on me to a frequently sullen touch-me-not. I rather miss the old 'vibes.'

'Frequent observations regarding moods appear in my notes. A broad overview suggests a rather intransigent sense of depression, often deep and accompanied by distressingly stubborn periods of apathy. Whether or not a causal link exists between my physical and psychological state would be difficult to determine. Certainly there have been a number of external problems increasing pressure on my bruised psyche. During periods of apathy, various self-destructive behaviors emerge. Not only do I fail to accomplish any substantive work, but I also neglect my appearance and my surroundings, ignore people, and phenomenologically experience a curious detachment from myself and objective reality. This I call the S.O.S. (shores of schizophrenia) status. Eventually, the mood passes and my behaviors cease to be perfunctory and resume purposiveness.

'Physiological effects of the estrogen pill in previous administrations, including nausea, weight gain, and splotched skin, have not occurred. Apart from some wrinkles and a few stray gray locks that might have emerged regardless, no adverse changes appear on the surface.'

EXPOSITION

Anyone meeting either of these two women would, without question, take for granted that they are heterosexual females. Nothing about their appearance, conversation, or behavior would show otherwise, either on first acquaintance or after a long and intimate friendship. Their cases absolutely contradict the proposition that their erotic orientation might be predicted as that of a male, on the basis of their 46,XY (male) chromosomal sex. Chromosomes, per se, obviously do not correlate with sexuoerotic orientation. Chromosomes are not, however, synonymous with heredity, nor with genes and the complete read-out of the genetic code. Hypothetically, it is possible that the missing genetic material presumed to be responsible for the dysgenetic gonads in Swyer's syndrome includes not only the TDF gene, but also another gene or genes responsible for masculine heterosexual orientation. To sustain this hypothesis, it would be necessary to make the unlikely assumption that the same genetic material is missing in the quite different syndrome of Case #2 — in which the TDF gene is clearly not missing, insofar as the testicles were very well formed and capable of being stimulated to function hormonally. On the basis of Case #2 alone, one might propose the hypothesis that biologically inert LH, or another substance riding in tandem with it (presumably a neurochemical), by being inert, allows a masculine heterosexual orientation to be suppressed, and a feminine heterosexual orientation to take its place. In the present state of knowledge, such a hypothesis is in the realm of science fiction, where it cannot be tested. Conjecturally, however, even in the realm of science fiction, it is less likely that the two different syndromes here represented would happen to share the same gene defect than that they would share a neurochemical anomaly that demasculinizes the sexually dimorphic differentiation of the brain in prenatal or neonatal life. If such an anomaly should indeed exist, it might in some way be associated not only with demasculinization of the external genitalia, but also with their representation as female organs in the body schema of the cerebral cortex. The feminine body schema would then be eligible for representation in the cognitional body image. Proprioceptively and tactually, the genital body image may, as in the case of phantom limbs, be independent of the visual body image. If so, then it could be a direct source of a diffuse awareness of feminine sexuoeroticism. (In Case #2, it may have been just such a diffuse feminine awareness that the patient was afraid of losing by consenting to removal of her testes.)

All of the foregoing speculation regarding an as yet unidentified, prenatally demasculinizing, neurochemical anomaly would apply not only to the two syndromes of this chapter, but also to other syndromes of prenatal demasculinization. It need not be conceptualized as an absolute determinant of erotosexual femininity, but rather as a contributory determinant only, originating in prenatal life. Subsequently, in postnatal life, it would be subject to the superimposition of other influences, notably those that commonly go under the name of social learning, and which enter the brain through the senses. The overriding power of these postnatal influences is made evident in other chapters in this book in which the pair is discordant for rearing and sexuoerotic orientation, though concordant for diagnosis and prenatal differentiation.

It is counterintuitive, at least in the thinking of many people, to be confronted with the evidence that the hormones of puberty, if they masculinize the body, do not automatically masculinize the mind. Case #2 of this chapter provides just such counterintuitive evidence. The flood of masculinized hormone that was released from the testicles, after they were stimulated by a test dose of gonadotropin, did not induce a masculinized sexuoeroticism. The woman did not find herself becoming attracted toward other women, nor falling in love as a lesbian. Instead, she continued to be attracted to men, and perhaps was a bit more passionate. Her case alone would not prove the point, but there are many others, some in the chapters of this book, that demonstrate that sexuoerotic passion as homosexual or heterosexual is completely independent of the male/female dimorphism of the hormones of puberty.

Whereas Case #2 is the first reported one of its kind and cannot, therefore, be compared with others, the same does not apply to Case #1, which can be compared with other cases of Swyer's syndrome. In endocrine and gynecological texts and research reports, it is always taken for granted that patients with Swyer's syndrome are women who come to attention because they are estrogen deficient and infertile, and not because of gender confusion. Even in the absence of a statistical survey, it is evident that the prevalence of lesbianism is negligible. The same applies in the case of Turner's syndrome which also is characterized by streak gonads, but differs chromosomally from Swyer's syndrome in having no Y chromosome. The typical chromosome count is 45,X, or some variant thereof, such as the mosaic 45,X/46,XX, in which some cells of the body have one X chromosome, and some two. There are 133 cases classified in the diagnostic file of the Psychohormonal Research Unit at Johns Hopkins listed under Turner's syndrome. None of these cases is cross-classified under either gender incongruity in childhood, or in adulthood under bisexuality or homosexuality (lesbianism). I have received only one professional consultation request regarding a woman with Turner's syndrome who was also a lesbian in sexuoerotic orientation.

Quite apart from formulations pertaining to sexuoerotic orientation, the patient with Swyer's syndrome in this chapter illustrates the way in which symptoms of a secondary syndrome may mask their derivation from a primary syndrome. As a teenager and college student, the patient appeared not to be perturbed by her primary syndrome, namely the failure of feminine maturation. There were no derivative symptoms of depression, anxiety, worry, or concern about her body and its functioning. Instead, derivative phobic symptoms manifested themselves – phobia of anesthesia, and phobia of riding elevators – which, by interfering with hospitalization, virtually ruled out the possibility not only of effective treatment but also of being given a dire and perhaps fatal prognosis. By the ordinary rules of rational thinking, it was not possible to recognize that there existed a derivative relationship between the primary syndrome of maturational failure and the secondary syndrome of phobic avoidance. This type of disjunction between a primary syndrome of birth defective sex organs and a secondary syndrome of behavioral pathology that derives from it is not unique to this case. Nor is it unique to syndromes of birth defect of the sex organs. It is likely to occur whenever a primary syndrome carries with it a stigmatizing and traumatic effect which becomes, metaphorically, an unspeakable monster in the patient's life. Unspeakable means, literally, not to be spoken about. The secondary syndrome masks or seals off that which is traumatizing and unspeakable. Thus it may be interpreted as a posttraumatic stress reaction. Unless the derivative connection between the primary syndrome and the secondary syndrome is recognized, there is little hope of effective therapeutic intervention for the relief of the secondary symptoms.

The unspeakable monster in the life of each of these patients harbored the classified secret of being sex-chromatin negative and chromosomally 46,XY. The patient with Swyer's syndrome might have been coercively entrapped into being publicly traumatized by this cytogenetic information had her career in sports progressed to the level of international competition. The International Olympics Committee requires that women competitors undergo proof of sex by submitting to the sex-chromatin test (*Lancet,* Sept. 19, 1987, 667–668; Nov. 28, 1987, 1265–1266). This requirement was first imposed in 1968, ostensibly in response to imposturing by males who had once taken estrogen so as to grow breasts and have the appearance of being female competitors in female events. Too prudish to require a straightforward physical examination of the genitalia, the Olympics Committee demurely opted for the modesty of the sex-chromatin test instead.

The argument that males have an unfair advantage over females in athletic competition is based on the difference in muscle mass and strength that is a direct effect of the male sex hormone, testosterone — known in sports medicine as an anabolic steriod. In the case of the two 46,XY women presented in this chapter, there was no spontaneous secretion of testosterone. Therefore, in competition with women of similar height and weight, they had no special athletic advantage. By contrast, they would have been hormonally handicapped in competition with males. After they began taking the female hormone, estrogen, they were hormonally on an equal footing with other women, despite their negative sex-chromatin status. In effect, the result of the sex-chromatin test would be telling a lie about their right to compete as women. It is a test that does not legitimately apply to 46,XY women. For them, it should be abolished. Measurement of the level of estrogen circulating in the bloodstream would be more legitimate and more fair. It would also be not traumatizing.

BIBLIOGRAPHY

Espiner, E.A., Veale, A.M.O., Sands, V.E. and Fitzgerald, P.H. (1970) Familial syndrome of streak gonads and normal male karyotype in five phenotypic females. *New Engl. J. Med.,* 283:6–11.

Money, J. (1968) *Sex Errors of the Body: Dilemmas, Education, Counseling.* Baltimore, Johns Hopkins University Press.

Page, D.C., Mosher, R., Simpson, E.M., Fisher, E.M.C., Mardon, G., Pollack, J., McGillivray, B., de la Chapelle, A. and Brown, L.G., (1987) The sex-determining region of the human Y chromosome encodes a finger protein. *Cell,* 51:1091–1104.

Park, I.J., Burnett, L.S., Jones, H.W. Jr., Migeon, C.J. and Blizzard, R.M. (1976) A case of male pseudohermaphroditism associated with elevated LH, normal FSH and low testosterone possibly due to the secretion of an abnormal LH molecule. *Acta Endocrinol.,* 83:173–181.

Swyer, G.I.M. (1955) Male hermaphroditism: A hitherto undescribed form. *Br. Med. J.,* 2:709–712.

CHAPTER 5

Concordant for 46,XY hermaphroditism, discordant for rearing and gender-identity/role: two matched pairs divergent for virilizing and feminizing puberty

SYNOPSIS

There is in this chapter not one matched pair of cases, but two. All four cases were classified as concordant on the criterion of intersexual birth defect of the sex organs which was generically diagnosed as male hermaphroditism – 40 years ago there were no diagnostic tests for specifying subtypes. Fetal differentiation of the external genitalia had been so demasculinized that the announced sex at birth was female. In infancy, the sex was reannounced as male in two cases, whereas in the other two cases it was not. Thus, from infancy onward, the rearing was in two cases as a boy, and in two as a girl. In childhood, the two boys underwent repeated masculinizing genital surgery to construct a urethra in the clitoridean penis and to bring the testes into the descended position in the reconstructed scrotum. There was no corresponding attempt at feminizing surgery in the two being reared as girls, because their genital anomaly did not generate special attention.

At the age of onset of hormonal puberty, one of the boys and one of the girls began to masculinize under the influence of the hormones secreted by their own testicles. The cells of their bodies were androgen sensitive and responded to the male hormone secreted by the testicles. The other boy and the other girl began not to masculinize, but to feminize. The cells of their bodies were androgen insensitive. They rejected the male hormone secreted by the testicles and responded only to the lesser amount of female hormone normally secreted by the testicles.

The girl whose body masculinized at puberty and the boy whose body feminized were traumatized by the sexological failure of their bodies to agree with, respectively, the femininity and masculinity of their minds. It proved clinically simple to arrest the girl's hormonal masculinization and to substitute not only hormonal feminization, but surgical feminization of the genitalia also. Clinically, nothing was simple

TABLE 1
Diagnostic and somatic data

	Pair I		Pair II	
	Status female WV	Status male MV	Status female WF	Status male MF
Sex assigned at birth	female	female	female	female
Sex reannouncement	none	$5^1/_2$ months	none	6 months
Age first seen at PRU	10:11	3:7	13:6	0:6
Chromosomal sex	male	male	male	male
Gonadal sex	male	male	male	male
Phallus[a]	3.5 cm × 1.5 cm hypospadiac with chordee	3.2 cm × 1.5 cm hypospadiac with chordee	6 cm × 2 cm hypospadiac with chordee	1.5 cm × 0.7 cm hypospadiac with chordee
Urethral opening	into urogenital sinus	into urogenital sinus	female position	into urogenital sinus
Labia/bifid scrotum	labia majora well formed, slightly fused	labia majora well formed, slightly fused	labia majora well formed, no fusion	labia majora well formed, no fusion
Vaginal opening	into urogenital sinus	into urogenital sinus	female position	into urogenital sinus
Vaginal canal	blind vaginal pouch	blind vaginal pouch	blind vaginal pouch	blind vaginal pouch
Internal genitalia	bilateral high-inguinal testes; well developed wolffian structures	bilateral inguinal testes; well developed wolffian structures	bilateral high-inguinal testes; well developed wolffian structures	bilateral testes above the inguinal ring; well developed wolffian structures
Puberty prior to hormonal therapy	virilizing	virilizing	feminizing	feminizing
Age and surgery	11 gonadectomy and phallectomy 21 vaginoplasty	$3^1/_2$ release of chordee 5 excise vaginal pouch, construct perineal urethra 7 hypospadiac repair	$13^1/_2$ phallectomy gonadectomy 23 vaginoplasty 29 minor surgery to repair constricting bands at vaginal opening	7 release chordee; excise vaginal pouch $10^1/_2$ hypospadiac repair 13 hypospadiac repair 13 mastectomy 16 repair urethral fistula
Hormonal substitution therapy	Premarin 1.25 mg/day	none	Premarin 1.25 mg/day	testosterone enanthate, irregular compliance
Response to hormone	good	not applicable	good	negligible

TABLE 1 (continued)

	Pair I		Pair II	
	Status female WV	Status male MV	Status female WF	Status male MF
Age at last follow-up	33	26	37	31
Years of follow-up	22	23	24	30
Diagnosis	male hermaphrodite with hypospadias	male hermaphrodite with hypospadias; 17-ketosteroid reductase deficiency conjectured	male hermaphrodite; partial androgen insensitivity	male hermaphrodite; partial androgen insensitivity

[a]All data on the external and internal genitalia apply to the age when first seen.

for the boy who feminized, for his body was genetically programed to be unresponsive to all attempts at hormonal virilization. There was no technology in plastic surgery that could enlarge his penis as an erectile organ. Apart from having a normal adult height, he was condemned to having, in adulthood, a perpetually juvenile appearance which denied him the social and vocational status of a man.

These four cases in two matched pairs illustrate that the masculinity/femininity variable of gender-identity/role overall, and in the sexuoeroticism of the lovemap, is a product in part of prenatal determinants, but with a heavy overlay of postnatal determinants. The similarities and differences of the four cases are summarized in Tables 1–4. In the narratives that expand the data of these tables, the codes are WV, woman who had begun virilizing at puberty: MV, man who had begun virilizing at puberty; WF, woman who had begun feminizing at puberty; and MF, man who had begun feminizing at puberty.

The four case narratives examine the vicissitudes experienced by the four people growing up to womanhood and manhood. The duration of the follow-up period ranged from 19 to 30 years. The age at initial contact ranged from 5 months to 13½ years, and at final follow-up from 29 to 33 years.

For each patient there had been kept a consolidated psychohormonal case file. It is composed of endocrine and other clinical notes, surgical reports, laboratory data, social and school reports, psychological test records, and transcripts of taped interviews with patients and parents, and observational notes. Each file has been indexed and abstracted in conformity with the conceptual design and rubrics of this fourfold report.

TABLE 2
Gender-identity/role and its erotosexualism

	Pair I		Pair II	
	Status female WV	Status male MV	Status female WF	Status male MF
Sex announced at birth	female	female	female	female
Postneonatal sex of rearing	female	male	female	male
G-I/R status self-defined	female	male	female	male
Clothing and grooming	female	male	female	male
Heterosexual dating	age 12	age 12	age 14	Age 17
Going steady	age 13	age 20	age 17	age 17
Falling in love (self-report)	age 19	age 20	age 17	age 17
Marriage	age 26	age 21	age 17; 23	age 22
Sexual partner(s)	exclusively heterosexual	exclusively heterosexual	exclusively heterosexual	exclusively heterosexual
Age kiss and pet only	12–24	16–21	16–17	17–18
Age at first intercourse	24	21	17	18
Coital imagery	present and heterosexual	present and heterosexual	none reported	none reported
Erectile phallus	clitoridectomy	yes	clitoridectomy	yes
Vaginal adequacy	good	n/a	good	n/a
Oral sex	yes	yes	no	yes
Frequency of intercourse	5–7 per week	daily	2–3 per month	infrequent
Erotic initiator	initiates erotic activity 40–50% of the time	initiates erotic activity 90% of the time	initiates erotic activity 20% of the time	varies greatly; erotic inertia and apathy
Orgasm	spasmic peak	spasmic peak	relief of tension	relief of tension
Orgasmic fluid	vaginal lubrication	no ejaculate	vaginal dryness	no ejaculate
Masturbation	frequent in childhood, negligible now	daily in adolescence, sporadic now	none in childhood, sporadic now	occasionally in childhood and adulthood
Erotic zones				
Penis	n/a	yes	n/a	yes
Clitoris[a]	yes	n/a	no	n/a
Vagina	yes	n/a	yes	n/a
Breast	yes	yes	yes	yes
Erotic stimuli	touch and nude pictures	touch, nude pictures, sounds, and smells	touch and romantic stories	touch and stag movies

n/a, not applicable.
[a]The zone from which the enlarged clitoridean phallus had been surgically removed.

TABLE 3
Education, occupation and group status

	Pair I		Pair II	
	Status female WV	Status male MV	Status female WF	Status male MF
Recreational energy history	vigorous, outdoor, competitive	vigorous, outdoor, competitive	sporadic, relaxing, non-competitive	sporadic, relaxing, non-competitive
Dominance fighting history	starts fights, fights in self-defense	starts fights, fights in self-defense	gets along with everyone	fights only in self-defense
Social leader	organizer and leader	organizer and leader	group participant	group participant; recent self-isolation
Bonding with own parents	positive	positive	resolved teenage discord with mother	extreme ambivalence; loyalty vs. reproach
Behavior disorder	none	marital discord	none	incapacitating bouts of inertia and apathy; failed marriage
Education	high school; secretarial; Assoc. Arts	high school; BS; professional school graduate	high school; beauty school	high school; BA; MA (incomplete)
Occupation	managerial business partnership with husband	professional	managerial business partnership with husband	student dropout; volunteer service
IQ test	WISC	WISC	WAIS	WISC WAIS
Verbal IQ	104	143	112	135 128
Performance IQ	113	125	103	97 99
Full IQ	109	138	109	119 116

MATCHED-PAIR CASES WV & MV, AND CASES WF & MF

Diagnostic and clinical biographies

Case WV According to family tradition, the genital anomaly in this patient passed unnoticed in infancy and childhood. She was declared and reared as a daughter. There was an older brother, aged 2, and seven years later a younger brother. The first signs of puberty, pubic hair and huskiness of the voice, were recorded by a local pediatrician who first saw the child when she was aged 10:6 years. He gave to the parents a diagnosis of cryptorchidism and hypospadias in a male, and advised that

TABLE 4
Data on parentalism

	Pair I		Pair II	
	Status female WV	Status male MV	Status female WF	Status male MF
Parentalism in childhood play	boys' play, help with domestic chores, babysitting	boys' play, no interest in domestic chores	maternalism in doll play	boys' play, no interest in domestic chores
Parentalism after childhood	adoption	donor insemination	changed to pro from con	no interest
Fantasy of pregnancy	yes	no spontaneous mention	negative report	no spontaneous mention
Attitude toward baby care	enjoys infant	enjoys infant	changed to pro from con	no interest
Attitude toward infertility	disappointed/accepted	denied/resigned	accepted	conflicted

the child should be a boy. Unable to reconcile the contradictions involved in having their daughter changed to a son, the parents insisted on a second opinion, and were referred to Johns Hopkins. They had correctly surmised that the external genitalia were not suitable for masculinizing surgery, and they knew that their daughter had never seemed to them to be like a boy.

When initially examined at Johns Hopkins in the pediatric endocrine clinic, the girl's age was 10:11 years, the height age 12:6 years, and the bone age 12 years. Somatically she was in the first phase of a virilizing puberty, ahead of her 13 year old, prepubertal brother. The genital examination revealed a clitoridean phallus, 3.5 cm × 1.5 cm, labia minora and labia majora (Fig. 5.1). There was no external vaginal opening. The single external opening, 1 mm × 3 mm, was interpreted as a urethral orifice rather than a large urogenital sinus, but internally it did connect with a small vaginal pouch as well as with the bladder. Gonadal masses could be slid down into the inguinal canals, but not as far down as the bilateral labia majora (also referred to as a bifid scrotum). The laboratory results showed 17-ketosteroids, estrogen, and FSH (follicle stimulating hormone) to be not remarkable in relation to puberty of virilizing onset. The sex chromatin was negative (male).

A consensus was reached in favor of preserving the child's psychosocial and psychosexual status as a female, rather than dictating a change of sex by edict and attempting masculinizing genital surgery. Accordingly, a program of surgical and hormonal feminization was designed.

The first step in surgical feminization included exteriorization of the vaginal orifice, gonadectomy, and cosmetic penoclitoral revision, and was completed at age 11

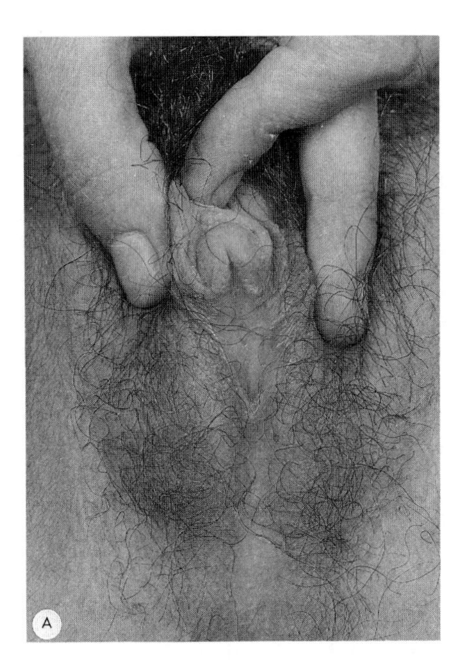

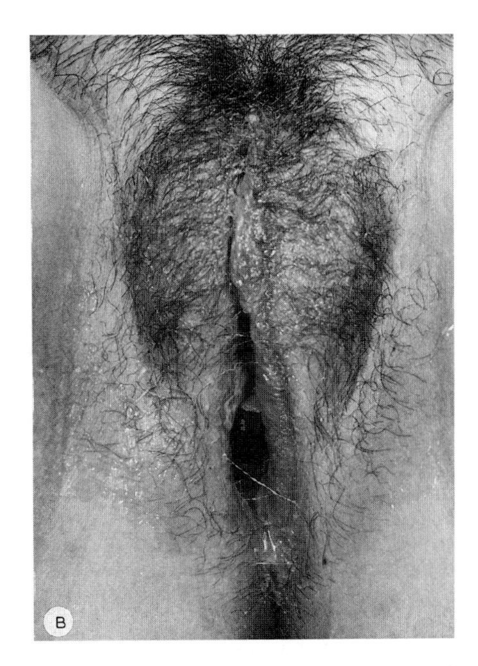

Fig. 5.1. Case WV. Appearance of the external genitalia (A) before surgery at age 11, and (B) at age 20, nine years after surgery.

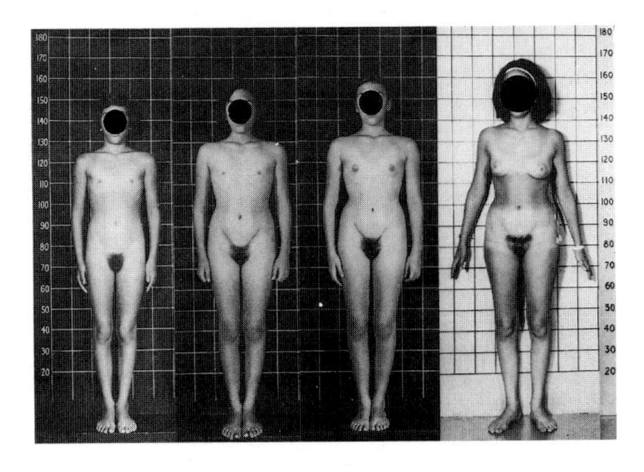

Fig. 5.2. Case WV. Four stages of hormonal feminization of the body habitus, ages 11, 12, 13, and 20.

(Fig. 5.1). Histologically, the gonads proved to be testicular. Their removal prevented further pubertal virilization and allowed the overall morphologic sex to differentiate very satisfactorily as womanly. There was no menarche, as the internal mullerian organs were absent. The second stage of surgical feminization, vaginoplasty to lengthen the existing cavity of the vagina, was performed when the patient was 21 years old.

The onset of estrogen therapy to induce pubertal feminization was postponed until age 12:5, so as to allow increased statural growth. During the interval the patient retained a prepubertal body configuration. At age 12:5, the height was 157.7 cm (62 in.; average for age (female norm) 58.7) and the weight was 49.2 kg (108.5 lb; average for age (female norm) 85.1). After the first eight months of hormone therapy (1.25 mg Premarin daily), breast development was evident. One year later, the brassiere size was 32A. The voice had ceased deepening. Examined at age 14, the vaginal mucosa was well estrinized, and the labia majora appeared well formed. The vagina was 6 cm deep and 1.5 cm in diameter at the opening.

At age 16, the body had continued to feminize (Fig. 5.2). The brassiere size was 34B. The height measured 165.4 cm (65 in., which is average for an adult female), and the weight was 49.2 kg (108.5 lb; average for age (female norm) 110.9). In adulthood, hormone substitution therapy (Premarin, 1.25 mg daily) was maintained under the supervision of a local physician.

Case MV This patient was the first of three siblings. The brother was $1^{1}/_{2}$ years younger, and the sister 15 years younger. The neonatal declaration of the patient's sex was revised at age 5 days from boy to girl, because of the appearance of the genitalia. When the baby was five months old, an aunt who was a pediatric nurse suspected a diagnosis of male hypospadias and urged a diagnostic re-evaluation. Four local specialists agreed that the child was a male with a hypospadiac penis (not an enlarged clitoris), and a bifid scrotum (not two labia majora). Therefore, surgery was performed to release the chordee which kept the phallic organ bound into the clitoridean position, and to permit it to dangle more. The parents were told to wait until age 7–9 years for further operations. Townspeople reacted to the change in sexual status by referring to the child as a 'morphodite'. The family physician, to prevent the difficulties of the boy's being unable to stand to urinate, referred the family to Johns Hopkins when the child was 3:6 years old. The earlier diagnosis was confirmed. The clinical technique of chromosome counting had not yet been developed. The sex-chromatin test (Barr and Bertram, 1949) was done by Murray Barr at the University of Western Ontario and was negative, that is to say, male in type. The gonads were palpable in the inguinal canals. They could be pushed down into the labioscrotum, but they retracted back into the canals. Later in life they descended permanently. By age 12, when puberty began, they proved to be androgen-secreting, for pubertal development was virilizing (Fig. 5.3). Their infertility was not confirmed until many years later when donor insemination was required for parenthood.

Two months after his 5th birthday, the child was admitted for the second operation for genital reconstruction, namely, excision of the vaginal pouch from the urogenital sinus, and closure of the sinus to form a perineal urethral opening. Another operation would be needed to construct a penile urethra, and the recommendation at age

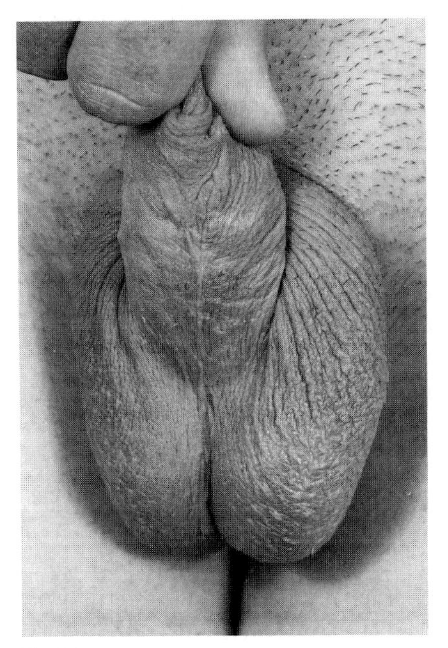

Fig. 5.3

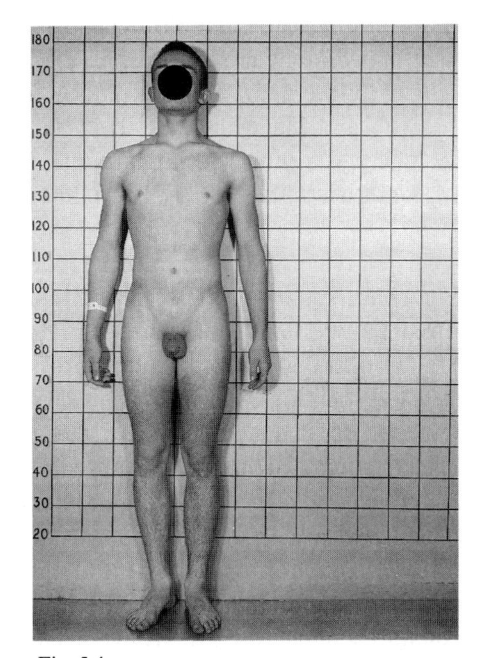

Fig. 5.4

Fig. 5.3. Case MV. Appearance of the external genitalia at age 16, after masculinizing surgery in three stages completed at age 7:6.

Fig. 5.4. Case MV. Appearance of the body habitus at age 16.

5 was to wait until the penis had enlarged at a later age. The possibility of using testosterone ointment to enlarge the penis was not mentioned. The penis was still small when the next operation was done at 7:6 years of age (Fig. 5.4). The result was that the boy was thenceforth able to stand to urinate. There were, however, recurrent acute attacks of urinary tract infection, involving inflammation of the bladder, testes, and scrotum, and requiring persistent antibiotic therapy.

At age 12, the boy returned for a pubertal-age checkup. Spontaneous masculinizing puberty was under way. The muscle tone was good. The voice had begun lowering. The height was 150 cm (59 in.; average for age 57). The weight was 39.2 kg (86.5 lb; average for age 78.3). Axillary and pubic hair were present. The once bifid scrotum was unusual in being bilobed in appearance, but was normally wrinkled and vascularized. The testes were firm and properly descended. They measured approximately 3 cm × 1.5 cm. The stretched length of the penis (also its erectile length) was 6 cm (2.4 in.). By age 14:3 years, the penile dimensions were 8 cm × 3 cm (3.25 in. × 1.3 in.), which is well below average. The testes measured 4 cm × 1.7 cm, which is slightly below average, and were softer than expected. The prostate gland was small. Hair was abundant in the axillae and in the pubic area. Beard fuzz had begun to appear. The height was 169 cm (66.5.; height age 13 years), and the weight 58.6 kg (129 lb; average for height 125). In adulthood, the patient returned for help with the problem of infertility. He was vehemently phobic of genital exposure and could not undergo a genital examination. He considered his penis too small. His general appearance was that of a well developed, well virilized man of normal build and stature.

Case WF This patient was an only child. Her birth was preceded by one and followed by two first-trimester miscarriages. There was vague uncertainty about the baby's sex at the time of birth, because the clitoris appeared slightly too large. When the child was 5 years old, a pediatrician in a major medical center said that nothing could be decided until puberty. The parents lived with a silent foreboding as to whether their daughter would develop normally. At age 12, the girl had what was construed as a first, very light menstrual bleeding and then, for several months, 'disappearance of the menstrual periods.' They consulted a local physician. Recognizing that the reproductive organs were atypical, he made a referral to Johns Hopkins. The patient was 13:6 years old. On physical examination, the body proportions were found to be moderately androgynous, but the muscle and fat distribution were feminine, and the breasts well developed (12 cm × 8 cm) (Fig. 5.5). There were a few hairs in the abdominal midline, over the sternum and over the sacral area. Pubic and axillary hair were normal for an adolescent girl. The height was 160 cm (63 in.; height age 16 years). The weight was 58 kg (128 lb; average for height (female norm) 111). The clitoridean structure was covered with much redundant skin (Fig. 5.6). Its stretched length varied according to the examiner: 6 cm × 2 cm and 8 cm × 2 cm. The urethral and vaginal orifices were separate. The vaginal mucosa was estrinized. The vagina had a diameter of 1.5 cm, and a depth of 3 cm. The sex-chromatin test was negative, as typical for a male. The 17-ketosteroids (18.9 mg/24 h) were elevated (female norm) and within normal limits (male norm). FSH was within normal limits. Estrogen was not measured, the method available at that time being undependable.

Three months later, exploratory surgery (laparotomy) revealed that the internal mullerian structures had failed to differentiate, and that the wolffian structures had differentiated. The gonads were removed from the inguinal canals and proved to be testicular in structure. The clitoris was cosmetically revised (Fig. 5.6).

The combined endocrine and surgical findings established the diagnosis as androgen-insensitivity syndrome of male hermaphroditism, incomplete type.

Vaginoplasty to lengthen the vaginal canal was postponed until age 16, when the body was fully grown. Postsurgically, over the course of years, constricting bands developed near the mouth of the vagina. They required minor surgery for their release, which was done at age 29.

The rationale for gonadectomy at age 13 was primarily as a prophylaxis against possible malignancy in later life. Since it removed the source of endogenous estrogen,

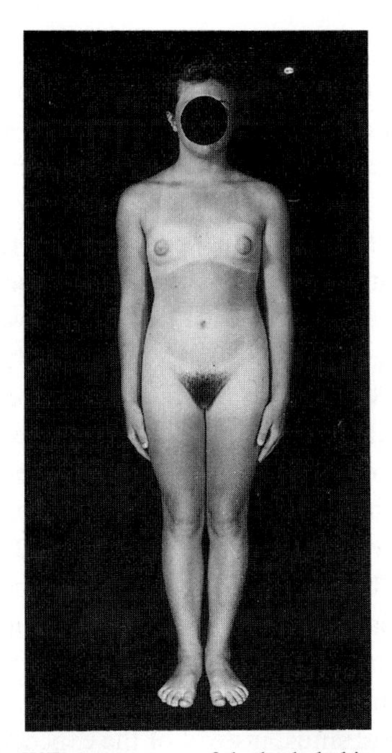

Fig. 5.5. Case WF. Appearance of the body habitus at age 13:6.

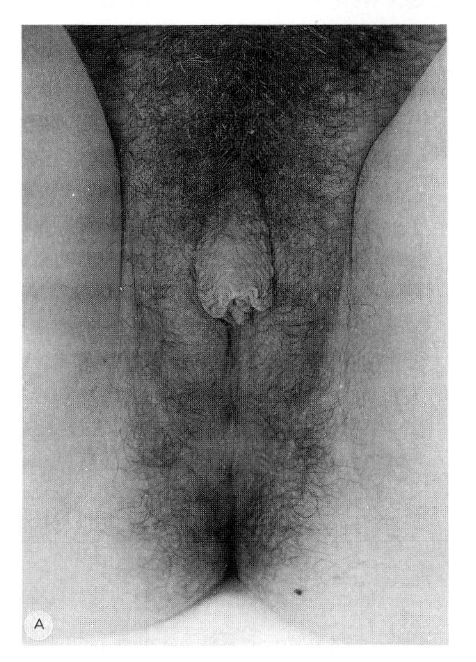

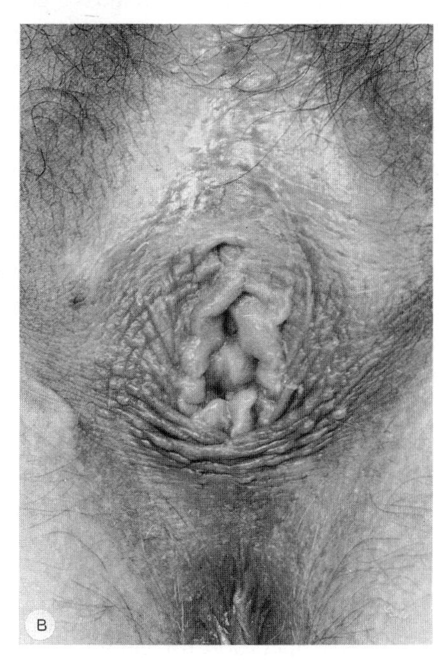

Fig. 5.6. Case WF. Appearance of the external genitalia (A) at age 13:6 and (B) in adulthood, 15 years after surgery.

hormonal replacement therapy (Premarin, 1.25 mg/day) was instituted and has since been maintained with complete success. As an adult, the patient is an attractive woman with a normal-appearing body and stature.

Case MF When this baby was born, the product of the mother's first and only pregnancy, the maternal grandmother alone insisted that the genitalia needed investigation. A doctor and nurse in the obstetrical clinic both were evasive. Not until three and a half months later did the parents find a physician who admitted that the vagina was not normally developed, and that the clitoris was a little enlarged. Another physician recommended that they 'wait until the child is fourteen and the true sex will show itself.' Not one to live in limbo, the mother pressed for further expertise. Two months later, when the baby was first seen at Johns Hopkins, she had a girl's name and the first reports were written using the pronouns she and her. A month later the pronouns had changed to he and him, the name had changed to that of a boy, and the sex had been reannounced on the authority of the findings of an exploratory operation. On physical examination, it had not been possible to demonstrate any masculine organs; but the genital examination disclosed that there was only a single opening, that of a urogenital sinus, instead of the expected two separate orifices, one urinary, and the other vaginal. No gonadal masses were palpable in the inguinal canals. Only under anesthesia could these masses be forced down to the position of the internal inguinal ring. They were located at the time of surgical exploration, biopsied, and released so that they could be brought down into the labia majora and sutured there. Microscopically, the biopsy was read as immature testis. Subsequently, the right testis remained in place, partially descended, but the left retracted into the abdominal cavity. The surgical exploration revealed a small vaginal pouch, opening internally into the urethral outlet. There were no mullerian structures, and the wolffian structures were poorly differentiated.

Surgical masculinization of the external genitalia was postponed until the penis grew bigger. In an experimental attempt to anticipate the effect of puberty in promoting phallic growth, a 5 mg dose of methyltestosterone was given daily by mouth. Three months later, there was a moot possibility of increased genital pigmentation, and of an increase in phallic length from 1.5 cm to 2.0 cm. The dose of methyltestosterone was increased to 10 mg daily. After another 17 months, the size of the phallus may have increased to 2.7 cm × 1.0 cm, unless the change was an artifact of measurement. Contrary to expectation no sexual hair had grown in response to the hormone. The concept of androgen insensitivity, scarcely recognized at the time, here was making itself evident – a bad omen for the boy's future development.

The first operation for genital reconstruction was done at age 7, with release of the clitoridean phallus from the chordee (actually the labia minora) that prevented it from dangling freely, and excision of the vaginal pouch. The next operation was at age 10:6 for first stage construction of a urethra (Fig. 5.7). The second stage was completed two years later, at age 12:6. The penis now was equipped with a urethra,

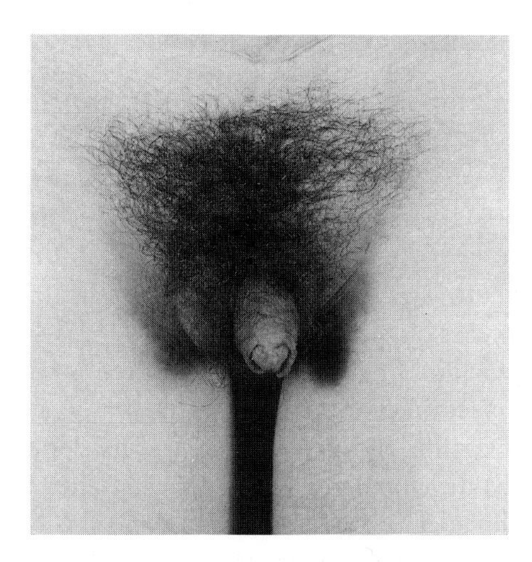

Fig. 5.7. Case MF. Appearance of the external genitalia at age 11:2.

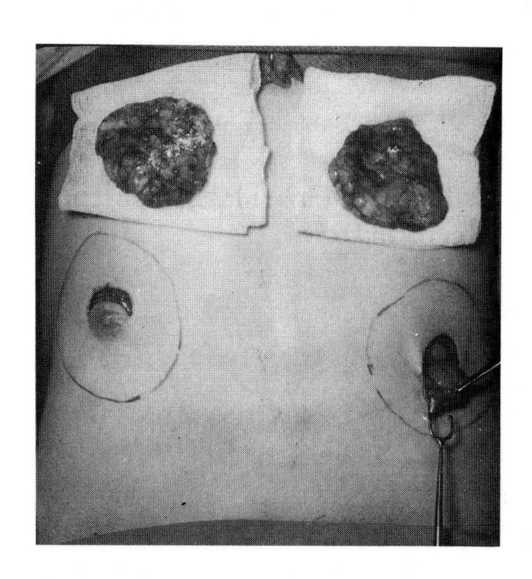

Fig. 5.8

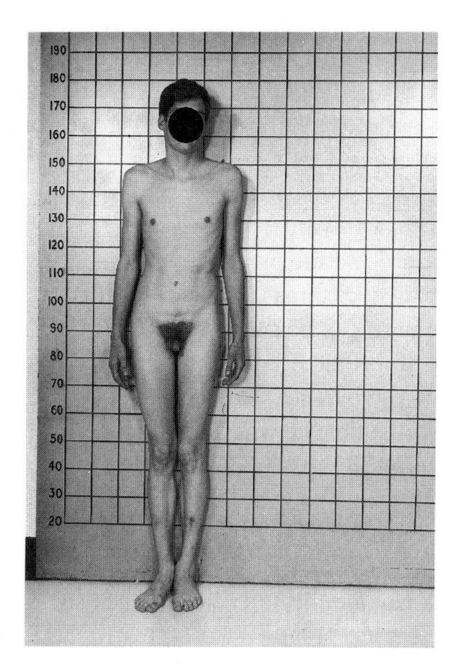

Fig. 5.9

Fig. 5.8. Case MF. Glandular breast tissue surgically removed at age 14:3.

Fig. 5.9. Case MF. Appearance of the body habitus at age 17:3.

but since it measured only 5 cm × 1.7 cm, it was too short to facilitate urination in the standing position. In any case, a fistula developed from which urine leaked. It did not finally yield to surgical repair until age 16.

The onset of puberty was first evident by age 12. It was not a virilizing puberty, but feminizing. The breasts were beginning to enlarge – yet another sign of the androgen-insensitivity syndrome (or the incomplete form of it). In two years they had become so large and stigmatizing that plastic surgery (mastectomy) was necessary (Fig. 5.8). The boy was 14:3 years old.

For the preceding year and a half he had been given a high dose of testosterone enanthate, 250 mg by injection every 10 days, in the first of a series of all-out efforts, in vain, to promote virilization. By age 17:3 there was moderate growth of pubic hair, but no prostate or beard development. The penis's stretched length, which is also its erectile length, was optimistically measured at 7 cm × 2 cm. Already at this age, there had begun a chronically recurrent urinary tract infection, secondary to postsurgical anatomical anomalies. Urinary backflow induced severe edema and cystitis.

Laboratory tests performed when the patient was 23 years of age indicated a 46,XY chromosome count; a normal production of testicular testosterone; and also the intracellular protein-binding deficit that prevents target cell response to the hormone. The height and weight at this time were 173 cm (68.2 in.) and 52.1 kg (115 lb), respectively. The appearance was of tall and gangling stature, and juvenile features, so that most people grossly underestimated the man's age and stigmatized him as a 16 year old (Fig. 5.9).

When dihydrotestosterone became therapeutically available, the patient was 26. Since it held the promise of being more effective than testosterone, of which the patient had tried massive doses (400 mg/week of intramuscular testosterone enanthate, at one stage), he contracted to try it. It came in the form of a lotion to apply to the face to promote beard growth. For months on end, however, he reacted to it with phobic avoidance.

By the time he was 31 years old, he had discovered that his problem of taking testosterone disappeared when he had pellets implanted subcutaneously every half year. At this time he was able, on the advice of a local urologist, to undergo surgery for the removal of the left, undescended gonad. The right, descended one was left in place.

A few months later he returned for re-evaluation of plasma hormonal levels: testosterone, 1410 ng/100 ml (greatly elevated), dihydrotestosterone, 72 ng/100 ml; LH, 40 mIU/ml; FSH, 44 mIU/ml. These findings confirmed the diagnosis of partial androgen insensitivity.

Lovemap biographies

Table 2 provides a tabular overview of the four cases on a range of variables pertaining to overall G-I/R and adult sexuoerotic functioning. From inspection of the table, it can be seen that, whereas all four patients had been neonatally announcted as girls, the two who were reannounced as boys, at the latest by age six months, subsequently diverged in G-I/R differentiation from the two who remained unreannounced and were known always as girls. The two girls grew up to have the social status of women, and the two boys the social status of men, irrespective of the personal trials and tribulations that were or were not involved in maintaining gender status. None had the status of being transexual, transvestite, gynemimetic, andromimetic, androgynous, bisexual, or homosexual – using the ordinary criteria by which these statuses are defined.

The lower portion of Table 2 deals with details of genital erotosexualism and pair-bonding. Inspection of this part of the table discloses a disparity less on the criterion of rearing than of diagnosis. There is a degree of parallelism between the androgen-responsive Pair I and the androgen-insensitive Pair II that transcends their respective divergency of status as man or woman. For example, both the man and the woman of Pair I reported having coital imagery, experiencing orgasm as a spasmic peak, and having had frequent juvenile masturbation and adult coitus, whereas those of Pair II reported paucity of coital imagery, of experiencing orgasm as a progressive relief of tension rather than a sudden and spasmic climax, and of having had a history of infrequent masturbation and coitus at any age. MF, the androgen-insensitive patient with a man's status conformed less to the masculine stereotype than WF, his matched-pair counterpart, conformed to the feminine stereotype. Both androgen-responsive members of Pair I conformed, respectively, to the masculine and feminine stereotypes.

Tables 3 and 4 permit comparison of the two pairs on various nonerotosexual components of gender dimorphic behavior related to work, play, social interaction, and parenting. Inspection of Table 3 discloses that the two pairs differ kinetically, with the androgen-responsive pair, regardless of upbringing, being more inclined to vigorous and competitive recreational energy expenditure, dominance assertion, and being a leader. By contrast, the androgen-insensitive pair, again regardless of upbringing, were more inclined to relaxational recreation, unassertively getting along with others, and following a leader. This contrast between the two pairs is consistent with observations, based on other human and monkey syndromes of hermaphroditism, regarding the behavioral effects of both prenatal and postnatal androgenization (reviewed in Money and Ehrhardt, 1972; Money, 1980, 1988).

Table 3 discloses also that the two individuals living as men had behavioral difficulties in the role of husband which their matched counterparts did not have in the role of wife. In MV, the marital problem was one of arbitrary authoritarianism in domestic affairs, and of what may best be called a counterphobic fantasy of hyperphilia. It appeared to be a problem of manageable dimensions. By contrast, in MF, the dimensions of the marital problem became unmanageable. The patient became progressively incapacitated with inertia and apathy, and at times unjustly and depressively vindictive toward those closest to him. Eventually, he left his wife and took up a self-sabotaging type of existence from which he eventually graduated to return to his home of origin, shortly after his father died. The chief precipitant of his despair and decline was almost certainly the total failure of his body to virilize, and its capability of fulfilling, at best, only the role of a eunuch.

Case WV At the time when this child's sexual status as a girl was first questioned, a few months after her 10th birthday, the father, a man in his late thirties, was recovering from a heart attack. Because of his health, the family had within the preceding six months relocated in a distant state. The daughter had then developed a tic-like mannerism of the face, neck and shoulder, as if tossing back hair that obstructed her vision. Since a shorter haircut brought about no change, the mother took the child to a pediatrician, requesting treatment. She was aware at this time of a huskiness in her daughter's voice, but respecting the child's modesty, had not seen the beginning of pubic hair, nor the protuberant enlargement of the clitoris which the prediatrician soon would demonstrate to her. 'He worked with the clitoris,' she recalled, 'and showed me the penis standing up. He frowned, as if to say he couldn't believe I hadn't seen it before ... and at the bottom of the left side he pushed down a testicle. ...' With the mother thus starkly confronted with the concept of having not a daughter, but a son, the pediatrician suggested calling the child Rob instead of Roberta. When he did so, himself, the child immediately replied: 'Don't call me that; that's a boy's name.'

'There was no cause for me to doubt that she was a girl when she was little,' the mother said. 'It was just clear cut.' But, searching her memory, she found 'one or two items a little bit different. ... When she was four of five we were out in the country, the boys and the girls running around just in their drawers, and they'd urinate, so she'd take her pants off and do likewise, like a boy. ... I said to her, you're a girl, you should be sitting down, not standing and urinating.'

'At the age of two years, it would have been so much easier to change. She would have gone as a boy. Now that she's playing with girls, polishing her nails, and thrilled about lipstick, it's going to be very hard for her. She wants her hair long, in a permanent.' The father had had a fantasy of himself taking a gun to the doctor if he robbed him of his daughter. 'Probably because I want her to be a girl so badly,' he said, 'I wouldn't listen to him.'

More realistically, both parents agreed with what the father later put into words: 'I always want to have two doctors' opinions before I do any surgery. Another thing, the pediatrician said there is a chance she could be made a girl by giving her cortisone, female hormone. [The pediatrician was referring to CVAH, see Chapters 10, 11]. I don't care if she would be sterile and not be able to have children, or anything. Not for the fact that I want her to be a girl – I just don't want her to be all messed up.'

Both parents assimilated elementary notions of the homologous nature of the male and female genitalia and of the surgical possibility of feminization. But they explicitly refrained, the mother later disclosed, from saying things that might imply that they were trying too much to influence the decision. Nonetheless, they had wanted a second opinion. Therefore, their pediatrician had referred them to the country's leading expert on syndromes of genital ambiguity, Lawson Wilkins of Johns Hopkins.

At the time of the initial examination at Johns Hopkins the parents were taken aback when they overheard a telephone call in which a urologist said that the child had 'male features.' Along with other physicians, they categorically disagreed. They knew that their daughter resembled other females in the maternal line. Her features were full rather than delicate in contour, and in marked contrast with the more finely chiseled features or her older brother. As a baby she had been unprepossessing in appearance. Her older brother always got comments on his good looks, and she on her good behavior – the little sister they had both wanted for the older brother. During the summer of her fourth year the girl wakened from bad dreams, crying, and asking her parents for reassurance that she was not an ugly girl, but a pretty one. Thenceforth the mother made a point of not making comparative comments about the appearance and behavior of the two children.

By the time of the hospital visit, the mother evaluated herself as equally partial to all three children, and reproached her husband for idolizing the younger son. He responded by pointing out that the older son's reading, modeling, and constructing interests made his wife more partial to him than to the daughter, who was a tomboy by comparison. By tomboy, they meant athletic – wrestling with her older brother ('beating him up,' she boasted); and playing cowboys and Indians with him and his friends (she was always 'a cowgirl,' she said, 'and a rough one'). She was good at throwing and catching a ball – superior to her brother, in fact – but she did not play with boys. 'I play with the girls,' she said. 'The boys are playing with their boyfriends.'

She had never been very interested in dolls, but was devoted to her little brother, and spent a lot of time with him. She had a same-aged best friend, Cynthia. Together they would discuss clothes and boys by the hour, and watch T.V., or go to the movies. They went also to a local youth center to socialize, square dance and, in the craft shop, make costume jewelry, placemats and other gift items. She dressed in the style currently fashionable for preteen girls. The dress code at school forbade jeans and cosmetics, both of which she liked to use at home and on weekends. At home she was like her brother in balking against doing household chores.

Her preference in movies was 'cowboy pictures and jungle pictures. ... I don't like love. There's too much kissing and everything.' In real life, if a boy got fresh with a girl, that meant 'the boy starts kissing the girl, and the girl doesn't like it; sometimes she'll smack him in the face.' She allowed that she had a boyfriend, and added, 'I like him, but not too much.'

A good thing about being a girl was having long hair (she wanted to let hers grow again) and being allowed to wear lipstick – and, 'well, you get babies when you're a lady.' Hypothetically, had she chosen which sex to be when a baby, she would have been a girl, the advantage being that 'I like it.' For a career, she thought that she might be 'a secretary or a nurse – mostly a nurse, to help people.' Since she would like to have a pony, a horse, or a dog, she allowed that a compatible career might be as a cowgirl. The inquiry then changed to dreaming.

'I dream that I'm a cowgirl – a secretary. I get married. Then we go on our honeymoon, on a boat ride. Then he'll kiss me. Then we'll go all around the United States ... California, Texas, Wyoming. Then two years after we're married, I get a baby. And my husband works. And I take the baby for walks. And sometimes I wake up.' Such knowledge as she had about getting a baby was from 'a book I had from the library ... something starts growing, and it keeps growing, and it was in something, and then when the time comes – I can't remember it.' She did not know the father's role.

Though feeling shy in talking about things sexual with either a woman or a man doctor, she established the sex difference as that the boy has 'a long one down there, a bit longer than the girl's.' She had not heard the term, penis – only pecker. She had seen her small brother sometimes walking around the house pulling on his. She lacked a conception of what another girl would look like when undressed. She had no word for the clitoris. She said that her own did sometimes 'get firm and hard,' and in response to direct questioning added, 'And I do feel like rubbing it, and I do rub it' but there was nothing said to indicate erotic sensuousness or orgasm.

'What *do* girls look like?' she asked ingenuously, and paved the way to disclosing that, of herself, 'I thought maybe it was different – my clitoris, I didn't know what it was. I just thought it was different, because of the way they examined me.' But before the onset of repeated examinations, she had been given no preparatory explanation of why she was currently seeing so many doctors for examinations, nor of why she had traveled to Johns Hopkins, even though she was nearly 11 years old. She realized that all her examinations had been from 'the stomach down,' and speculated that constipation might be the problem. She had heard the term, sexual development, and construed it to mean development of the genitalia. Her mother's earlier attempt to prepare her for breast development and menstruation had been an embarrassment for both of them, and the daughter brought it to a close by saying that she already 'knew about all that.' The gaps in her sexual knowledge, including sperm function, were eventually filled in by her 'talking doctor' at Johns Hopkins. This same doctor, after the decision in favor of surgical and hormonal feminization had been jointly agreed upon by all concerned, provided the girl with the information necessary to prepare her for a surgical admission. She followed it up postoperatively with more explanation, suitably worded, about the time of estrogen substitution therapy, breast development, lack of menstruation, motherhood by adoption instead of pregnancy, vaginoplasty at a later age, and sexual intercouse. The girl thought she certainly would not ever want to do that (intercourse), but the eventuality of a change was implicit in her questioning the doctor as to whether she herself did it.

The girl and her parents both needed to rehearse how to answer questions about the reason for having been in the hospital. They were satisfied with explaining the operation as a congenital hernia repair, which, in fact, it qualified as being.

At the time of discharge, all the members of the family were satisfied with what had been achieved. At home a dissenting voice was that of a local urologist who had misgivings about how nature would reveal itself. His misgivings failed to materalize, and the girl did, in fact, continue to develop among her preadolescent and adolescent female peers as one of them. For the next four years, she returned to Johns Hopkins every six to nine months and then, because her treatment on estrogen was completely successful, did not return for four years.

At the time of the first postoperative visit, the first noteworthy observation was that there had been no further progress of pubertal virilization, and maybe some regression, as a consequence of gonadectomy. The second was that the neck and shoulder and facial tic and completely disappeared. As usual, she contrasted greatly with her mother in being taciturn rather than talkative.

In the youth center in her home town, she had begun to stand out as a leader. She won a craft prize for the doll she submitted. The great social event had been a school dinner dance for which she had her first evening gown and corsage. There was an older boy who heard on the grapevine that she loved him. He sent back a message that he loved her too, but they did not meet.

Because her height age was not advanced, it was considered advantageous not to commence estrogen therapy until age $11^1/_2$. By this time, her nonchalance barely masked her eagerness to get breast development in synchrony with her age-mates. Her mother considered her too young to date. Therefore, her ruse was to take her young brother to the movies and to bribe him not to tell that she and her girlfriends there met their boyfriends. They paired off together, holding hands. There were also occasional walks together, and a few furtive kisses – until her own boyfriend moved out of town.

Her breasts grew well under the influence of estrogen therapy in the intervening seven months before the next clinic visit. A sign of her pride in them was that, in the fashion of the day, she wore her cardigan sweater backwards so as to show them more effectively. Husky breaking of the voice had totally disappeared, and so had the hot flashes that had been sporadic but severe before the onset of estrogen therapy. The girl and her mother were in dispute regarding more freedom in dating, less maternal supervision, and occasional permission to stay out until 11 p.m. The mother was contemptuous of the changes that had taken place for girls in teenaged culture since she, herself, had been a teenager – particularly the new idea of going steady at so young an age. For a while, the girl's compromise was to have a boyfriend in the military, with whom she corresponded, but had not met. He was 'crazy about her.'

The patient was 20 years old when, at her own request, she returned for vaginoplasty. On the day of admission for surgery, she had a long interview of two and a half hours which she summarized on tape, beginning with words of wisdom that she might have for a younger person coming into the clinic for the first time with a condition like her own. 'I've been through a heck of a lot' she said, 'and I've never been too downhearted. I've always had plenty of faith that everything will work out okay.' She concurred that her condition had been something of a hurdle to her in her boyfriend relationship, but not in school achievement (she had graduated from high school at age 17), job achievement, or family life. At age 10 she had had such a close relationship with her older brother that she had been able to tell him about her infertility, with the result that they'd been even more close ever since. Her younger brother was still much too young for confidentiality. With her parents, she said, 'We just didn't talk about anything really confidential like that. My mother still thinks I should wait for this operation until six months before marriage. I think "she has trust in me," but she would disapprove of me going out and having sexual relations with a fella.'

The girl did not perceive that her condition had been a hurdle in her social life, 'I look the same as other girls; and I've never been so intimate with anyone that it has given me any trouble'. She described herself socially as 'just quiet, easy to get along with among my friends, my closer friends. Once I've got myself going I'll have a good time, but amongst a group I'm quiet, and it takes me a while to get to know a person. ... I started dating – I was going steady at thirteen.' By age 20, there was one boyfriend over whom she'd had many heartaches. He seemed to be unable to make up his mind on marriage. She had decided she would wait for the ring before having sexual intercouse for the first time – 'because of him,' she said, 'it's just so much to him. ... I don't think it would be wrong, but well, so far I couldn't.' Because of her condition, she had held back even on heavy petting, she allowed.

She was disinclined to make a connection between that and her boyfriend's indecision regarding the proposal of marriage which she so much wanted. 'I really don't think that's it. He's just afraid to assume the responsibility of having a family and supporting it. I don't know why he hasn't given me more of himself, more consideration and things like that, instead of holding back. As strong as I do feel about him, I'd do anything for him, really – give up anything for him. But I don't want to have to give up anything for him, I just want to do more for him.'

To resolve the uncertainty, she fell back on the tactic of testing him by leaving him. She told him that during her absence for six weeks, including the period in the hospital, she would not write to him. 'I told him I just wasn't going to, and for once I'm not going to break down. I told him that what I want, I want. It's not a game to me. I want a lot.'

The parents perceived this boyfriend as being too self-oriented. By contrast, there was another young man whom their daughter failed even to mention. He was making himself sick over her. He had timed his vacation to coincide with her hospitalization. He spent many hours sitting at her bedside – all to no avail. She did not care for him and could give him none of the response that, in his lovesickness, he craved.

By contrast, her response to the boy whom she herself craved was reflexive. 'If he just touches me, I tingle,' she said, 'whereas another person, another fella, could do anything he wants, I think, and it wouldn't phase me in the least, really. But when *he* blows in my ear and kisses my neck, I go wild. Of course there are other parts of the body, too – my breasts, especially around the nipple, and uh, I guess around the vagina, and the sides of the clitoris – do I or don't I have a clitoris?' The reply was: 'To answer your question very directly, you have an amputated clitoris. You get feelings, as you do, out of the stump and covering skin. This I've discovered from most of the girls I've talked to who have your type of medical and surgical history. You don't lose the feeling.' 'No, I don't at all,' she laughed – 'on the contrary, really.' She considered that the feeling was both deep, and from the skin surface – 'that whole area.' Then she added a qualification: 'Well at one point right over here it's numb – in my groin, right about the join of my leg to my body. I spoke with my surgeon today and he said that, during the last operation, he cut the nerve which sometimes happens in operations like that. Sometimes, he said, the feeling comes back after a while, but it's been ten years now since my operation. Just that one little area, that's all.'

The response to a query about genital feeling that became powerful and overwhelming was: 'Like I want to burst? Yes ... just from him touching me and kissing me – holding me close to him, like I'd be touching his body.' She did not wake up from sleep with this sort of feeling. 'I could be lying in bed thinking about him, and I'd feel like I want to hold him so close, so tight, right then and there – I get that urge, but I'm not excited, I would say. I just always want to be with him.'

'I never stimulate myself with my own hands,' she said. 'I've never even wanted to. And with a fella, I've always just pushed him away from touching' the genitals. 'I guess you can say I'm frustrated.' The feeling of being ready to burst down below always was associated with having her boyfriend close, and did not occur on other occasions.

Erotic arousal through the skin senses, and by the lover, took precedence over visual stimulation. 'I've seen Elvis Presley on the screen,' she laughed. 'I don't know, it – something tingles inside of me. I wouldn't say it's sexually aroused. I'd never scream and go wild over him – you know how some girls do. But I guess I could reach out and grab him, but I don't think I've ever gotten excited. On the street I can just look and admire someone and say, well, he's good looking, but they don't do anything to me ... just a thought in my head, a wild thought,' she laughed, 'that's all it is.' She had no history of having been erotically aroused from narrative material. 'I'm a romanticist,' she said. 'I like it when the couple end up living happily ever after.'

'I laugh,' she said, when asked what sort of things happened in her mind when she saw a picture of a sexy-looking pin-up girl. 'I mean, like they have these girls' shows, and other girls go to it. I could never go to it. You'll never catch me at one because I'm just not interested in other girls, not even to look at them like that. Not me. Seeing a girl undress doesn't entertain me at all. Not at all. I just can't see it either,' she said, referring to what men find in watching the same shows, 'like a guy sees a pretty girl – how he can undress her in his mind and how he can be so easily aroused and all that. I think just these degenerates go to see these shows, just to be aroused because they can't get it any other way.' If handsome male bodies showed themselves off in shows for women, 'it wouldn't interest me,' she said. 'It really wouldn't. Not even if they had Elvis up there.'

About menstruation, she said, 'Well, many women have said it's a pain, and they wish they didn't have it. Well, I don't know – I'm sorry that I can't have children.'

The operation for vaginoplasty took place without postoperative complications. It was difficult for the patient to adapt to the insertion of dilators, but she and a hospital room-mate who had undergone the same operation were mutually supportive, and the patient always could count on her supportive parents. There were no subsequent problems of vaginal constriction. Hormonal substitution was maintained on Premarin 1.25 mg daily, which eventually was supervised locally.

Twelve years passed before contact was re-established for a psychologic follow-up, which was done by long-distance telephone. At this time she had been married for seven years. Her husband was the man with whom she had been in love when she talked of him at the time of vaginal surgery. For the first three years after the operation she had had other men friends, and then for two years kept an exclusive relationship with her future husband. During this period of their courtship, their sexual relationship progressed from kissing and heavy petting to include coitus, and encompassed also reciprocal oral sex. These are the activities that have also comprised their sexual life in marriage. Penovaginal coital thrusting alone had proved insufficient to release an orgasm, whereas oral or digital stimulation of the clitoral region by the husband was effective. Vaginal dryness was sporadically a problem, no more than once a week, the usual frequency of intercourse being nightly. The area of numbness formerly described had persisted; and to be touched there was unpleasant.

The husband knew before marriage that parenthood would be by adoption. After a waiting period of two years, they had, only six months prior to the phone call, taken into adoption a six week old boy. In order to be a full-time mother, the wife took a leave of absence from the family business. She 'loved it,' having so much time with the baby whom she described as a good, happy baby who sleeps all night, eats well, and is good-natured and relaxed. He accompanied her on visits and errands, and also when she and her husband went cycling, swimming, or to other sports activities – tennis, running, and her husband's football games, and riding dirt bikes. They visited also on a frequent basis with members of the family who lived in the area. They had close-knit ties with the extended family on both sides. The baby had a cousin, the youngest son of his mother's older brother, only a few days younger than himself – the 'twin grandsons.' Her husband was as delighted with being a parent as she was. She laughed off his typical male aversion to dirty diapers and spitting up. He had maintained the family business during his wife's motherhood leave. They had jointly established their own business as consultants in public relations, for which they each had special training, and were doing well. They planned on adopting a second child, by preference a daughter, in a couple of years.

Case MV When this boy was 12, he returned to the hospital for the first time in five years. He remembered that prior to surgery at age 7 'I never went to the lavatory because it looked kind of funny everyday going into the little booths to sit down. So I just stayed in the room and said I didn't have to go.'

He said he had begun to grow sexual hair at age 11, and that when his penis erected it was about two or three inches long – about the length of his middle finger. Its size had not worried him. He had not noticed feelings in his penis, and thought that playing with it would be silly. He was not aware of ejaculatory discharge. He said he didn't know anything about how human beings reproduce except that a baby comes from a woman. 'I think I'm too young,' he said, and indicated that the topic had not been dealt with at home, nor at school. 'I don't think it's too easy to talk about it,' he said. He listened to information without much responsiveness.

Regarding girlfriends, he allowed that he had one at school, but hadn't told her yet. Fourteen, he thought would be the right age to begin dating. There had been no love dreams, and he guessed a boy 'would feel kind of funny having a boyfriend in a love dream.' He hadn't thought too much about marriage: 'I think I want to be, but I'm not quite sure yet.' Professional education, in the footsteps of this father, had priority in his planning for the future.

The next clinic visit took place when the boy was 14. In the taped summary of the interview, he began by saying that the 'talk about sex life had been the most important. Just from having my operation, I didn't know if I would be able to have intercourse successfully.' His concern was not with the shape or size of his penis, but with whether it would 'be able to function properly and produce semen ... if I wanted to get my wife pregnant.' He lacked decisive evidence regarding semen, and could report having observed only a transparent, viscous discharge that oozed rather than ejaculated from his penis. He did not endorse masturbation, but also did not negate it as a test for ejaculation. He could not recall having had any wet dreams. The penis was able to erect and sometimes did so in response to reading, pictures, or from watching pretty girls which, in school, sometimes required hiding his crotch behind a book.

Socially, he regarded himself as very shy with girls, but expected he would 'grow out of it. I got enough courage to ask a girl for a date. I can again, too.' About sexual intercourse before marriage, he said: 'I'm not going to, because I don't think it is right,' and he would not be able to condone it in his friends. He accepted family planning because 'it would be hard to support a large family,' but he did not agree with birth control for unmarried teenagers to enable them to have sex without the risk of pregnancy.

He had read about homosexuality in a sex information book his father had bought for him and his brother. If homosexuality entered into a friendship, 'I might try to help them – suggest they go to a doctor,' he said, 'but I don't think you could help it really that much.'

The boy had a strongly positive relationship with his father who accompanied him to the clinic. In the sexual guidance of his sons, the father wanted to avoid anything even remotely similar to the error his own mother had made with him when he first began having wet dreams. To hide the evidence, he had hidden his soiled pajamas. When his mother discovered them, she chastised him for his evil habit, and as a punishment grounded him for the summer, and cancelled his Boy Scouts' camping trip. Years later, the family doctor disclosed that it was on his advice that the mother had made a weeping apology to her son. The doctor had also censured the boy's father for neglecting his duty to his son.

As it turned out, the father's resolve not to repeat the error under which he had suffered would be put to the test in a couple of years when, at age 16, the patient had to request his dad for medical help. He had lost a piece of bead-chain while using it as a masturbation accessory in the large cavity of the reconstructed urethra. It had become entrapped and could not be retrieved. The father properly responded to the problem as a medical emergency and not as a moral delinquency, and it was readily resolved without subsequent complications. Recalling the incident years later, the son said it was 'probably one of the best times in my life, in the fact that it brought me very close to my father; and I was really surprised. It was very hard to go in and tell him, that morning, you know, and the way he handled it brought me very close.'

The interview with the boy at age 16 added little to the information he had given two years earlier. He was no longer averse to masturbation. The frequency was more or less daily. The imagery was orthodoxly heterosexual with foreplay and coitus with himself on top. The use of the piece of bead-chain had been atypical. It was associated perhaps with exploring internal responsivity, for the grafted skin that formed the urethra on the shaft of the penis lacked sensation and was numb.

Penis size and appearance had not proved to be a problem of stigmatization in the locker room at school, nor had the sites where skin for grafting had been removed from the thighs. He estimated the erected length of his penis at about $3^1/_2$ inches.

There was an eight year interval before the next interview in person. On this occasion he was accompanied by his wife of three and a half years, in preparation for donor insemination. For a year they had been trying for a pregnancy. He had considered the possibility of 'weak ejaculations, very small in amount [the prostate was indeed small]; and also very little penetration.' Finally, a fertility check gave a verdict of azoospermia.

In accordance with routine procedure, the husband and wife were given individual interviews followed by a concluding joint interview. Their reports were discrepant. His was the report of a man more turned-on to sex than his wife was – more often, more orgasmic, more variety in foreplay, including oral sex, more positions, more pleasure – whereas she was more into romance and upper-body sex than genital stimulation. However, their disputes were about money as well as sex, he said.

The wife's report was one with less emphasis on sexual compatibility and personal sexual satisfaction than on income and career compatibility, the sacrifice of her professional career to his – working only to support the marriage while he completed his education. Though each had different working premises for the stress they engendered in one another, each believed that a baby would be the solution.

A positive premise of the marriage was that neither partner's erotosexual imagery and fantasy excluded the spouse. Husband and wife each had had imagery of having sex with another partner, but not obsessively so. That 'is about as kinky as it goes;' he said, 'no sadomasochistic tendencies or anything, or homosexual, that I can think of ... with her has been my only sexual experience ... usually I don't see another partner or fantasize as such. I'm just involved in what I'm doing with her.' For her, the chief problem had been one of being turned off to penises in general, as exemplified especially in oral sex to which her husband was greatly responsive, both ways. She had not reached the full peak of orgasm.

The concealed proposition in the erotosexual syllogism of this marital relationship was that, provided her negation of genital satisfaction could be displaced onto financial discontent, he could maintain his myth of macho erotosexual prowess, irrespective of the inadequacy of the organ of prowess.

The dynamics of sex therapy are such that, on this first occasion of being able to discuss her sex life with someone who responded nonjudgmentally, the wife graduated to a newfound liberation as her husband had done 10 years earlier. She summed it up in a return visit three months later by saying, 'It was a real high for me. I guess maybe I was a repressed nympho.' She now initiated sex more often. She described orgasms as 'more violent and they last longer.' The best position for intercourse was with her on top, because it minimized what they joked about as 'the fall-out game' which made him very discouraged.

With the wife's newfound coital exuberance paradoxically his 'fall-out' changed from being a game – what had been fun now became failure and a threat. The return to the clinic for donor insemination metamorphosed for the husband into another threat. Reversing this own former fantasy of adultery, he had a fantasy of her going out to get a man who could get her pregnant. He became irritable and jittery. He defined the clinic visit as being exclusively for his wife's sake. He made it known to her that he would not under any circumstances submit to the follow-up physical examination, which he had formerly agreed to – and indeed he did not. Such highly charged defiance was totally out of keeping with his past record. One conjecture is that he was threatened not only by revealing the size of his penis, but also by what might be revealed to him as a consequence of the physical examination. Perhaps he knew more than he had ever disclosed about his neonatally declared status as a female and had a phobic fantasy, or terror, that a physical examination would confirm the existence of hidden female evidence. Should there be any credibility to this conjecture, then the episode at age 16, of losing a piece of bead-chain in the urethra, may be construed as having been the improvised equivalent of an exploratory culdoscopy.

The end of a saga coincides with the end of its chronicling. There may one day be more to add to this story, and a chance of ascertaining how much the patient either knows or suspects about his neonatal history. In the meantime, he has not taken up the option of further visits.

Donor insemination failed at first, but eventually a son was born. 'This baby has worked miracles for us,' his mother wrote when he was two years old.

Clinics and hospitals are good places when needed, but better still when not. They are expensive and dangerous places that may harbor infections and false doctrines. They cure and make healthy, but they also stigmatize and invalidize. They dictate people's lives as also they democratize them. The possible harm of imposed follow-up has not been researched. There may well be a time when in a patient's life the healthiest policy is to let well enough alone.

Case WF When her parents first let her know that they were concerned about her sexual status she was 13:6 years old, and their concern was expressed not as failure to menstruate but as disappearance of periods. The girl and her mother both had interpreted a blood stain six months earlier as a period. She shared her parents' concern that there had been no recurrence. She did not have their foreboding based on clitoral hypertrophy. 'I didn't think too much about it,' she said, 'until they mentioned coming up here,' to the clinic.

As a young child, she had never seen another girl, even a baby, naked, and had not inspected the female genitalia. Later, she had seen other girls taking a shower, and they had seen her, but there were no comments or comparisons regarding genital appearance. She did not see herself as being particularly unusual. She reported no awareness of clitoral erection and erotic sensitivity, nor of masturbatory stimulation. By contrast, she reconized nipple sensitivity and protrusion. She dated the onset of breast development at around age 12, which kept her in synchrony with many of her age-mates. During the preceding year, her class at school had been shown a film on the birth of a baby. At home there had been no sex education from either parent. The mother had attempted to bring up the subject but quit, because the daughter would change the subject. The girl had an impoverished knowledge of reproductive anatomy, and did not know of the shallowness (3–4 cm) of her own vagina. The psychiatrist who interviewed her used diagrams to explain reproduction. The conversation included the matter of clitoromegaly. The girl indicated that she would like to look more like other girls. She asked if there was a nonsurgical way of shrinking her clitoris. If not, then she would want surgery.

The history of the nongenital aspects of her G-I/R development was that of a girl. As a 13 year old, she listed her recreational interests thus: 'I just like to play around outside, play with my dolls, ride my bicycle, play basketball ... play with animals, raise fish ... I like to collect stamps and pictures of movie stars ... after school, I watch television, do my homework, and during the summer go over and play softball.' The basketball and softball teams were for girls. One of her team-mates was voted queen of the school, which was one reason why she definitely did not think of herself as a tomboy. Her preferred movies were comedies and westerns. Mystery stories took precedence over love stories except for, on television, a program on modern romances which she and her mother watched every week.

In an interview at age 13, she contemplated an inquiry on whether boys and girls like being what they are, and responded in favor of being a girl, beause 'you get nice clothes.' She had always dressed according to the fashion in vogue for girls of her age. She could not think of any special advantages of being a girl instead of a boy, but concluded, nonetheless, 'I think I'd rather be a girl.' It required more thinking before she justified her preference with, 'You don't have to work as hard,' though she conceded that housework and raising children was work – but she wanted to do it, anyway. She would want 'about two' children. With her girl friends, whenever they had played house with their dolls, she had always played the role of the mother.

At age 13, her career plan included being a veterinarian, because of her attachment to animals – 'but I wanted to be a nurse first,' she said. Her future plans included marriage at the age of 25–26, though she projected no detailed wedding fantasy, and at the time had no important boyfriends in her life. Her ideal boyfriend would be cute (maybe like Elvis Presley), and would have a good job. He would also have to be 'one that doesn't run around.' Like the other girls in the peer group she belonged with, she had not yet begun dating.

The girl herself, her parents, and all of her professional consultants agreed on exploratory and corrective feminizing surgery for which she was appropriately counseled. As happens all too often in clinical psychohormonal follow-up, there was no administrative coordinator to synchronize the scheduling of follow-up consultations to coincide with the patient's return for a surgical admission. Thus three months later, the admission for surgical exploration and feminization took place without concurrent interviews and psychohormonal follow-up.

The same thing happened again 10 years later when, at the age of 23, the patient returned from a far distant state to be admitted for vaginoplasty. She had just been divorced after a marriage of six years. She attributed the failure of the marriage to overall incompatibility, and not to a specific coital problem with her vagina. In fact, however, her vagina proved to be too shallow, with a depth of only 4 cm, despite the dilatation effected by coitus. Belatedly, and in anticipation of remarriage, she requested the surgery for vaginal lengthening that she had, at age 13, been told would be advisable some time after pubertal maturity at around age 16, in preparation for establishing a sex life with a copulatory partner.

Postoperatively, the extended vagina was adequate for penile penetration and readily could be dilated with a form, as prescribed. Within 12 months, however, it developed an obstruction, a tightly constrictive band of scar tissue, about four inches within the vaginal canal. It was at the site where the skin grafted extension had been added to the apex of the original shallow vagina. The constriction prevented full penetration of either dilator or penis. The husband had not complained. He said his erected penis was very firm, about six inches in length, and adequately accommodated vaginally in coitus. Nonetheless, he endorsed his wife's decision to return to the hospital for a surgical re-evaluation.

This decision came after they had been married five years and was prompted by a long-distance phone call from the psychohormonal research unit at Johns Hopkins, as part of a program of long-term follow-up. The husband accompanied his wife. Each gave an interview on their sexual life as a couple.

At this time the patient continued to be maintained hormonally on estrogen replacement therapy (Premarin, 1.25 mg/day), without which she experienced 'nervous spells and flushes.' She had not made any inferences regarding a possible relationship between estrogen and erotocism or genital function. Her hormone prescription had never included, as well as estrogen, either progestin or a minimal dose of androgen.

In addition to the coital problem of the intravaginal band of constriction, there was typically a coital problem of vaginal dryness also, corrected with a vasoline lubricant. This may have been a problem of too short a period of preliminary stimulation, as the prolonged stimulation of romantic reading or movie watching 'is about the only time that I do' produce natural lubrication. The husband's estimate of the duration of foreplay was two or three minutes of kissing, caressing, and finger play. For the wife, the breasts were erotically sensitive. The vulva had 'had a lot more feeling' before the onset of vaginal constriction and the pain associated with it. The site of the former clitoridectomy was not erotically sensitive, nor was the region of the urethral meatus. Both partners agreed that they were not turned on to oral sex.

Estimating the frequency of sexual intercouse, both agreed that the time-table was variable, depending on mood and work schedule. The wife's long-range estimate was once or twice a week. They both felt very compatible with one another. 'Whenever I've been in the mood, she's always been in the mood,' the husband said, 'and vice versa. Sometimes I'll go to bed with the idea of just going to sleep ... and she's kind of, you know, in the mood, so we'll have relations.' Overall, however, the husband would have liked to have sex more often.

The wife, in one interview, evaluated herself as not overly interested in sex. She had always been indifferent to masturbation. She rated her first marriage as a mistake, but the second one as totally successful, sex life included. She and her second husband both had had only one other partner as a criterion of comparison, and very little exposure to explicit materials on erotic technique. One book, on the sensuous woman, she had set aside. 'I felt like it was going a bit too far,' she said. Apart from the standard missionary position for sexual intercouse, she had tried only the female superior condition. Questioned about having experienced orgasm, she reponded with, 'No, I'm not sure. Before I started having this problem [of vaginal constriction] I did have more feeling like that – what I suppose was an orgasm.' The husband's verdict was, 'I really don't know. At times I'd ask her, and she'd say yes.' Usually he maintained penovaginal thrusting for five to ten minutes before ejaculating. For his wife the experience was, according to the sum total of reported evidence, erotically pleasurable but anorgasmic.

The wife had no history of erotic imagery in dreams, and no experience corresponding to the ejaculation of a wet dream. Her imagery and ideation of erotic turn-on did not involve tastes or smells, nor sound apart from words with romantic meaning. Visually, she was not turned on by explicitly erotic images either in print or real life, as compared with touch or tactile imagery. However, in the manner commonly reported by women of her age and time, she was able to be aroused by romantic stories or movies in which 'sex is insinuated, not actual.' If erotically aroused by a romantic movie (neither she nor her husband had seen an explicitly erotic movie), then she might put herself into the scenes. As for the male lead, she guessed, 'I would send him off with a woman in the movie,' and then replace him with her own husband. Her husband himself would have been more likely to have had, in the manner of masculine fantasy, an affair with the female lead.

Neither in fantasy nor actual life had the woman been in a bisexual or lesbian relationship. 'I don't think I've ever had any tendencies like that,' she said. 'It seems very distasteful.' Her husband shared this reaction, but like his wife had an easy acceptance of gay people whom they encountered through their work.

The husband's knowledge of his wife's diagnosis had always been in terms of infertility, and by implication rather than dramatic diagnostic disclosure. He himself had grown up with a surfeit of infant care, for his mother was a professional foster mother. He had an older sister who, like himself, for many years elected against parenthood. Eventually, however, he and his wife underwent a change in favor of adoptive parenthood, and they had initiated an adoption procedure. His wife had forfeited her erstwhile devotion to three poodle dogs in favor of a child. One measure of her self-concept as wife and mother was her serious concern as to her status regarding a Pap smear.

There were very few people to whom she had ever confided anything regarding her diagnostic self-knowledge. Her parents had been totally hush-hush when she was a teenager. 'I felt like that I was really a freak, in a way, for a while,' she said, 'until I began to understand, now, that there is a lot more to it.' Thereafter, she kept her medical story a privileged communication. When she came to the hospital for vaginal dilation (an outpatient procedure of minor surgery, so it transpired), she told her business patrons only that it was a female problem. They asked no further questions. They had no reason to do so.

Case MF In this case, the boy grew up unacquainted with the fact that the original civil and religious records of his birth had been destroyed and replaced with new ones. Only the closest family members had at the time been informed of the change. The parents relocated out of state. Subsequently only two former friends remarked on their recall of the baby as a girl, but without further ado accepted the correction.

When the psychohormonal study unit was first formed, the boy was 4:6 years old, at which age his language competence was superior (Stanford-Binet IQ, 123). His parents said that within the preceding two months he had been trying to stand to urinate, and that he had learned how not to expose his genitals. For example he bathed himself, unaided, if a baby-sitter was left with him, and when outside at play with friends he always came into the bathroom to urinate.

By the time of his next visit, a month after his 6th birthday, it became apparent that he had conceptualized his urinary problem in terms of not yet having been able, like other boys, to learn to stand to urinate. He knew one boy who was only 4 who had learned already. His explanation for his own failure was that his penis was too small, not that the urinary meatus opened in the wrong place. It was from an improvised demonstration with rubber tubing that he learned for the first time that other boys can stand to urinate because they have a penile urethra. The primacy of smallness in the conceptualization of his urinary problem persisted throughout childhood. After puberty began shortly before his 12th birthday, the first sign of sexual development that he listed was: 'Like, getting larger around your penis.'

The urologist had waited for pubertal enlargement of the penis before, at age 12:6, undertaking the final stage of surgically constructing a penile urethra. With the operation completed, the boy contemplated a query as to whether now having a penile urethra would make a difference in his life. He came up with the reply: 'I don't think much ... probably, if I can use my fly, it would.' In actuality, it still was more expeditious for him to sit to urinate, his penis being too small to manipulate so as to keep his pants dry while standing to urinate.

The main advantage of having a penile urethra the boy specified as: 'I can have a family.' He was aware of sexual intercouse, through the progression of sexual information that, beginning at age 6, had been given on his biennial checkup visits and interviews. It was in keeping with his parochial school religious education that he referred to the sexual use of the penis as having a family. The same education had shaped his response to masturbation, which he recognized as associated with erection from age 7 onwards, when he said, 'I try to quit it. I wonder when it will get bigger, and how long it's going to take before they can do it' – the operation. 'Just lately,' he said at age 12, 'I try not to do it,' and after the operation he wondered whether masturbation would be injurious. He was able to live with the discrepancy between his moral standards and his actual practices, and was not inhibited in talking about either. The talk, however, was with his doctors rather than his parents, even on matters of medical history and prognosis. The mother accepted the responsibility of answering sexual questions if he asked them, she said, but the boy did not take on the reciprocal responsibility of asking them. He had, according to his mother, the same talent as his father for accepting the status quo and, when nothing more could be done, going to sleep.

His genital problem did not preclude his participation in boyhood activities like going to camp or locker-room changing. He was confident that he could find ways of not exposing his genitalia. His sports included swimming, bicycle riding, basketball, baseball, and football. He did a short free-association exercise at age 10, and one response was: 'Life, swimming and fishing, football – that's about all.' He did a lot of reading, with a specialty interest in U.S. Indian history and ethnography.

When he was 7, he said he would rather play with boys than girls. 'Sometimes girls are just too sissy ... you know sometimes boys like to do rough stuff ... like jumping over a wall or something ... once in a while I'll play with girls, not always ... I like Colleen. We play together.' Colleen was 6. They would watch her dog race with other dogs, or watch her little sisters dancing on a stage she made for them. He did not, during childhood, have a girlfriend. There were no childhood reports of either romantic or erotic rehearsal play.

If the prognosis at age five and a half months had passed the test of veracity, then the surgical repair of the penis at age 12:6 would have marked the triumph of success, and the patient would have been released from the tyranny of doctors and doctoring, healthy and intact. With the knowledge of hindsight, it can be said that the prognosis had been based on conjecture, and was wrong. Its wrongness became evident when swelling of the breasts was one of the earliest signs of puberty. At first it was hoped, in vain, that it would prove to be nothing more than adolescent minor gynecomastia, a self-limiting, self-correcting minor growth of mammary tissue that occurs in perhaps as many as 25% of boys an the onset of puberty. The falseness of this hope was fully evident when the boy was admitted to the urological service again, at age 13:3, for additional penile surgery. This unpredicted operation was needed to repair a fistula that had opened up, allowing urinary leakage from the base of the urethra, where the skin graft had failed to heal properly – an all too common complication of hypospadiac repair.

Though his breasts had continued to grow, and though he had held back from swimming on several occasions during the preceding summer, the boy himself had his usual phlegmatic approach to his medical problems. His mother was far from phlegmatic, however, and so were his doctors. The mother guessed, correctly, that the breasts were a sign of masculine failure, and that it would have been better had the sex never been reannounced. With recent advances in the classification of syndromes of hermaphroditism to guide them, the doctors were aware that they were confronted with the syndrome of partial androgen insensitivity, called at the time testicular feminization, and also known as Reifenstein's syndrome. The prognosis was permanent inability of the body to virilize, sterility, and an increased risk of testicular malignancy.

There was absolutely no evidence that the patient could undergo a sex reassignment so as to live in the body of a woman. The possibility was not, in fact, even mooted. Instead, a continued plan of attempted masculinization was put together: repair of the urethral fistula; plastic surgery, mastectomy, to flatten the chest; hormonal treatment with a very high dosage of testosterone; and postponement of a decision regarding surgical castration and cosmetic implantation of testicular prostheses.

The boy himself was willing to go along with the idea of eventually losing his testicles, if there were no other alternative, but clearly relieved at the possibility of being able to keep them, regardless of the prognosis of their almost certain infertility. He absorbed the information about infertility in his usual imperturbable way as yet another blow of the hammer of fate, and likewise the added new information of parenthood by way of donor insemination from the sperm bank.

He elected to postpone a decision on mastectomy on the off chance that high-dosage testosterone might shrink the breasts and make the operation unnecessary. He reaffirmed his decision in favor of repairing the urethral fistula, and had the operation two days later. The next postoperative return to the clinic was in four months, at which time his age was 13:8.

During the intervening four months, he had been receiving 250 mg of testosterone enanthate by intramuscular injection every 10 days. Its virilizing effect had been negligible. Though the breasts had not reduced in size, nor had they increased, possibly because endogenous secretion of estrogen as well as androgen from the testes had been suppressed by the exogenous, injected hormone. There had not been, in the boy's own estimation, any increase in the frequency of erections (one or more a day), masturbation, or erotic imagery. Masturbation frequency was about three times a week, for five minutes or longer each time. There had been accompanying fantasy for about a year. In the fantasy, 'there's a girl, and there's nobody else around ... and I'm in it ... usually she is somebody I'd recognize.' The imagery would include kissing, hugging, and sexual intercouse with 'me on top of her.' As the fantasy began, the two might be either dressed or undressed. Though he did not look closely at the girl, genitally, nor she at him, he did not conceal his genitals. In the fantasy, he did not watch or observe himself. 'I just see whatever I'd see if I was really there,' and his penis was 'like I'd be when all the operations were done, but still you could see that I'd been operated on,' except that his penis appeared bigger. 'It's like you're just right – you had scars, but never thought anything of it, or something.' By contrast, in the imagery of the fantasy, the chest was not represented as either flat or with gynecomastia.

Neither at the culmination of a masturbation fantasy nor at any other time was there an ejaculation of fluid. 'It's hard to describe what does happen,' he said, 'just, uh, a squeezing ... like grasping, sort of – I don't know exactly how to put it.' He allowed that, with eyes closed, he could imagine that something was actually being squirted out, as he knew to be normal in males. On a subsequent occasion, he described it as a 'rushing sensation that comes and goes in a matter of seconds, and afterwards you just don't have the urge to masturbate.'

In the absence of ejaculation, there was no history of actual wet dreams, but in sleep there had been some sexy dreams. Their erotic imagery was similar to that of the masturbation fantasies, but without the feelings of culmination.

Saying that 'I don't dream too much,' he could not recall other dreams, 'except for a nightmare, once. I dreamed I came here and they were going to take off the testes ... just sort of a fouled up dream. I dreamed that nobody had signed the operating permit, and still they tried to operate, and chased me all over the place. They couldn't catch me. Nobody had signed an operating permit, but still somehow they had got me down to the operating room. About a month ago I dreamed it.'

Prior to recounting this dream, he had projected himself into a future occasion of having sexual intercouse, and said he would 'feel sort of funny if they weren't there, I guess.' Without them, he would want to have silicone replacements, soft to the feel. With prompting, he thought about when, and how much he would tell about his medical status to a future sexual partner. The recommendation was to take disclosures only after the relationship or love affair was rather well established.

He did not yet have a girlfriend, but of one girl he said: 'She likes me, but I don't know her that well. She's not exactly my girlfriend.' In the parochial system in which he was being educated, he identified the beginning of the dating age as coincident with the first year of high school, one year ahead, when he would be 14.

A supplement to the foregoing erotosexual information was obtained after an interval of seven months, at which time the boy was 14:3 years of age. In considering the sensory channels of erotic arousal, he said that he could respond to nude pictures of females, as in *Playboy* magazine – but he would feel funny if his parents found him with a copy of the magazine. Whether or not he might be aroused by a woman, as in girl-watching, would 'depend on, like you're just sitting, and your imagination takes rein.' Then he might daydream, with imagery progressing to sexual intercouse. He had not experienced arousal from other varieties of erotic image – no brutality or sadism, no masochism, no animals, no members of his own sex, no threesomes, nothing unusual or worrisome, and nothing freakish.

He was at the hospital on this occasion for a five-day admission for mastectomy. Permission to do a testicular biopsy was requested, and verbally agreed to, but when it came time to give signed permission he didn't sign.

He had continued taking testosterone enanthate, 250 mg intramuscularly every 10 days, with negligble effect on virilization. He had pubic and axillary hair, but no development of body and facial hair. The pitch of the voice had not lowered. There was no palpable prostate gland.

There was still a persistence of urinary leakage where the former repair of the urethral fistula had been unsuccessful. Rather than face another surgical admission, he tolerated the urinary problem for another two years. He had just turned 16 when the returned for a further attempt at urethral repair, which proved to be successful.

At this time his pubertal status had not changed, except that his voice might have deepened a little. It transpired that he had given up on hormone injections and had had none in the 21 month interval since his previous clinic visit. Without the hormone, genital and erotosexual function remained the same as with it.

Socially he had progressed to occasional social dating. 'I don't think I'm the most popular person,' he said, 'but I think I get enough girls, though.' Whereas he believed he had had no trouble in being accepted for his age, he had made what appeared to be a trade-off between school grades (settling for a C average) and social participation as a regular guy, despite his handicap. Because he had been separated from his friends by being relocated in a high IQ group in parochial school, he and a friend arranged to fail so that they could be transferred to the public high school. A possible confounding variable, however, was his difficulty with math and praxic reasoning on the basis of a lower nonverbal than verbal intelligence which had been in evidence since age 7. At that age his WISC IQs had been Verbal, 135, Performance, 97, and Full 119. Nine years later, the corresponding WAIS IQs were 128, 99, and 116; and the IQ equivalent scores for the special factors, namely, Verbal, Numerical and Praxic reasoning were 138, 108, and 89, respectively. The basis of such wide discrepancy is unknown, but may well share the same origin as the genital anomaly.

The specific difficulty with praxis showed up in the poor reproductions of the Bender-Gestalt figures, and in the human figure drawings done at this time. But the human figures were also ominous, drawn with strokes that were fragmented and exploding apart. Added to the history of academic underachievement, hormonal noncompliance, and postponement of surgery, they suggested, to quote the report written at the time, 'that some of the outward calm is a freeze, with a rigidity not so permanently stable as it may seem.'

In the ensuing year there were no major changes. By age 17:3, he had a steady girlfriend. His genital deformity, and the fact that his penis was barely three inches long when erect, had held him back with regard to receiving genital petting. He did not hold back in petting her.

In appearance he was a quite handsome looking boy, well-dressed in the Ivy League style. There were no aspects of his appearance or manner that would evoke the term, feminine; but it was difficult, on appearance alone, not to underestimate his age, and hence not to talk down to him. The problem of being beardless and too juvenile in appearance became increasingly malignant to his development as an adult. When he was 18, and about to graduate from high school, the mother was sufficiently concerned to request a follow-up appointment. She wrote, 'I detect some concern because his complexion is so smooth. He is one of the youngest looking boys in his class and he is not one of the youngest in years.'

When he arrived for his appointment, his dress and conversation were clearly those of a college freshman, not of a high-school boy, but he looked too boyish for his age and achievement. Never one to dwell on his problems, he said of the failure of testosterone treatments, 'I haven't been really satisfied, but I haven't been dissatisfied to the point where it's ever bothered me. It's annoyed me from time to time, but as far as constantly thinking of it, or allowing it to hold be back from anything too much, it hasn't bothered me ... looking younger than I am hasn't bothered me too much.' The more important order of business was his first steady relationship with a girlfriend.

They had been going together for about six months before intercourse finally happened. About being in love, he said, 'I don't know. It's hard to say. I feel very attached to Cindy, but I don't know how to – I don't really know how to define love. I don't know actually if you could say I was in love with her.'

In obedience to the demands of the sexual taboo, they accepted the traditional restrictions on their teenaged right to unmarried sexual privacy, and so had had few opportunities when there wasn't 'always the matter of tension, because you're afraid of somebody coming home.' He had found that the best coital position for him was when he was lying on his back with her straddling him, though he thought that she enjoyed it best when he was on top of her. In this male superior position, he guessed he could get his penis inside the vaginal opening to a depth of about one inch. She had, he estimated, reached a climax on about five percent of occasions. With him, it had been more often, but always without a discharge of fluid. 'It builds up kind of gradually until late towards the end,' he said, 'and then it rushes to a point where it drops off.' The most important thing about whether he got aroused or not was 'having her feel aroused, too.' She had not ever made reference to the size or appearance of his genitalia. He had not mentioned sterility to her, and on some occasions she had been in a panic because her period was a day late.

Insofar as he had criteria of comparison, he said, 'I don't find, you know, too much difference between me and anybody else, you know – like fellows would be sitting around having a bullshit session, and I never find any great difference.' He rated himself as not the athletic type, and not aggressive – in fact not assertive enough when people talked down to him. As compared with his friend, Nick, if he'd pick up a girl, he said, he'd 'do everything else except [expose his penis], for the reason I'd be afraid of being embarrassed' [by its size].

With respect to masturbation fantasy and erotic dream imagery, there had been no change. A picture of a sexy looking male, as compared with a female, 'only makes me feel uncomfortable,' he said. Before he was old enough to get a driving license, he had frequently hitch-hiked. 'It's a kind of rough neighborhood,' he said, 'a lot of weirdos, a lot of various types of kooks, I don't mean strictly homosexuals ... but it happened four or five times ... and I told him, aw, shut up, just leave me, just let me out of the car. And he always did.' He said he recognized the type 'because they'd wear the kind of clothes that kids would wear, and drive the kind of car that would impress a kid, and play the radio on a station that caters to kids – and then they'd start with, What did you do tonight? I was out with my buddy. Went out with a broad? You know, something like that.'

He saw his girlfriend two or three times a week. 'If it was up to her,' he said, 'I'd be seeing her seven nights a week,' and that he found too demanding. 'Some things,' he said, 'you just can't sit down and shoot the breeze with a girl. ... I don't know, I'm not sure, but I sort of have the feeling that it is – that our relationship isn't going to go on. I don't know.' His uncertainty was prophetic. Some months later, when she announced her decision to date other men, he decided to declare his problem of infertility. Her reaction was 'one of shock.' She quit having sexual intercourse with him, and in two weeks withdrew from their relationship forever. His reaction was that, 'regardless of how much I ever loved a person, I'd never be able to marry.'

Fifteen months later he wrote, 'I have some news that you will probably be glad to hear. Six months ago I met a girl whom I love. I explained my problem to her.

She says that it doesn't change the way that she feels at all. I've felt guilty, so this past weekend I forced myself to tell her about my being sterile. She said that she loved me, and would be very happy with adopted children. I couldn't believe that I had been so wrong in what I had felt for quite a while. To tell you the truth, my being sterile has always made me feel that my ever having a truly normal or happy life with family and children was a pipe dream ...'

In the same letter, naively oblivious of the hypothesis of a connection between sex life and career life, he wrote: 'The second thing that is bothering me is school. I know I have the intelligence to do well in school, and I know that I'm not lazy. What I do wonder is this. Some people, regardless of their native intelligence are natural students. I'm dissatisfied with school. Often I feel that I would be much happier working. The main thing which keeps me from quitting school is that I'm afraid of hurting my parents. ... I can get more satisfaction from washing dishes or laying bricks than I can from studying. ... At school I can get more interested in the work a janitor is doing, and respect him more than the Ph.D. who is giving the lecture. ... Do you think I can plan for marriage and do you think I should continue school?'

Symbolically, this question signified that a janitor is a male who has reached maturity as a male, a wage-earner, and a sexual partner or husband. By contrast, a Ph.D. instructor represented the patient himself as a perpetual student, unable to mature somatically and become either a wage-earner, or sexual partner. The symbolic discrepancy between the two was prophetically foreboding, and permanently without resolution.

Three months later he wrote about the virilizing failure of hormone treatments in past years: 'I don't think I can impress on you just how desperate I am about the situation, and how willing I am to try any possible treatment.' In response, he was told that there had been no new hormonal discoveries that would benefit him, but that he and his fiancee together might find value from a joint interview. The ensuing silence lasted for three years.

It was broken by a letter sent to the patient requesting a progress report. His mother replied. In the interim he had married, two months before his 22nd birthday. His wife was the same girl with whom he'd been going steady. At the same time he had quit college for a job as an itinerant salesman. Customers discredited him because he looked too juvenile. He took an apprentice job, at which he excelled, as a dog cosmetologist, clipping the hair of show dogs in preparation for contest exhibition. Psychodynamically, this career complemented perfectly his own deficiency of facial and body hirsutism, but it came to a premature end when his tutor died quite unexpectedly with cancer. Thereafter he drifted through a prolonged period of unemployability and pathologically depressive inability to cope with life either maritally, vocationally, or academically.

When he was next seen again for an interview in person his age was 23:4, and he had been married for 18 months. The appointment had been for him and his wife, but he came alone. In the interview on erotosexual functioning nothing essentially new was added, except that he was now married. What he had to say about his relationship as a husband with his wife was positive, erotosexually and otherwise, except with respect to the unresolved issue of donor insemination or adoption as the way in which they might establish a family. The de facto resolution of that issue was being taken care of covertly by reason of his financial status as a vocational and educational dropout. Resentfully, he depended in part on assistance from his parents. The insouciance and careless optimism which had been so valuable to him in encountering the surgical problems of his childhood were now no longer as serviceable as they had been. One danger signal emerged during the course of the interview, namely, a report of bad dreams and episodes of sleepwalking that occurred when he was at home, asleep in bed with his wife. Another danger signal was the chaos of his answers to a routine sentence completion (Sacks) test. In confirmation of what his father had conjectured, he disclosed that he had earlier been too shame-faced to admit that, with the instigation of a friend, he had when younger tried to prove his masculinity with a prostitute. Consequently, he had been picked up by the police.

As a sequel to this visit, he agreed to resume treatment with testosterone enanthate, 400 mg every three weeks, the injections to be given by his wife who had had training as an assistant nurse. 'When I didn't see any results, I said to hell with it and quit,' he had reported. In addition to noncompliance because the treatment was ineffectual, there was also a phobic inability to attempt treatment. He had been supplied with a free sample of the new and commercially unobtainable hormone, dihydrotestosterone. It was in a cream base for direct application to the face, in the hope of encouraging beard growth. He was unable to take the jar down from his bureau and use it. It stood there, 'accusing' him, he said, 'a rebuke and a reproach' to his manhood, every morning.

His age at the time of the clinic visit when he reported the foregoing information was 24:4. He was accompanied by his wife on this occasion. Each gave a separate interview in which they downplayed such difficulties as existed in their relationship together, in anticipation that things would be better. The husband had made preparations to complete his college degree. The pros and cons of adoption versus donor insemination had been given further airing.

The wife regarded her husband's testosterone injections as having increased their coital frequency from once or twice to three or four times a week, 'and even now it is still more frequent than before the shots, even though he hasn't had them for a few months.' The strength and maintenance of erection of the penis had always been adequate to stimulate her to orgasm. She had no knowledge of whether testosterone might or might not have affected private masturbation nor, indeed, if it had occurred. The husband's own estimate was 'a few times a week ... in the shower.' There was never an ejaculate, and it did not displace coitus of which his estimate was 'once or twice a week, maybe three times.'

Most of what he reported about erotosexual functioning reaffirmed what he had reported in earlier years. His orgasm he described as a 'feeling of very intense pleasure. My penis thrusts or contracts rhythmically. Also some feeling, like around the anus, of tightening,' without ejaculation. Like his wife, he enjoyed oral sex, but his preference was to have the climax 'through intercourse.' The easiest way for his wife to get him aroused would be by body contact, then, 'once I'm aroused I enjoy looking at her.'

He had had more opportunity to see explicit erotic movies as he became older. He was accompanied by his wife only once. It was at a drive-in, where she had the privacy of being in the car and not seen. 'I don't know as I really fantasized too much, while I was watching them,' he said. 'I think I was more watching them, or probably was more projecting myself, you know, into the part of the man in the movie ... like, boy, that looks like fun! I wish that I had the opportunity to do, you know, what he's doing with her – more that sort of thing, than actually being on the screen. ... I'd fantasize being home with my wife rather than being with another girl, or woman.' Though he could imagine having sex with another partner he had not in fact done so – 'I'd have to be pretty sure I wouldn't be caught.'

He recalled 'once, just only once' having had an erotic dream that included orgasm – 'real orgasm, physical not just mental,' but as always nonejaculatory.

He had a real-life test regarding homosexuality on his next trip to Baltimore. Around the corner from his hotel he visited a bar frequented by bar girls. They were soliciting drinks and money, so he left and went into a coffee shop next door. It turned out to be a hangout for gay people and impersonators. Two young men sat with him and asked if he were gay. Though he said no, and had no interest in them sexually, they stayed and talked with him about the entertainers, some of whom were transvestites or transexuals. A couple of hours later, at closing time, patrons were pairing off to go home together; and he went alone to his hotel; and pursued his experience no further.

To keep procrastinating and postponing is to ensure never having to confront hope defeated. This is the principle that explains the paralysis of behavior that increasingly overtook this man as he progressively confronted the stark realization that he would never have the appearance of virile maturity, and never have a big enough penis. At college, this paralysis made him a failure as he tried to resume his studies so as to graduate into the world of vocational maturity. In the clinic, it made noncompliance an obligation, so that the ultimate failure of the treatment to produce mature virility would never be absolutely confirmed. At home, it made soporific intoxicants like alcohol and sometimes tetrahydrocannabinol seem to hold time in check until virility could be achieved.

None of the sneaky strategy of hope about to be defeated worked well enough, of course, to prevent hope's ultimate defeat. It never does. In seven months, the patient requested an emergency appointment to talk again. 'Mainly,' he said, 'I have been very depressed, and finding it very difficult to just carry on the necessary things of day to day life. I guess the main reason would be that it's really becoming a discouraging hangup looking so much younger than I actually am. Things that should be done from day to day, whether they have to do with your job, school, or what, I'll put them off, and keep putting them off. ... I feel somewhat that perhaps I've come down to the bottom, and started on a bit of an upswing, just by the fact that I finally admitted to myself – and this is the first trip that I've ever made to Baltimore in my life, without my parents dragging me down by the heels. Last year I was pestered into it. This is the first time I've ever said, well, I'd better do something.'

There was a brave optimism in these statements, and one that, judging by a letter two weeks later, had at least a short-lived carry through. He had found help in the concept that he might adopt tactics, including even the gimmick of a theatrical moustache, to enforce people to react to him according to his actual age But he had not found the internal resource with which to implement any changes – not even to persevere with testosterone injections. The happiest outcome had been in his relations with his wife (whose personal interview had been by long-distance telephone). He told her that when he was little and in a hospital bed he would tell himself he had to hold back from everybody. Now with her he felt, 'I want us to be so close.'

At work, his official excuse for an unscheduled absence had been either 'personal business' or 'a death in the family,' both of them evasions that did not bode well for facing self-truth. In three weeks, he was in contact again, by phone, taking up once more the possibility of a prescription for an antidepressant, and of a local referral for therapy on at least a weekly basis. Another month later, he was incapacitated, sitting alone in the dark, hitting his head and saying: 'I'd like to die, but I haven't got the guts to kill myself.' He was in treatment for depression, and on medication with amitriptyline (Elavil).

The ensuing years were marked by ups and downs of depression, of progress in college, of family relationships, of marital compatibility, and sex life, and of compliance in treatment. A new trial of therapy to induce virilization with both testosterone and dihydrotestosterone failed to have any effect after two years. He was now 27 years old, and early adolescent in appearance. In the interview at this time, he introduced the topic of his neonatal history.

'I was aware and had been,' he said, 'from about the time I was thirteen, that I had been assigned female gender when I was born. I had decided just never to talk about it before, until now. ... I kind of put a cap on it, and cut the conversation off when my parents tried to talk about it a couple of years later. ... I was in my father's drawer and I came across, not my birth certificate, but my baptismal certificate, which had stated on it that this child has been found to be a boy and his name is changed to John. ... I don't think I felt very shocked – I think if anything a little bit of suspicions confirmed.' Whereas he might have thought of himself as some sort of a sex-change freak, he recalled it as more likely 'that I sort of talked myself into the idea that doing that would be sort of dramatizing the situation. ... I think I've always felt that the best thing was just sort of to really ruminate on it for, you know, a good long time. ... The idea of being better off changed to a girl? Yes, it crossed my mind, but I think as quickly as it crossed it, I just put it in the category of dramatizing things because I think actually, like just socially the type of life to lead, I'd be much happier being a male, and I just couldn't picture – you know, I just feel more comfortable as a male. I mean I feel uncomfortable sometimes as a not very masculine looking male. But I think I would rather be as I am than be a feminine looking woman, because I just don't think I'd like being a woman. ...

'It puts me in a funny position. You hear, you know, things that sound sort of arbitrarily masculine and feminine, and you realize that – at times I might enjoy doing things that are considered strictly masculine. And yet, I think how close I came to – I mean, let's say that my parents were other people and I was taken to a real quack somewhere, and he said the best thing to do is some other course of action. If that had been followed, I probably would have been imprinted as a female. It just makes me, it makes it very difficult, except for, like, sexual functioning, for me to look at a lot of things as being, like, masculine habits and feminine. You know, like hobbies or pastimes.'

Would it have been better to have enforced discussion of the birth records in teenage? 'I don't think so,' he said. 'I'm kind of glad that I discovered it under a situation ... where I was able to absorb it and do with it in my own mind what I wanted to. I didn't feel forced into discussing it, because nobody knew how much I knew. I don't know how you determine at what point or stage a person should be told.'

He contemplated the proposition that implications of his early status as a female might have had something to do with the spells of depression he had been through. 'No,' he said, 'though I think that at the time – there's always an element of paranoia, and having knowledge that nobody was disclosing to me sort of reinforced the paranoia a little. Do you follow my reasoning? I don't think it was the knowledge of my sex assignment at birth that upset me. I think it was, sort of, the way people were reacting to me as a male. But, having that knowledge. ... I felt like I didn't know where my psychiatrist was headed, or what he was trying to get to ... and then I just felt anger at myself for my own appearance, and the way that – like coping with it was not, you know, a good way to cope with it.'

Six months after the foregoing interview, and far away from home at graduate school, he was coping more effectively. He wrote: 'I'm still taking my hormones [testosterone and dihydrotestosterone], and I don't look a day older to have a face any rougher than a baby's bottom, but it just isn't an issue any longer. I still have an awful lot of devils to deflate, but it is nice to have one out of the way. ... I've had an affair which has somehow changed my feelings about size, etc. I can't really sort the whole thing out, right now ... we are really talking seriously and comfortably about a child for the first time. ... I used to think that maturity would be no longer having strange feelings and finding yourself in frightening, confusing situations ... now I see it as the ability to take those feelings and situations and do something with them. Somehow, all of that makes me feel as though I'm ready to take on the responsibility of a kid.' Paternity had always been important to him, but he had never shown spontaneous interest in baby and child care and parentalism.

Within a year, his wife had divorced him and left town with one of his friends. It may have been that their relationship was one of those that, like the nurse-patient relationship, come to an end when the partner who was being ministered to becomes able to cope more self-sufficiently. The husband's self-sufficiency had indeed increased, and included greater erotosexual self-confidence as a sequel to having extended his sexual experience to include a different partner. It gave him a criterion of comparison. His wife, he came to realize, had always been 'inhibited ... she just never seemed to enjoy sex ... and I always thought this must be my fault because of my penis.'

The divorce did not, of course, help in exorcising his 'demon of sexual insecurity' nor any other sources of insecurity. Nonetheless, he was able to write a year later, 'Though I'm still subject to depressions, I deal with them much better than in the past. They don't paralyze me as they once did.' He established himself professionally in one of the helping professions. Erotically and sexually he was able to cope with the 'initial gut reaction' of blaming the size of his penis when a relationship broke up, even though he knew there were other, logical reasons, and he expected eventually to establish another long-term relationship. He coped with the trauma of his father's early death, following which, for family reasons, he changed his place of residence. It was while en route through Baltimore that he had stopped at the hospital for the most recent follow-up visit in person. As a gift, he brought with him one of his houseplants, a tall, single-stemmed, thorny cactus. It served as a silent satire of the medical promise, since he was an infant, that he would one day be a hirsute man with a full-sized penis.

EXPOSITION

It is self-evident from these four cases that the XY pair of chromosomes domiciled in every cell, every brain cell included, does not, per se, inherently obligate the organism to differentiate as a male in G-I/R (gender-identity/role).

In all four cases, the exact chain of biochemical command through which the XY genetic code governed incomplete differentiation as a male (Koo et al., 1977; Silvers and Wachtel, 1977; Eicher et al., 1979; Engel et al., 1980) cannot be ascertained retrospectively. Nor could it have been ascertained anterospectively, case by case, even supposing there were no technical obstacles, for the very act of intervention would have been inimical to fetal survival. It can be inferred that the embryonic testes did secrete their mullerian inhibiting hormone (Josso, 1984; Josso et al., 1977), for in all four cases the uterus and fallopian tubes did not differentiate.

It can be inferred also, in the two cases of partial androgen insensitivity, that target cell unreactivity to embryonic and fetal hormones was responsible for absence of prostatic and seminal vesicular differentiation, and for the clitoridization of the penis and lack of midline perineal fusion. In the other two cases, those of androgen responsivity, the parallel defect of the external genitalia represented not a target cell insensitivity to hormones, but an insufficiency of testicular androgen, or a biochemical deficit in hormone synthesis, so that the necessary androgenizing steroid (possibly dihydrotestosterone) was missing.

In the two androgen-responsive cases, the effect of this missing steroid on the brain and its sexually dimorphic differentiation is not simply speculative, but subject to great uncertainty. Animal experimental studies implicate estradiol, which is normally considered a feminizing hormone, as being the masculinizing brain hormone. Thus, there is at least the possibility that when the synthesis of testicular steroid fails, the fetal brain gets the estradiol it needs from either the placenta or the maternal blood flow.

The brain is a multiple organ. What happens in one part under the influence of hormones, prenatally, does not necessarily parallel what happens in another. Nothing may be assumed. Rather, by means of painstaking inquiry, experiment and observation, it is necessary to develop a catalogue of items of behavior (to coin a much-needed term, behaviorons) that are differentiated in the brain as sexually dimorphic, in greater or lesser degree.

In human beings, the only behavior that is sex-dimorphic, absolutely and irreducibly, is in men, impregnation, and in women, menstruation, gestation, and lactation. Other sex-dimorphic behavior that comes under the influence of prenatal hormones may in subhuman mammals be absolute and irreducibly sex-different, but in human beings it is best conceptualized as being threshold-dimorphic only, and otherwise, sex-shared. That is to say, the behavior may be elicited in either sex, but more readily or more frequently in one sex than the other.

Urinary posture is a good example. In sheep (Clarke, 1977; Short and Clarke, undated), as in many other mammals, it is irreducibly sex-dimorphic, and is determined prenatally by the level of testosterone in the fetal bloodstream during brain differentiation. In human beings, by contrast, either sex may sit to urinate, and either may stand or squat. The design of clothing and urinary fixtures makes standing more convenient than squatting for males, and vice versa for females. At the age of attaining continence (which adults mistakenly attribute to training) small children may attempt both urinary postures, at times, but usually one more often than the other. That is to say, one of the two postures has a lower or easier threshold to surmount in order to manifest itself and be put into practice.

There are some children who were masculinized in utero, but not enough to get a penile urethra, who yield rather slowly to the convenience of sitting to urinate. In the case of WV, the mother made special mention of this phenomenon. The same children may also have prolonged persistence of bed-wetting, which suggests an additional anomaly or imperfection of the brain's urinary governance.

There does not yet exist a definitive list of behavior in humankind that is sex-shared/threshold-dimorphic on the basis of prenatal hormonal brain programing. Nine possibilities are listed in Money (1980), as follows.

Nine parameters of sex-shared/threshold-dimorphic behavior

- General kinesis-activity and the expenditure of energy, especially in outdoor, athletic, and team-sport activities
- Competitive rivalry and assertiveness for higher rank in the dominance hierarchy of childhood
- Roaming and territory or boundary mapping or marking
- Defense against intruders and predators
- Guarding and defense of the young
- Nesting or homemaking
- Parental care of the young, including doll play
- Sexual mounting and thrusting versus spreading and containing
- Erotic dependence on visual stimulus versus tactual stimulus arousal

The governance exercised by prenatal sex hormones in the differentiation of sexual dimorphism centrally in the brain and peripherally in the sex organs belongs, according to the traditional academic point of view, in a different universe of discourse, namely biology, than the governance of what happens after birth. The latter is traditionally assigned to the social or psychological universe of discourse where it is by many people considered to be totally divorced from biology, and akin to the occult, 'spookology,' or the spiritual.

In fact, of course, the brain is a biological organ. Its cells function with supreme indifference to the developmental origin of the program that governs them, whether it be in the genetic code, the molecular chemistry and physics of the embryo, the invasion of viral or toxic particles, the trauma of intrusive injury, or the stimuli imprinted through the special senses from either interoceptive or exteroceptive sources. The latter include visible waves entering the brain through the eyes, audible waves through the ears, touch transmitted through the skin, and smell and taste.

Just as the organism as a whole is both separate from, and in symbiosis with its surroundings, so also is the brain. It assimilates programing from its surrounding interoceptive and exteroceptive environment. What it assimilates may become just as permanently and effectively programed into the brain as that for which the genetic code is responsible. Environmentics is every bit as important as genetics to an understanding of brain function. For example, genetics alone, without environmentics, cannot explain native language; and one's native language becomes as imbedded into the brain's program as if it were put there by genetics.

Assimilation of native language serves very well to exemplify the parallel phenomenon of the assimilation of G-I/R on the basis of exposure postnatally to other people.

In all four of the cases under discussion, the baby was, at the time of birth, considered to be a girl on the basis of the feminine appearance of the genitalia. Two continued to have the status of girls throughout childhood, one (WV) without, and one (WF) with parental doubt as to the prognosis. The other two (MV and MF) became the subject of diagnostic doubt and, before they were six months of age, were formally reannounced as boys, and subsequently continued throughout childhood to have the status of boys.

The two who were reared with the status of girls were, in childhood, emancipated from medical and surgical intrusions on their well-being, and saw doctors on the same basis as did their playmates and siblings. The two who were reassigned and reared with the status of boys were, in childhood, entrapped into what they experienced subjectively as the status of being the medical and surgical victims of defective genitals. Without at least the first stages of corrective surgery as a token of things to come, their genitalia would have lied first to their parents, and subsequently to them, as to the correctness of their sexual status as boys. The only available alternative statuses would then be either that of a girl or, in the concepts of folk knowledge, that of a freak belonging to neither sex.

Comparison of the four cases shows that pubertal hormonal feminization posed no problems. The Eve principle, which is the Adam principle in reverse, signifies that, in the cells of hormonal target organs, feminizing has developmental priority over masculinizing. That is to say, masculinization can be prevented, or held in abeyance, and feminization can be substituted more dependably than vice versa – at least on the basis of today's endocrine skill. The great stumbling block to masculinization is the intracellular phenomenon of androgen insensitivity, which may be either complete or partial in degree. There are some family pedigrees of male hermaphroditism in which androgen insensitivity is not present, and some in which it is present. Each newborn hermaphroditic baby with a history of androgen insensitivity in the family tree can almost certainly be assumed to have inherited the defect. In sporadic cases of male hermaphroditism, where there is no family history as a guide, it is necessary, as one learns all too well from the case of MF, not to make an untested assumption of androgen responsivity.

At puberty, the hormonal changes of the body tell either the truth or a lie about a developing child's status as a boy or a girl. Only one of the two girls in this study (WV) was confronted with this lie. Though she made no report in words that something was amiss, there was a nonverbal body signal, namely, a tic-like movement of tossing back her girl's long hair so that it did not fall across her eyes. This symptom went into spontaneous remission after hormonal feminization was therapeutically ensured. The other girl, who feminized spontaneously, had only the concern of not menstruating. It was less threatening to her than to her parents. It confronted them with their inchoate fear, suppressed since her infancy, that their child would develop as a homosexual.

The contrast between the two girls is, by the design of the study, paralleled in the two boys. The one of them (MF) who feminized at puberty, and for whom all heroics of bodily masculinization were a total failure, should be an exemplar for all of medicine. Henceforth, it should be categorically a malpractice, in the case of a birth defect of the sex organs, to predict developmental prognosis on the basis of chromosomal and gonadal status alone. Henceforth, no boy should be required to undergo so much suffering and failure, and in manhood be condemned to the irreversible status of a hormonal eunuch with a micropenis, simply to satisfy professional hubris. The enormity of the audited costs of so great an error, long-term psychiatric costs included, is indefensible – to say nothing of the unaudited personal costs.

For practical purposes, in the delivery room, the configuration of the external genitals is a prime guide. If the genitals look predominantly female and have a hypospadiac organ small enough to be a clitoris, and not large enough to be reconstructed as a penis, then the sexual status is unquestionably female, no matter what tests are subsequently performed, and no matter what their results.

The problematic decision belongs to the case in which the phallic organ is not so small as to be left intact as a clitoris, but on the borderline of being large enough to serve as a penis. The larger it is, the more likely that it was, and will remain androgen responsive. To test for androgen responsiveness versus insensitivity one applies a testosterone ointment (0.2% testosterone propionate in a water-miscible stearin-lanolin cream base) to the genital area on a daily basis for four to six weeks. If the tissue is androgen responsive, it will begin to mimic the changes of puberty and become more reddish in color, with enlargement of the penis, and growth of silky pubic hairs. The changes will cease when the ointment is discontinued. By extending or sporadically resuming the period of treatment, the penis can be induced to enlarge further, but every millimeter gained in infancy is borrowed from the growth that would otherwise taken place at puberty. In adulthood, the penis is disproportionately small again.

Surgical reconstruction of hermaphroditic genitalia as for a female may be evaluated according to three criteria: cosmetic, copulatory, and orgasmic. In the cases here represented, there was no cosmetic impediment to feminine genital reconstruction. There was one postsurgical copulatory impediment, namely, the development of a band of vaginal constriction in Case WF. It was correctable by minor surgery.

In some cases vaginal lengthening can be achieved without surgery by means of dilation. In some instances, dilation has been effected by gentle coital pressure of the penis, frequently, and over a period of months. In other cases, self-dilation with progressively sized dilators has been effective. Concurrent sexological counseling has commonly been necessary to prevent noncompliance among those for whom compliance means complicity with the forbidden sin of masturbation or the invasion of the body's internal privacy by a noxious object.

The effect of surgical reconstruction on orgasm in each of the four cases is reported and compared in the preceding section. The responsibility of deciding on the surgical removal of erotically responsive genital tissue is always a grave one, regardless of whether the decision is made for reasons of ritual, as in circumcision, for reasons of pathology, as in malignancy, or for reasons of cosmetics as well as coital function, as in the surgical repair of a hermaphroditic birth defect of the sex organs. In the latter case, feminizing surgery poses the particular problem of the loss of the erotic organ, the clitoridean phallus or the phallic clitoris.

Since cosmetic surgery is elective, neither the parents nor, when older, the patients need be under pressure to consent to it. Since, however, it is pathologically stigmatizing to grow up as a girl with a genital organ that looks like a penis, it is, in effect, semantically dishonest to characterize surgical feminization only as cosmetic and elective. In the present day and age, there are in our society no institutionalized roles of gynandry or androgyny provided for people with sex organs that look ambiguous and that, uncorrected, have neither male nor female coital capability.

Feminizing surgery in infancy precludes the possibility of before-and-after reporting, and hence of ever knowing whether the operation changes either the quality or the quantity of erotic feeling, including orgasm. One knows only that, later in life, those who have undergone surgical feminization in infancy typically are able to report the experience of genital erotic feeling that intensifies to a maximum, and then relaxes.

Those who undergo genital surgery later in life so that they can give a before-and-after report, though few in number, typically indicate a change in the focal location of erotic feeling and possibly in its quality also, and a change in the timing of the build-up to maximum intensity. In men, even loss of part or all of the penile shaft typically does not abolish the orgasm as experienced preoperatively (Money, 1961; Mathews et al., 1979; Money and Mathews, 1982; Money and Davison, 1983). In male-to-female transexuals who undergo surgical sex reassignment, erotic sensitivity associated with the corpora of the penis is lost, whereas that associated with the skin of the penis may or may not be, depending on the surgical technique employed. In the preferred technique, the skin of the penis is not severed from its blood and nerve supply, and is used to line the vaginal cavity. Regardless of surgical technique employed, male-to-female transexuals vary in their postoperative reports of orgasm ranging from what amounts descriptively to a warm glow throughout, to a climactic peak that resembles, but is different from the preoperative orgasm. In either instance, there is for the male-to-female transexual great subjective satisfaction in being able to satisfy the male partner.

In many cases of feminizing surgery in hermaphroditism, postsurgical eroticism is focalized genitally in and around the urogenital sinus and vagina. Especially in those cases in which there is no urogenital sinus, in order to maximize the preservation of erotic feeling there should be minimal surgical excision of the clitorophallus – for example, by dividing it and imbedding each half in one of the reconstructed labia majora. Conservation of the clitorophallus is highly recommended in all cases, including those with a urogenital sinus. The chief contraindication, that the imbedded tissue may become painfully erected, does not apply when estrogen is prescribed.

Cosmetically, the results of masculine genital reconstruction in hermaphroditism are variable. In the case of MV they were not good, and in the case of MF, very definitely bad. Functionally, there are two criteria of the success of surgery, urinary and copulatory. Postsurgical complications from urinary infections following urethral reconstruction are a hazard, and one which may persist, off and on, indefinitely, as in the case of MF. Urinary strictures may also develop. Postsurgical copulatory complications relate closely to the size of the penis. Penis size cannot be enlarged surgically. A penis that is too small and inadequately erectile is simply incompatible with intromission. This was to some degree a problem in the case of MV, and to a far more serious degree in the case of MF.

Judged on either the cosmetic or the functional criterion, masculinizing genital reconstruction requires more surgical admissions than does feminization, and is more subject to failure, namely the opening up of urethral fistulas that require additional surgical repair. A boy may be confronted with as many as a dozen different surgical admissions in the years of his boyhood, and each one may for him be the equivalent of a trip to a purgatory of separation and abuse.

For the surgeon, by contrast, each of his patient's admissions is for him the equivalent of a musician's appearance in Carnegie Hall – a virtuoso's challenge to perform. It is cynical to carry the analogy too far. Nonetheless, it is only fair to keep in mind, in accordance with the principles of the sociology of knowledge, that there is more meaning and significance to customs and practices than what the orthodox justifications and explanations of them lay claim to. In the health professions, dedication is an example: it is all too easy for dedication to patients, on whose behalf one makes decisions, to become dedication to what one decides for them. Patients then become victims of one's professional self-aggrandisement. Pride then insists that one's own doctrines, methods, and decisions are beyond criticism and revision. Perhaps because emancipation from the sexual taboo is no further advanced in hospitals and medical schools than in the rest of society, professional pride, often only partially informed, still is the basis on which many decisions about hermaphroditism are made.

In three of these four cases (MV, WF and MF) the parents had information about their baby's sexual anomaly and (in MV and MF) sex reannouncement which they classified as top secret. They were young parents in their twenties at the time of their baby's birth. They had heard about a birth defect of the sex organs, if at all, only in the context of 'morphodites,' half-men, half-women, in the freak show at the circus. In tune with the sexually taboo-ridden society of which they were a part, they took immediate refuge in the defense of secrecy, for which they found ample support among those professionals who were as prone as they were to the authoritarianism of prescribing without explaining. Their rationale was to 'spare the child' by relegating the topics of sexual anomaly and sex reassignment to the category of the unspeakable – the unmentionable skeleton in the closet of family ghosts.

Nearly four decades later, and with the knowledge of hindsight, it is obvious that total destruction of all evidence is more fantasy than fact. Families keep documents, birth gifts, pink or blue mementos, and early snapshots. Hospitals keep records in the original name with pretreatment clinical photographs attached. The only safe assumption is that there are no medical secrets, and that sooner or later a child will be old enough to uncover the family's unspeakable secret about her/him – and maintain it as his or her own unspeakable secret.

The idea is to have a birth defect of the sex organs managed so correctly from the moment of delivery onwards that secrecy will not become an issue. Failing that, the ideal is to intervene early, giving the parents guidance that will emancipate them from being secretive with their child and its siblings, while following a policy of guarding privacy within the family (Money et al., 1969). If this maneuver fails, then the clinician may be prevented from serving the best interests of the growing child because of the litigious self-righteousness of the parents in refusing to give informed consent. In the majority of instances, however, the parents can be enlisted as collaborators who help themselves simultaneously as they help their child by ridding themselves of the deception that alienates them from one another. The earlier in a child's life that this is achieved, with appropriate conceptualization and wording, the better. The longer the unspeakable secret of the child's genital anomaly and/or sex reannouncement is kept secret, the more dramatic and traumatic is the occasion of its disclosure. It is not too difficult for a child to grow up always knowing that he or she was born with 'unfinished sex organs,' or with the vagina 'covered over,' and always knowing that the correction still needed is anticipated. Nor is it difficult to know that in such an instance, parents and doctors may be temporarily unsure of whether the baby is unfinished as a boy or as a girl. By contrast it is traumatically difficult to discover secret documents that inform you that once you were not the sex you now are. You wonder about the operations to make you what you are. You are completely alone. Not even your parents can be trusted. They must not know that you know. You are a spy – an alien in your own home.

What should be the therapeutic policy of the professional who has complete access to the patient's 'forbidden' medical history and who suspects that the patient, in teenage or later, already has knowledge of it, but is intellectually too paralyzed to talk about it? For 20 years or more, my policy was to be attentive and to wait. Let only fools rush in, prematurely, where angels fear to tread. They would do more harm than good. Now, on the basis of information from various patients, I have come to the conclusion that the harm is in waiting too long. It is the doctor's responsibility to prepare the parents and the child to open the forbidden topic to discussion – progressively and not precipitously, and with ample opportunity for therapeutic follow-up. The Parable Technique is an effective technique for making a breakthrough. The parable is a story of how a person, who the doctor thinks may have been in the same predicament that the patient may presently be in, made the breakthrough in talking of the forbidden topic. The breakthrough may have occurred in a dream. A parable requires no confession. It simply opens for discussion a hitherto unspoken topic.

The forbidden topic of one's past history, insofar as it remains forbidden, invades one's present history destructively. It becomes a forbidden topic in the relationship between lovers or spouses, and between parents and children, stepchildren, or adopted children. It may become a menace to one's own health and well-being, if one's closest relatives by marriage or in the next generation do not know enough of one's medical past to give an accurate history to a physician in case of a medical emergency.

Not only is the hazard of the forbidden topic in one's past history represented in the four cases here under discussion. Other hazards that may impinge on the differentiation of G-I/R in hermaphroditism are also represented in diverse permutations and combinations, including incongruity between chromosomal status, prenatal hormonal status, and/or neonatal diagnostic status; the history of sex announcement or reannouncement; the quality and frequency of counseling available to the parents and subsequently to the growing child; disparity between pubertal hormonalization, civil status, and the efficacy of therapeutic endocrine intervention; the postsurgical coital efficacy of their genitals; and the degree of psychopathology that afflicted them in adulthood.

Despite their diversity, all four maintained a G-I/R consistent with the sex in which they had been reared since early infancy, and toward which the clinical endeavors for their habilitation had been directed. Not one of the four would be recognized or diagnosed at any clinic or by any specialist as having either a partial or a complete syndrome of gender transposition or gender dysphoria. As conventionally defined, the gender transpositions of transexualism, transvestophilia, homosexuality or bisexuality would not apply.

Nonetheless, the four cases are of great significance with respect to the variables of gender transposition, and the theory of how gender transposition might originate and be defined – the theory of homosexuality included. They show that postnatal history can assimilate prenatal history, and then redirect the differentiating G-I/R, and redetermine its outcome. In these cases, the postnatal history included gender-dimorphic intervention, surgically and hormonally, with variable success. It included also gender-dimorphic intervention by way of rearing according to the conventions of being boy or girl.

The two people who were habilitated as women show, with absolute certainty, that neither heterosexuality nor homosexuality (nor transexualism or transvestism) can be defined on the chromosomal and gonadal criteria alone. Thus it would be absurd to say that the husbands of WV and WF were homosexual on the criterion of their wives' chromosomal and gonadal status, for the two women had a female morphology and they functioned sexuoerotically as females with a feminine G-I/R. There are many other cases and different diagnoses of hermaphroditism that demonstrate the same logic.

Pubertal hormonalizing was incongruous with gender status in the cases of WV and MF, thus demonstrating that masculinizing and feminizing pubertal hormones, per se, cannot be held directly responsible for either heterosexuality, homosexuality, transexualism, or transvestism. Here again these two cases do not stand alone. Many other hermaphroditic cases reinforce the same conclusion. Some hermaphrodites do, however, request a reassignment of sex.

Such requests are more likely to be honored when they are made by XY hermaphrodites who at, or after, puberty request a female-to-male reassignment. Some such individuals progress, longer than in the case of WV, into a nonfeminizing puberty without corrective feminizing measures available to them. Their parents and physicians, biased by their nonfeminizing, nonmenstruating status, are further biased if the gonadal status can be established as male. In such a case, physicians consent to, and indeed even impose reassignment as male. The XY hermaphrodite who requests male-to-female reassignment is less likely to meet with success. The only XY hermaphrodites who encounter no difficulty in being permitted female status are those with complete female body morphology and the complete androgen-insensitivity syndrome.

By contrast, an XX hermaphrodite who has a male body configuration, even including a complete penis, is seldom allowed the option of having the status of male. The existence of ovaries and a uterus governs the decision that the status will be female. There is a celebrated Johns Hopkins case from before the era of corticosteroid therapy, quoted in Jones (1979). The patient was a fully virilized adrenogenital XX hermaphrodite whose phallus had been reconstructed as a penis. He took revenge on his surgeon by poisoning himself with mercuric chloride and dying a lingering death. The surgeon had refused to provide the church with medical authorization for marriage as a man, even though the patient had spent his entire life as a man in a body that was irreversibly masculinized and was genitally competent for intercourse as a man.

Quite apart from personal diffidence in making intimate sexual disclosures, self-incrimination is a serious legal obstacle to the advancement of the present type of research on G-I/R, and to the advancement of sexological science in general. Though states do not have a uniform code of sexual offenses, they all make it virtually impossible for a virtuous citizen not to be a sexual criminal. For example, according to the current (1976) revision of the Annotated Code of Maryland (Article XXVII, Sec. 554, Unnatural and Perverted Sexual Practices) it is a crime for anyone, even spouses, to engage in oral sex. The penalty is either imprisonment for 10 years or a fine of 1000 dollars, or both.

It is quite possible that the 30–40 years spanned by these case reports will prove to have been a historically unique period in sexological history, and that it will not be legally and bureaucratically permitted to accumulate such extended developmental sexological histories in so much detail again for quite some time to come.

BIBLIOGRAPHY

Barr, M.L. and Bertram, E.G. (1949) A morphological distinction between neurones of the male and female, and the behavior of the nucleolar satellite during accelerated nucleoprotein synthesis. *Nature*, 163:676–677.

Clarke, I.J. (1977) The sexual behavior of prenatally androgenized ewes observed in the field. *J. Reprod. Fertil.*, 49:311–315.

Eicher, W., Spoljar, M., Cleve, H., Murken, J.-D., Richter, K. and Stangel-Rutkowski, S. (1979) H-Y antigen in trans-sexuality. *Lancet*, ii:1137-1138.

Engel, W., Pfäfflin, F. and Wiedeking, C. (1980) H-Y antigen in transsexuality, and how to explain testis differentiation H-Y antigen-negative males and ovary differentiation in H-Y antigen-positive females. *Hum. Genet.*, 55:315–319.

Jones, H.W. Jr. (1979) A long look at the adrenogenital syndrome. *Johns Hopkins Med. J.*, 145:143–149.
Josso, N. (1984) Hormone anti-müllérienne. In: *Médicine de la Reproduction Masculine* (G. Schaison, P. Bouchard, J. Mahoudeau and F. Labrie, Eds.), pp. 7–14. Paris, Flammarion.
Josso, N., Picard, J.-Y. and Tran, D. (1977) The antimüllerian hormone. *Recent Prog. Hormone Res.*, 33:117–167.
Koo, G.C., Wachtel, S.S., Krupen-Brown, K., Mittl, L.R., Genel, M., Breg, W.R., Rosenthal, I.M., Borgaonkar, D.S., Miller, D.A., Tantravahi, R., Schreck, R.R., Erlanger, B.F. and Miller, O.J. (1977) Mapping the locus of the H-Y gene on the human Y chromosome. *Science*, 198:940–942.
Mathews, D., Robinson, S., Mazur, F. and Money, J. (1979) Counseling after resection of the penis. *Am. Family Physician,* 19: 127–128.
Money, J. (1961) Components of eroticism in man: II. The orgasm and genital somesthesia. *J. Nerv. Ment. Dis.*, 132:289–297.
Money, J. (1980) *Love and Love Sickness: The Science of Sex, Gender Difference, and Pair-bonding*. Baltimore, Johns Hopkins University Press.
Money, J. (1988) *Gay, Straight and In-Between: The Sexology of Erotic Orientation.* New York, Oxford University Press.
Money, J. and Davison, J. (1983) Adult penile circumcision: Erotosexual and cosmetic sequelae. *J. Sex Res.*, 19:289–292.
Money, J. and Ehrhardt, A.A. (1972) *Man and Woman, Boy and Girl: The Differentiation and Dimorphism of Gender Identity from Conception to Maturity*. Baltimore, Johns Hopkins University Press.
Money, J. and Mathews, D. (1982) Prenatal exposure to virilizing progestins: An adult follow-up study of twelve women. *Arch. Behav.*, 11:73–83.
Money, J., Potter, R. and Stoll, C.S. (1969) Sex reannouncement in hereditary sex deformity: Psychology and sociology of habilitation. *Soc. Sci. Med.*, 3:207–216.
Short, R.V. and Clarke, I.J. (n.d.) *Masculinization of the Female Sheep*. Distributed by R.V. Short, MRC Reproductive Biology Unit, 2 Forrest Road, Edinburgh EHl 2QW, Scotland, U.K.
Silvers, WK. and Wachtel, S.S. (1977) H-Y antigen: Behavior and function. *Science*, 195:956–960.
Tjio, J.H. and Levan, A. (1956) The chromosome number of man. *Hereditas*, 42:1–6.

CHAPTER 6

Lovemaps discordant for transexualism in two siblings concordant for 46,XY hermaphroditism and female sex assignment, one reassigned in adulthood as a man

SYNOPSIS

The matched pair of this chapter are siblings. Neonatally, they were concordant for sex of assignment which was female, on the basis of their female genital appearance at birth. They were concordant also for having subsequently been given a diagnosis of male hermaphroditism on the criteria of 46,XY chromosomal sex; male gonadal sex with undescended, defective testes; lack of internal female reproductive organs; and masculinizing hormonal puberty (with normal testicular testosterone synthesis as judged by plasma testosterone and gonadotropins, and by peripheral 5α-reductase, and receptor affinity for dihydrotestosterone).

The pediatric histories of the two children were discordant. The hermaphroditic status of the older child was established at age 5, as a result of which sex reassignment as a boy was recommended. The recommendation was not implemented but put on hold. The parents procrastinated, and three years elapsed before the child was again evaluated at a major referral center. The decision there was that the external genitalia could not be surgically masculinized, and that the child should be kept in follow-up to ascertain whether, at the expected age of puberty, hormonal intervention would or would not be necessary.

The hermaphroditic status of the younger child was not established until age 13, at which time it became evident to the mother that, as in the case of the older sibling, the early signs of puberty were not feminizing, but masculinizing. At this time, the clinical history of each sibling was concordant, insofar as each elected to be hormonally and surgically habilitated as a woman. Postsurgically, their histories diverged and became discordant. In desperation, the older sibling, at age 23, self-initiated female-to-male sex reassignment by living in society as a male. Masculinizing hormonal sex reassignment to live as a man soon followed. Surgical sex reassignment of the genitalia was technologically dubious, as well as unfundable. The younger sibling had no inclination toward sex reassignment. She continued to live hormonally and socially as a female, in conformity with stereotypes of femininity typical of her community. She was not beleaguered by erotosexual turmoil and despair, as was her sex-reassigned, female-to-male brother.

In view of the concordance of these two siblings on the criteria of chromosomal sex, congenital anatomical sex, and pubertal hormonal sex, the discordance of their adult gender-identity/roles as predominantly masculine and feminine, respectively, can most likely be attributed to a history of covert discordance in their rearing and family interactions. The older sibling's masculine hermaphroditic status, first discovered when she was 5, compromised her status as a daughter, as became evident in the special relationship she had with her father, and the male-bonded activities they shared together. By contrast, the younger sibling's masculine hermaphroditic status remained unascertained for the first 13 years of her life. During these years, her parents did not have an image of her, as they did of the older sibling, as a daughter who was not an authentic female, but a male hermaphrodite masquerading as a daughter. Moreover, when the younger child's diagnosis was eventually ascertained at age 13:8, the diagnosis did not have a long-term repercussion on the father-daughter relationship. Afflicted with arthritis, psoriasis, leukemia, and the stigma of having fathered two hermaphroditic children, he died from suicide by carbon monoxide poisoning when she was 15.

CASE #1: OLDER SIBLING

Diagnostic and clinical biography

When the patient was first referred to the Johns Hopkins Hospital for evaluation in its Pediatric Endocrine Clinic and Psychohormonal Research Unit (PRU), she was 14:3 years old. She did not know why the referral had been made, except for her mother's explanation, namely, 'that she'd have to have the glands taken out, and to have hormones that would help her to develop more like other girls.' The only recall that she had of hospitalization and surgery at age 5 was: 'All I knew was that I was in and out of the hospital.'

The child had been born when the mother was 18 and the father 21 years old. Though the mother had not considered the baby, her first child, defective at birth, in retrospect, she allowed that the clitoris might have been considered large. Not until the baby was aged 2 did she become concerned that the clitoris 'looked kind of large for a girl.' She waited until the child was 5 before she took her to a local physician. He referred the child for an exploratory laparotomy. The operation established the absence of fallopian tubes and uterus, and that the internal gonads were not ovarian but testicular in structure. 'They wanted me to have her operation done, and change her over into a boy,' the mother recalled, 'and I wouldn't do it.' Her decision was influenced by the female appearance of the child's external genitals, and by family and community acceptance of her as a girl, together with the child's own compliance in being accepted as a girl. In addition, the mother was overwhelmed by her husband's history of being crippled with arthritis and psoriasis, and the family's economic dependence on welfare. Three years elapsed, until the child was 8 years old, before the local physician's referral to the nearest major medical center was acted upon. There a buccal smear test was sex-chromatin negative, as for a male, thus confirming the earlier surgical evidence of male hermaphroditism. The diagnosis was considered most likely to be male hermaphroditism of the androgen-insensitivity, testicular-feminizing type. The recommended treatment in this syndrome is that the gonads should not be removed prepubertally, as they induce pubertal feminization spontaneously. Hormonal intervention is not needed. The clinical recommendation was that the child return to be evaluated again at age 11. She did not return, however, until age 13:10, ostensibly because the family was emotionally and financially stricken by the fact that the father had developed leukemia, in addition to his other health problems. By this time, the child had reached the age of puberty and was not feminizing, but virilizing (Fig. 6.1).

On physical examination, the height measured 171 cm (67.25 in.) and the weight 75 kg (165 lb). The external genitals comprised a clitoridean phallus, 5 cm in length, hooded with a redundant foreskin and chordee instead of either labia minora or urethral tube (Fig. 6.2). The urethral meatus opened not in the well-formed glans, but at the base of the clitoridean phallus, in approximately the female position, in a urogenital sinus situated in the cleft between the wrinkled, scrotalized labia majora. With a probe, the vagina could be identified as opening into this urogenital sinus. It was blind, and only 4 cm deep. The gonads, palpated in the inguinal canals, measured 2×1 cm. Pubertal hair growth was evident on the limbs, face (daily shaving), axillae and pubis (Tanner stage 4). There was mild facial acne. Breast development was masculine, in keeping with the masculine body habitus. The voice had deepened. The chromosomal karyotype was 46,XY.

Because there was a conflict of philosophy among the specialists working on the case as to whether the child should be rehabilitated as a girl or be subjected to an enforced sex reassignment, an additional opinion was sought at Johns Hopkins

→

Fig. 6.1. Case #1. Appearance of the body habitus at age 14:3, prior to treatment.

Fig. 6.2. Case #1. Appearance of the external genitalia at age 14:3, prior to surgical feminization.

Fig. 6.3. Case #1. Appearance of the body habitus at age 25:0, showing the effect of hormonal feminization.

Fig. 6.4. Case #1. Appearance of the external genitalia at age 25:0, showing the effect of surgical feminization.

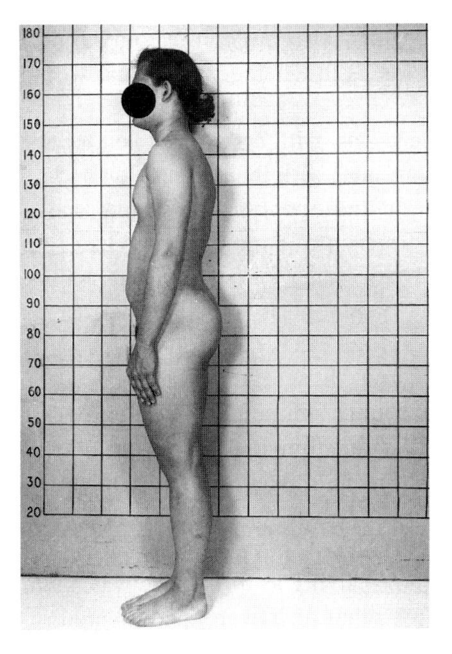

Fig. 6.1

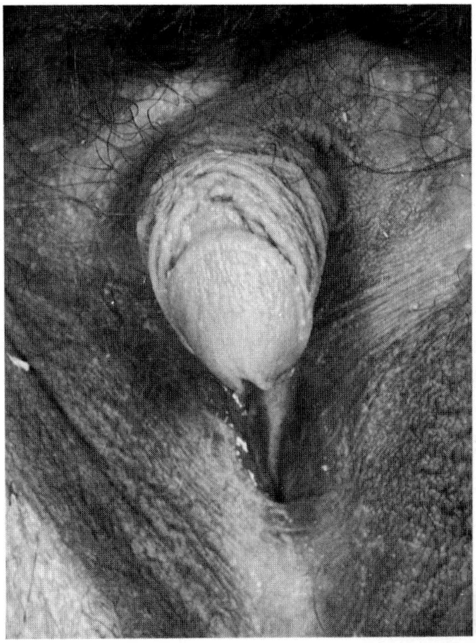

Fig. 6.2

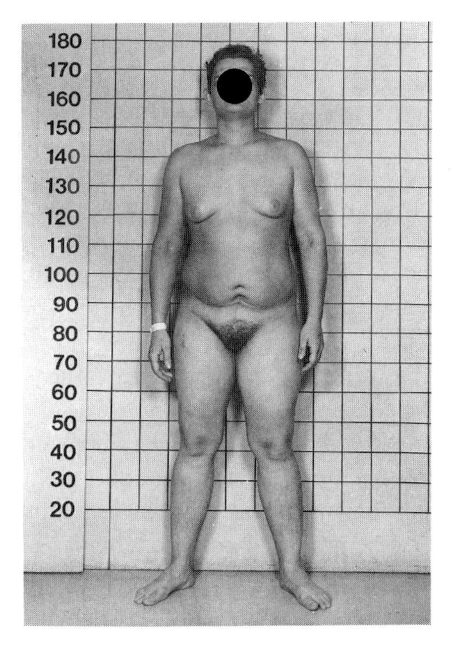

Fig. 6.3

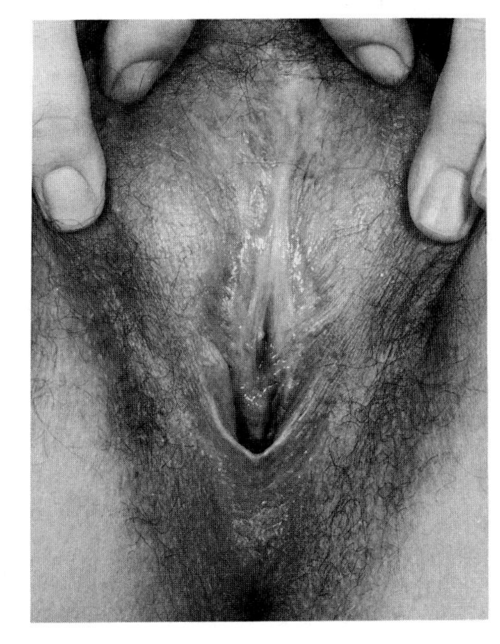

Fig. 6.4

where it was known that psychologic, not only cytogenetic, endocrine, and surgical considerations would be weighed in a joint decision as to whether the patient would live most successfully as a woman or a man.

The joint decision favored retention of female status, with appropriate endocrine and surgical rehabilitation. This decision was congruent with the patient's own ideas and self-prognosis which she discussed extensively and openly in taped interviews. Accordingly, she was admitted for feminizing surgery three months after her 14th birthday. The first-stage procedure comprised gonadectomy, removal of the phallus, and reconstruction of the labioscrotum into labia. Vaginoplasty would be performed later, as a stage-two procedure. Hormonal feminization was begun, and subsequently maintained on Premarin, 1.25 mg twice a day. Three years later, at age 17:5 years, the breasts were well developed, the body contour more feminine, and the body hair less wiry (Fig. 6.3). Vaginoplasty was completed at this time (Fig. 6.4). The patient returned home fitted with a vaginal form, to be worn until healing was complete, so as to prevent vaginal constriction.

At the age of 23, the patient underwent a sociolegal, self-initiated change of status from she to he. Two weeks before his 25th birthday, living as a man, he was again evaluated in the PRU (see below). He then disclosed that he had stopped taking Premarin at approximately age 18 and had not taken any hormone since. Treatment with testosterone enanthate, 200 mg every four weeks, was begun. The surgical consultant advised against an attempt at phalloplasty. Medical economics dictated a surgical referral for mastectomy in the home state. The mastectomy was performed with a satisfactory, flat-chested result.

Gender-coded social biography

By the time the patient was interviewed at age 14 she had had, as a girl, a lifetime's history of tomboyish play and recreational interests. 'Most of the stuff she was kind of tomboyish at, and then other things she was more girlish about,' the mother said. Like her similarly diagnosed younger sibling, she 'would rather be out playing ball, running through the fields or riding horses than sitting and playing with dolls or something like that. But then I can think back when I was a girl, and I was the same way,' the mother recalled, 'She was always trying to imitate her uncle, her father's youngest brother, and wanted to do everything he did.'

The girl herself listed her favorite activities as horseback riding, bicycle riding, and hiking. Of dolls, she said: 'Well, any doll we got, we took the head off and used it for a baseball.' She had no interest in babysitting. Whereas she rated herself as mildly attracted to the idea of marriage and parenthood, she projected both possibilities as far ahead as age 30.

Though she recognized no one in her family as being particularly close to her, she believed that she was her father's favorite. She would always do things together with him. 'My father taught me about guns and stuff when I was four or five years old.

He taught me how to clean them and take them apart and oil them. ...I've been hunting ever since I was seven. ...I used to go hunting without a license when I was ten years old. I could hit a woodchuck in the eye at about fifty yards, with a twenty-two.' As well as hunting and fishing and trapping, her dad taught her mechanics, like how to tear down and rebuild a standard transmission. She was able to help him do things that his arthritically crippled hands prevented him from doing alone.

Concerning clothing, the girl said of herself, 'Well, mostly I wear pants and shirts, unless we go somewhere, and then I wear a skirt and blouse.' After surgery at age 14 she became somewhat more dress-conscious and wore, for example, a blouse with puffy sleeves. Her general preference, however, was for unisex clothing, and for indifference to cosmetics and jewelry. These preferences were consistent with the international teenaged style and fashion of the day. They were favored by her friends. Though her physique and appearance were virilized sufficiently to elicit in the observer an expectation of masculine mannerisms, her body language conformed to neither an ultramasculine nor an ultrafeminine stereotype.

As a child, the girl had had no schooling until age 8. The mother attributed the delay to inadequate school transportation, but it was more likely procrastination secondary to not having had the genital anomaly attended to. School achievement was consistently substandard relative to IQ (Wechsler Intelligence Scale for Children, at age 14:3: Full IQ, 107; Verbal IQ, 104; Performance IQ, 108), except in art, music, and gym. Chronologically out of step with class-mates of her own grade level, the girl either kept to herself or, as a teenager, mixed socially with a group of older adolescents whom she characterized as 'bums, freeloaders, and hippies,' whose main interest was in 'cars, motorcycles, and driving fast.' She did not develop a close and confidential friendship. At age 16 the bond with her father was broken when his illness terminated in suicide at age 39.

After quitting school in 10th grade, at age 18 in the year after the surgical admission for vaginoplasty, she had no serious occupational prospects, and unemployment locally was high. She moved away from her family to live in the house of her widowed grandmother, helping her with heavy chores. She hoped to be able to get employment in a factory, or as a truck driver, and to save money to travel. She had a fantasy of getting a guitar and being a musician.

In the ensuing seven years, she led a personally and vocationally chaotic life, self-designated as a freak, on the fringe of society, unsettled, in self-imposed exile from her family, fleeing from relationships that were no longer tenable, being frequently dependent on others for support and twice failing at suicide by overdosing on illicit drugs, and once when 'I wrecked a car one time on purpose.' The social change to living as a man at age 23 did not bring sustained relief. A year later, a shattered love affair precipitated renewed contact with the clinic and a return for sexological re-evaluation and hormonal treatment as a man. Within the ensuing months, there was gradual evidence of some improvement in lifestyle, one mark of which was graduation with a high school equivalency diploma at age 25:6.

For the next two years, he found employment only sporadically. He was arrested on a charge of driving while intoxicated and spent three days in jail (see Lovemap biography). He was released on bail. Then, he and a cousin, both unemployed, went to the northwest in search of work. None was to be found. For a year he subsisted on welfare payments. He was arrested again on a drunken-driving charge and given an 18 month suspended sentence. Four months later, he had a near-fatal motorcycle crash, of which he wrote: 'I've had a lot of people tell me that I'm self-destructive. I'm not so sure that they're wrong. The accident was really strange, because I'm not so sure it was an accident. All I can really say is that I was kinda shocked when I came out of it. It was like I didn't have any control over what was going on. It was like I knew I was in danger, and I just pushed it a little harder. I left the road once, and pulled it back and opened the throttle up. The police said I was going over 70 miles an hour when I hit the pickup. They say they didn't know why I wasn't killed.'

He was incapacitated for seven months, after which he returned east and lived with his paternal grandfather. Then he established a mutually supportive family life with a woman and her 9 year old son. She was partially paralyzed and unable to use her vocal organs, having suffered brain damage in a recent automobile accident. She could communicate by using a computer. She had a disability pension.

He took out a student loan, which enabled him to go to a community college. He improved his grades to A's and B's during his second semester.

Lovemap biography

Though congenial and cooperative when interviewed at age 14, the girl was not given to being talkatively self-revealing. She could not specify exactly how she had conceptualized her sexual problem when younger. When her normal sister and brother were younger, she had seen them both naked, but had not closely examined their genitalia. She disclosed no evidence of having criteria by which to compare male and female genital anatomy with her own. She claimed no acquaintance with masturbation, nor with love dreams or sex dreams involving erection, ejaculation, or orgasm, though she had noticed that her genital organ would sometimes swell and get hard. Of sexual play with a partner, she said: 'No, I never did it.' Sex education at school had covered the basics of reproduction, but not intercourse and orgasm. 'At home,' she said, 'we never talk about it, or nothing.' The gaps in her knowledge were closed with the necessary information and explanation.

Using the Parable Technique, the interviewer told a very explicit story of how another patient like herself thought only of becoming more like a regular girl, whereas still another thought that nature had made a mistake and that she really should have been a boy. Initially the patient could not liken herself to either of these two girls. Eventually came the query: 'When you're thinking about that sexual part here between your legs, what answers do you come up with?' The reply was a question: 'You mean which side of the fence I'm going to be on?" She then added: "Mostly this one.

You know, go right straight through. Have it taken off.' She claimed that she had thought of the opposite possibility 'once in a while, but not very often.'

She had only once, while asleep, dreamed about her ambiguity. It happened after she returned home from the medical center at age 13, prior to first coming to Johns Hopkins. From the dream she 'woke up and I thought it was all over with, and I was laying in bed.' She could not specify what was all over with.

By way of testing the limits, the interviewer posed the possibility of an enforced change of sex. Her feeling toward the doctor, in such an exigency, she summed up as: 'Probably would have liked to wring his neck.'

On the subject of teenaged love and romantic feelings toward one's own sex or the opposite sex, she responded with: 'You mean which one I like best? I like the male.' She had not, however, had any romantic experiences with either sex, at that time.

When she returned to the clinic for extensive endocrine studies at age 15, she was more at ease in responding to sexual inquiry than she had been prior to gonadectomy and clitoridectomy seven months earlier. She disclosed that, presurgically, she had experienced erection, 'If I was reading a good dirty book, or something like that.' Postsurgically, by contrast, 'It's different, you know, ... I can get all warm and everything else. ... It used to be like I'd read books and it would go right through me. Now, I don't know, er, it's different. It used to be that I'd read books and wouldn't pay any attention to them. Now I get something more out of it.'

The books she referred to were sexy paperbacks, some erotically explicit ('spicy' she called them) and some less so. Some she had obtained on exchange from friends, some she had borrowed unofficially from her mother's room, and some she had purchased for herself. Before surgery, the genital feeling aroused from reading them had been accompanied by a small amount of fluid coming out from the urethra, which she described as being 'sticky and a watery color.' After surgery, the fluid 'was not coming out any more.' From what she said, it was not possible to construe that the genital feeling she experienced had been orgasm. She could not specify whether this feeling was the same after surgery as before. She described its postsurgical status thus: 'Well, you can say, like, when you lift up your hand, the muscles get tight. And that's what happens between your legs. Something like that.'

When the inquiry was switched from narrative to visual erotica, the patient's reply about pictures was: 'Well, it depends on what I'm looking at them for. Pictures don't do so much, really — there's nothing wrong with them. The human body is beautiful. There's nothing to compare with it, you know — like the human body has a brain of its own, and it reproduces, you know. Really, it doesn't turn me on, but it doesn't turn me off.'

When asked about her reaction to the onset of her breast development, she said: 'I could take it or leave it. ... It's nice, you know, but it hasn't been here real long.'

At age 17, when the patient was admitted for vaginoplasty, she gave an update on her sexuality. She had not had any romantic interests in any of the boys that she knew: 'I can live with them or without them.' Reflecting on the idea of bisexuality, she could not venture a rating for herself without trying it. She said: 'I think you should be able to try anything once.' She gave her opinion about lesbian love: 'If you're that way, it's okay, I guess. I mean, I don't put nobody down for doing it. I have a few friends, you know, and that's their thing. I believe if they want to do it, they can. ... I had one kiss me one time. ... I was pretty well looped, drinking a little with some of my friends. She just like walked over and put her arm around me and kissed me.'

The subject of erotic arousal from books or pictures, first explored two years earlier, came up again during this admission when she requested 'spicy books' to read while recovering from surgery. Her reaction to reading them, she explained, was 'that I like to learn different things and — like in books, you learn a lot.' Though she disclaimed any increase of feeling in the sex organs, and any urge to masturbate, while reading the 'spicy' content, she added: 'I think some of the stuff is pretty cool, you know, groovy. My sister, you know, she says I get a kind of wicked smile on my face. But other than smiling, I don't think they do much for me.'

There was a brief annotation in the patient's history of a phone call that she made at age 20, in which she disclosed that she had sampled homosexual and heterosexual experiences. There was no further elaboration on this disclosure. For the next three years no further contact was established.

Contact was reopened when the patient phoned four months before her 24th birthday. Still identifying herself by her female name, she said: 'I've come to the sad fact that I'm a man.' She was going through the bureaucratic process of having her name changed from female to male. For five months she had been living with a woman in a remote part of the woods. They were going to get married, except that 'there wouldn't be much sense to it,' because of the condition of the genitalia, even though she believed the 'thing' was getting bigger. Below the skin, it would erect an estimated two and a half inches in length. She wanted it released. She said she could 'get excited and all,' but not the same feeling as she used to get, presurgically.

A year later, the change to living as a male had been completed socially, and the biographical history as a female was in the process of being reconstrued as the biography of a man wrongly assigned, witness the following letter, written at age 24.

I'm not really good with words, and sometimes I have trouble getting a point across, or making something understandable to someone who doesn't know what I'm talking about or even much less care.

'I've always had trouble being truthful with people, not that I lie, but what I say may not be how I feel. Saying what you mean, and how you feel, to me are different somehow. And in not being able to say what I feel, I can't show it either. It's like having a force field around myself. People can't get in, and I won't or can't come out. It may sound funny to someone who doesn't know me or what I've been through, but really it's not. I've had it around me for so long that I can't get out even when I want to. And that's why I'm writing my feelings out on paper so that maybe once they start coming out that they'll keep on coming out all the time. But most of all it is so that someone else besides me knows how I feel.

'The first feeling I can remember having was confusion. Knowing that I was different, but not knowing how. And no one really taking the time to explain it to me. Anger was next, I think. Having everyone tell me I was a girl and knowing in myself they didn't know what they were talking about. But still insisting on it all the time. Telling me it's the way it's going to be whether I liked it or not, the decision had been made. Hate was the next, because having the doctor tell me I had a choice of being what I wanted and having my parents telling me I had no choice not only made the confusion I already had worse, but made me hate everyone, including myself. I hated my parents because they were telling me I had no choice, and hating the doctor because I had been told I had no choice and having him tell me I did. Knowing full well that no matter how I felt my parents' wishes would be done, no matter how I felt. And hating myself for going along with them. Not knowing really what to do one way or the other. I wished they would all just leave me alone. I said once, "It's like a nightmare. I just wish I could wake up and everything be OK."

'Well, I woke up all right, but when I did the nightmare was still there only worse. This next part will sound strange. I was glad. Not about what they had done to me, but because my parents finally had what they wanted or thought they did.

'I had another feeling when I woke up. Hate, only worse and more at myself than at anyone else. I felt like a part of my life had been flushed down the toilet. And for no more reason than because I knew when I woke up in that bed that the choice that had been made for me was wrong. But I couldn't go back and I couldn't say anything and I couldn't feel anything but hate. For myself, for not standing up for myself. For my parents, for making a choice which should have been mine and mine alone to make. For the surgeon for cutting me, and for the head doctor because I hoped he could see through me and hoped he knew how I really felt.

'The last one was the dumbest of all, though. I knew I should have just told the doctor what I wanted. But I was just a kid and had been told how it was going to be before I met him. And to tell the truth I didn't think it would make a difference one way or another. Only to make trouble in my family, and they already had all that they needed with me.

'I had another feeling when I woke up, too. I wished I was dead, a feeling that's stuck with me through the years. There was one other feeling too. I just stopped caring about everything. I just felt empty inside. I just didn't give a damn about anything. That's when I started using drugs, and I've used them ever since.

'I didn't have any friends. Somehow most everyone knew about me and people can be pretty cruel. Most people thought I was a freak and some even came right out and told me I was to my face. The worst part of that was I believed them because, in a way, I knew it was true. I didn't dare get close to anyone because I knew people talked. And most people couldn't handle being around me. The worst part about it was I couldn't handle it very well myself.

'But even after the surgery I couldn't accept the fact that I was a girl. And that really bothered me, because I picked up the name of queer. Most people couldn't understand why I liked girls as more than a friend. I'd messed around with a couple girls before the surgery and didn't see why I should stop. But the name queer bothered me a lot, so I thought I should see if I was. Since I'd been told I was a girl all my life, I decided to see if I was.

'From the time I was nineteen until I was twenty-two, I had sex with over twenty different guys. Young ones, old ones, it didn't matter, just anyone. I didn't like it, not even a little. I liked the attention, but not the sex. In those four years I tried to kill myself at least three different times. I can't understand how come I'm not dead really, because all three times I did it, I took enough shit to do the job. It just didn't do it?????????

'That's when I decided no more men. I just couldn't handle it. They made me sick. I knew then I wasn't a female and couldn't ever be again. If ever I was? I knew then I was a man and I needed a woman. I had a few, but they were bona fide lesbians, and they thought I was too much like a man for their taste, so I quit bothering with them. I started going after straight women, and I liked them somewhat better, but they had the idea I was a Lessie myself and couldn't stand the idea of being seen with me. They'd go to bed with me, but that was as far as it would go. For some I guess that would have been enough, but I just couldn't see sneaking around like that. And it hurt me to think they were ashamed to be seen with me.

'Then I met Shannon. She was everything I'd ever wanted. I told her pretty much the story of my life, and she accepted me for what I was, for a while. I should have known it wouldn't last. But I didn't. The first six months it was wonderful. But then I guess it caught up with her. I should have known it would because she'd been around. She needed more than what I could give her. She needed a cock, a real one. She tried, I guess, but the need was still there. We were together for one year and seven or eight months, but she started to run around after about the first six or seven months we were together. Maybe before that. I don't know.

'All I know is I need help, and bad, and I don't know where to turn. I don't care anymore. I'm just waiting to die. Something has got to happen, even if it's wrong. I know there's no guarantees in life, but a little something isn't asking too much, I hope. Just a little hope. I haven't worked in almost two years, and I've never been able to hold a job much longer than six months. Most of the time not even that long. Please say there's something that can be done to help me even if not physically then mentally.

'I've stopped hating. All I need now is help of some kind before I do something dumb. Please write back soon. I can't wait much longer. It's too much for me to handle anymore.'

Four weeks after the letter was received, two weeks before his 25th birthday, the patient was admitted for a four-day evaluation directed toward rehabilitation as a male. He had adopted a completely new name, Matthew, in place of Molly, and added a middle name, Justin, in recognition of his favorite uncle, his father's brother only five years older than himself.

Talking about childhood rearing and the sense of being a boy or a girl, he began with a query: 'I mean, how many fathers take their daughters out and show them how to grind valves, you know, or replace rings in a burnt-out engine, or to replace a crank shaft, or to make sure the engine was tuned up right? It was kind of strange, you know, and confusing. I was the first child. There was two years between me and my sister. But the stuff that was done with me was the things that a father would do with a son — like taking them fishing, or taking them to a bar, or just the general things. I mean, like, the old man would throw me a Playboy book and let me look at it, and shit like that. And all this time they were telling me I was a girl, and girls don't do this, and girls don't do that, or they do this, and they do that. At the same time I was hunting and fishing. My father bought me a twenty-two. I think I was seven years old, and he took me out and taught me how to shoot it. He taught me how to shoot a revolver when I was really kind of a small child, and taught me how to cast for bass, and trout fishing, and the things that most fathers would do with a son.' The sister, by contrast, 'was more or less ignored.'

He recalled his mother as being 'the disciplinarian.' The upshot of one disciplinary incident in teenage was that she hit the child — 'So I turned around and said, you stupid bitch, don't ever hit me again. And she hit me right square in the mouth with her fist, and knocked me right up against the wall, knocked all my teeth in front all loose. I went to the dentist and he asked me what happened. I told him. He looked at me, and looked at my mother, and he didn't say a word. He just went about stitching my jaw up. Like I told you before, her family and my dad's family, too, really aren't affectionate people. They don't caress or anything like that.'

In teenage, after his father was dead, he was not able to talk about his medical situation with his mother. 'If I started to say anything,' he said, 'my mother would tell me to shut up.' Years later, when he talked to her, 'all she could say was she was sorry, and that she knew she had made a mistake. But she said that my father told her — his words were: Jesus Christ, what's people going to say? That I can't even have normal children? That they're freaks?'

At the time the wrong decision was made, his recall of his own mistake was: 'When I came here, Dr. Money told me I had a choice, and to think about it. And when I mentioned it to my parents, I was told again, there's no goddamn reason to think about it, because it's going to be this way. So get used to it. I was confused, and I was mad as hell, too, at the same time. So like when doctors asked me questions, I would tell them — like Dr. Money, he really fucked my mind up when I first met him, man, because, like, he come out and asked me, you know, he says, do you jerk off? You know? No, I don't do that, man. Do you play with yourself? No, I would never think of doing anything like that, you know? It was just, it was all weird to me. It was a new experience, and I couldn't grasp it. ... I just gave up. I didn't give a fuck. There were a lot of things I didn't tell Dr. Money, because I was about scared to death. You know, I was only thirteen, fourteen years old."

As an adult, he was able to disclose what he had not dared disclose while living as a girl in teenage, namely that from around the age of puberty onward, as she, he had engaged in sexual experiences with girlfriends at school, and with a younger, genitally normal sister. 'I enjoyed it,' he said, 'but I felt guilty about it.' His penile organ would become erect but, in retrospect, he could not be sure of how large it would become. 'Maybe three inches,' he estimated. 'It took a lot of talking, but we had hand jobs, and stuff like that. The position of the penis, being bent, crooked, or however, prevented it from sexual intercourse. But at the same time there was orgasm. I did come. ... Wet dreams, I would have them just like any other young boy would have. I would dream about girls and stuff like that. And I would wake up either with a full erection or wet pants or — I had already come, or something. It was just a wet dream.'

Referring back to the age of 14 when, after surgery, estrogen substitution therapy was begun, he said: 'I didn't like what the pills were doing to my body. I just got soft, and my chest enlarged a little bit.' Secretly he quit taking the pills. This discontinuance of hormonal replacement therapy may have been a factor in his failure to achieve orgasm, despite erotic interest and arousal: 'I get a good feeling,' he said, referring to occasional masturbation, 'but I don't get an orgasm or anything.' The orgasm returned at age 25, after he had begun hormonal replacement therapy with testosterone. At that time, he wrote: 'I feel wonderful about myself, life, and everything in general. I've experienced quite a number of orgasms, all self-inflicted. They are getting more intense. I just love the shit out of it. At first I felt funny about masturbating. But I remembered all the times I made love and nothing happened, and soon forgot about any funny feelings I had. I think about how it is now, and kick myself in the ass for not doing anything sooner.'

He had not ever thought of himself as bisexual, even while living as a female. On the audiotape at age 24, he said: 'The only reason that I ever had sex with a man was because I had been told that I was a woman, a female, and it had been really pumped into me all my life that women fucked men, and men fucked women, and that's the way it's supposed to be. So, like as I got older, and after the surgery, for a long time I never did anything, you know, because I was still chasing after chicks, and I got the name of being a dyke, a lesbian, or whatever you want to call it. It really bothered me, because I knew I wasn't a lesbian. I know now that I'm not a lesbian any more than anyone else in my situation would be. But, at the same time, I was being told that I was a woman, and I was to fuck men. I had to find out for myself, because inside I felt as though I was a man in my body and in my mind. And I just went out and fucked a whole bunch of guys, just for my own curiosity, to find out if it was really as good as everybody said it was, and to find out if it was for me. ... And it wasn't. I didn't enjoy the sex act. I never got anything from it. ... It just sickened me. While I was fucking these dudes, I was fantasizing about chicks. ... The only thing I could ever say I enjoyed about the sex with men was, maybe, the attention. But it was only for a short time and then it was gone. I didn't care for it, not at all. It made me feel like a homosexual. That's really how it made me feel. It made me feel as though I was lowering myself to do something even though I was doing it to find out if I was right, or they were right. It still at the same time made me feel like I was belittling myself or dirtying myself, or something like that.' The males did not know about the actual medical details of their partner. 'They just wanted to get their guns off, and that was it. ... But a lot of them thought that I was really strange.'

Having re-established herself to himself, the patient was not bashful about making an approach to a woman. 'If you were willing, I would be willing,' he said. 'I don't have to be in love to have sex, but I enjoy it a whole lot more, if I am. You know, I can just go out and pick up a whore and get laid. But I don't enjoy it as much, and there isn't as much satisfaction in it for me as there is to make love to someone I'm in love with.'

After giving up on sex with men, he had sex with several different women, casual acquaintances, some of whom he met in bars, and some as friends who shared communal living in the hippie lifestyle. There was one relationship, living together with his friend, Shannon, that lasted for approximately two years. Its breakup when he was 24 precipitated his return to Johns Hopkins for the admission to ascertain how complete a female-to-male sex reassignment could be accomplished hormonally, even if not surgically.

The story of how he met Shannon is one not only of immediate attraction or love at first sight, but also of stereotypical masculine hyperbole: 'We were all sitting at the window smoking dope and really getting bombed. We were drinking and raising hell. She came walking around the corner and, like, I looked at her, and it was just like a clump of loveliness. And, like, most of the guys were sitting around the table, you know, saying that they were going to get head from her; they were all going to gang-bang her, and just fuck her to death, and all this shit, you know. And I just laughed you know. I told them, I said: You guys ain't going to have a chance, because it's going to be mine. You know, and they just laughed at me because, like, most of them had known me for quite a few years, and they all laughed at me and told me, aw sure, I know her and she ain't going to have nothing to do with you. Well, like, two days later we were in the sack, and we were that way for a long time.'

Looking back on their relationship, he said: 'We had a pretty good thing going on. We had a very active sex life. When we had sex, it was not one of those one-hour things. It could last from six o'clock in the morning until five o'clock the next morning. We would be in bed constantly. ... We would have her brother take the children. ... We counted, one day, and she had eighteen orgasms in twenty-four hours. ... I don't know if it was true, but she told me that I was the first person she ever had an orgasm with, the first time I made love to her. And that she had an orgasm every time and would still want more. ... I ordered a plastic penis, and I'd wear it when I made love to Shannon. She liked it. I'd mount her and, like, get down on her. The way it was made, I'd insert it into her and push, and it would push into the stub of my penis. That brought me up, but not as close as one time' when she used her finger.

'When the stub gets hard, it's like a real short penis. It's under the skin, but the tip of it feels like the head of a penis. There's an indentation in the center of it. My girlfriend, Shannon, she used to have long fingernails. There was a certain way she worked her nail down into the head of it and massaged it really carefully; and it would send, like shocks, you know, good shocks up in my stomach, through my chest, down in my legs. It would, it was rather pleasing. ... It was just like having sex for the very first time, because I'd never experienced it before. It was like, if it had gone on for a couple of minutes, an orgasm would have happened. It felt just like somebody reached inside me, and just grabbed my guts, and was shaking me up and down, you know. It was tremendous. ... It was like everything just blacked off, and the next thing I knew I was hyperventilating, you know, like hell. ... And afterwards, even though I had not come, it was like I was just drained.' Shannon tried to find the same spot on subsequent occasions, but in vain. 'She could get it, but she couldn't maintain it, or I couldn't maintain it. You could find it, but it wouldn't last. It wouldn't be there.

'She likes me going down on her, and that turns me on too. ... Sometimes, I would go down on her first, you know, to get her relaxed and stimulated at the same time. I had a small vibrator, and I had the dildo, which was larger. I would stimulate her clitoris with my mouth. Then, I would grease the vibrator and insert it into her anus, and slowly massage her with it while I was going down on her and insert the dildo. And I'd work it on her and she would come from it. ... It was an occasional thing.'

Once Shannon tried the vibrator in his anus. 'I don't see how anyone can enjoy it,' he said. 'It hurt like hell, and there wasn't anything sexy about it. ... I thought it was degrading,' By contrast, he found anolingus acceptable and stimulating. He pointedly omitted any reference to the region of the neovagina.

There had been half a dozen occasions when the two of them had investigated the erotic effects of bondage and discipline. 'I enjoy bondage a little bit myself. I don't like pain. I don't like to inflict pain. But a little bit of pain, you know, applied lightly to certain areas of the body, is stimulating. Not hard enough to bruise the skin or break it.'

The only other aspect of sexual activity that might qualify, even remotely, as kinky was that he was not excessively modest about not being seen while having sexual intercourse, provided those who might see were at a distance and did not intrude. One-on-one was his style.

With respect to conditions and stimuli of sexual arousal, he made the following summary: 'I can look at a girly book, right? A nude picture book, and it doesn't stimulate me just to look at the pictures. But at the same time to have a real live naked woman walk in the room, you know, that would stimulate me. And sex noises turn me on, you know. And talking dirty and shit like that when we're getting it on, you know, that turns me on.

'The sex smell — even the smell of, say, like a woman, say like her armpits or her cunt, or even the smell of her hair, or a perfume, can turn me on. You know, I'm surprised, myself, a lot of the times, like I'm walking through a corridor or somewhere, and a really nice looking woman walks by and I get a whiff of her perfume, and I'm just ready to jump. Taste? Yeah, I think a woman's cunt is one of the nicest tasting things I've ever tasted, along with, better than steak or milkshakes, you know.

'Touch? Right on the stub of the phallus is really sensitive, but my whole body is. I mean my ears, my arms, my legs, my back, my ass, everything, you know. I've always been that way. I mean, somebody can rub the back of my neck, you know, and in a little while I'm turned on.'

What about narratives? Things you read? 'Reading and dirty movies, not all dirty movies, but sex movies and sex books, I really get turned on from them. It's better if there's sound and thrashing and gasping, all that good shit.'

The positive effect of testosterone on orgasm, aforementioned, did not carry over to the rest of his life. It may possibly have had a negative effect in releasing daredevil recklessness and self-destructiveness, as evidenced by the drunken-driving charge, and its prison sequelae, already mentioned. His audiotaped account was as follows.

'Well, they busted me for driving while intoxicated, and no insurance. They put me in jail for three days. ... By the law of the United States of America, my birth certificate says female. And my biological, you know — sometimes I'm not even sure what the hell I am. They couldn't put me in with the women, you know. And they didn't want to put me in with men, so they put me in a neutral cell and left me in there. I was by myself; solitary confinement. ... When he looked at my birth certificate, you know, he didn't believe it. He thought I was lying. Then he looked at me and said, "Are you telling me that you don't have a cock or balls?" I said, "Yeah, that's what I'm trying to tell you." He said, "Oh, fuck. Get in there and strip." So I went in there and stripped. He walked in and, he was a big guy, and he looked, and he turned around and walked out and brought another guy in, an old guy, and he, the guy just went white and started shaking, you know. I scared him. He said he'd never run into anything like that in his life or ever heard about it. I laughed at him. ... Well, one guy called Albany, and he said in the State of New York, in the history of the State of New York, they have only ever had three other people in my situation, that were arrested, and they didn't know what to do with them people either.'

From this time onward, he was always the male in his sexual relationships. At age 26, he wrote: 'I lived with this woman and her husband. I learned a lesson there, let me tell you. It's one thing to fool around with a married woman. But it's another thing to live with her and her husband. Needless to say it only lasted about three months. My fault though, I had an attack of conscience. Just didn't like being his friend, and then having sex with his wife as soon as he was out of sight. The only thing that bothered him was that he wasn't invited to join. And she wasn't having anything to do with him. And the only thing wrong with me was, I didn't have any feelings for her, except I was horny. You know, something strange: I had more sex before I started on the hormone than I do now, and I sure do enjoy it a lot more now. I have a real hard time explaining to women why I don't have a dick. You'd be surprised how many women are turned off by it. Although I have run into one lady here in the mountains who is real understanding. But it's pretty hard to get around in sub-zero weather without any wheels, so I've only seen her twice. But I hope to see a lot more of her.'

The most recent follow-up was by long-distance telephone. He was 30 years old. With his present woman friend his sex life was good. They both like oral sex, he said, and added: 'Or if I climb on top and rub on her, I get off.' Instead of giving his own estimate of whether his partner would rate their relationship as bisexual, heterosexual, or homosexual, he said, 'Wait a minute and I'll ask her.' Severely disabled from an automobile accident, as aforementioned, she had to punch out the answer on a computer keyboard with one finger that had escaped paralysis: 'Bi to straight.' His own answer to the query was: 'I'll tell you, doc, I really don't know. Hey, there's a big difference between medical law and man's law. I have kind of a hard time with that part of it myself.'

The idea that he might consider himself a perfect bisexual match for a woman who was both bisexual and handicapped had been introduced five years earlier when he had been talking about his breakup with Shannon. In the guise of a request for information, it was actually a covert therapeutic proposal which, at the time, equipped him with a fresh alternative to self-stigmatization and failure as a man.

He agreed that he and his new partner might have found just the right degree of mutual dependency. Considering whether their relationship might work for a long time, his reply was: 'Who knows about that, doc? Fuck, I know straight people who are normal who don't get along for very long.'

With the knowledge of hindsight

The patient was requested at age 24 to offer ideas on what might be done in the future to prevent a repetition of the error that had been made in his own case when, 10 years earlier, it had been decided to go ahead with feminizing surgery and hormonal therapy, directed toward continued living as a female. His reply was: 'I don't really know what to say. Catch them when they're younger, and tell the parents to stay the hell out of the kid's life. If they're old enough to come in here and talk to you, I think they're old enough to know what they want without having the parents pressure them. ... I knew I was making the wrong choice. I just wanted to get it over with so people would leave me alone. ... Like you said, it had to be done with a limit on time. But it is something that should be talked in front of the parents and everything, because I know myself that it's really hard to say things especially about sex and stuff, you know, in front of your parents, to a doctor or anyone else.'

In effect, the patient had been a casualty of the sexual taboo, and the antisexualism of our society. As is evident in the section on lovemap biography, above, there was plenty that as a girl, she could have said about her experience of her masculinized genitalia in masturbation and in sex play with girls. The price of protecting herself against self-incrimination, then, escalated for an entire lifetime.

For this patient, and for the younger sibling as well, the ideal clinical history would have begun with a close inspection of the genitalia at the time of delivery, and a recognition that the vulva was not perfectly differentiated as female. The decision to announce the baby as a girl would have been made, as in fact was the case, on the criterion of the feminine configuration of the genitalia and their complete unsuitability for surgical masculinization and eventual male copulatory function. The diagnosis of male hermaphroditism would have warned of the absolute necessity of doing an androgen-sensitivity test, in order to predict whether puberty would be masculinizing or not. This test is done by applying testosterone ointment to the genitalia, daily, for a period of approximately a month. In this case, the cells of the genitalia would have shown themselves sensitive to the masculinizing hormone. The response would have been similar to that observed at the onset of puberty, namely swelling, wrinkling, and reddening of the genital area, and enlargement of the phallus. With the test completed, in order to guarantee against inadvertent masculinization at the age of puberty, the first, external stage of surgical feminization, including removal of the gonads, would have been undertaken. The second stage, construction of a vaginal cavity, would have been postponed until around middle teenage, when the body had grown to adult size.

With the foregoing method of case management, the parents are, from the beginning, oriented toward rearing a baby who is announced as a girl with a girl's name and referred to as she. They are not confronted with the crisis, as in this case they were when their child was 5 years old, of being told that their daughter is not a female, and should have her sex socially reannounced. The evidence of their senses, when they looked at the child's genitalia, told them otherwise. They dealt with the crisis by taking no immediate action, other than to know that they were rearing a daughter who, with the full weight of medical authority, had been pronounced male, not female. At this point, the bureaucracy of the health-care system became essentially immobilized. The child's future sexological welfare was subject to nosocomial neglect for three years.

When the clinical evaluation at a major referral center was finally accomplished, the child was 8. At that time, the test for androgen sensitivity was not done. Instead, parental compliancy was wrongly taken for granted, and it was assumed that future follow-up visits would give ample early warning of the onset of puberty and of whether it would be masculinizing (androgen sensitive) or feminizing (androgen insensitive). Because of the father's deteriorating health and the family's lack of financial resources, the child's medical follow-up at the time of puberty was progressively postponed until masculinization was well advanced. When evaluated at age 14, she was already sexuoerotically experienced, but too intimidated by the unpredictability of reprisals to be able to incriminate herself by giving a complete sexological history.

The momentum of the health-care system under public assistance did not, at this time, allow for the institution of the Two-Year, Real-Life Test (Money and Ambinder, 1978) prior to the irrevocable step of genital surgery. Had this test been implemented, then the girl would have had a two-year period in which to take female hormones (which would have suppressed her own testicular androgen) and to test her capacity to be habilitated socially as a teenaged girl. At the same time, had finances been provided, she would have been maintained in regular sexological counseling, at least once weekly, so as to gain confidence in disclosing the true biography of her lovemap development. The counseling sessions would have allowed time for a more extensive use of projective tests and inquiries that permit a patient to disclose clandestine, gender-related information in response to requests for drawings, modelings, and narratives, and dramatic acting.

CASE #2: YOUNGER SIBLING

Diagnostic and clinical biography

The initial contact of this sibling with Johns Hopkins was for evaluation prior to admission at age 13:8 years. It was timed to coincide with her older sister's second admission, at age 15:4 years, for postsurgical evaluation and one-year follow-up. The younger girl had noticed that her sister 'had hair on her lip, like I do.' Otherwise she had had no knowledge of her sister's condition, nor of her own, until three months earlier, when she had undergone a school medical examination. Her mother then told her that she was 'not developing,' that it 'had something to do with hormones,' that she was like her sister, and 'would probably have to take pills like your sister does, and might have to have the operation and have the glands taken out.'

The mother's own story was that, despite her attention to the genitalia of her second child, because of what she had suspected and then had had confirmed about the

---

Fig. 6.5. Case #2. Appearance of the body habitus at age 13:8, prior to treatment.

Fig. 6.6. Case #2. Appearance of the external genitalia at age 13:8, prior to surgical feminization.

Fig. 6.7. Case #2. Appearance of the body habitus at age 23:4, showing the effect of hormonal feminization.

Fig. 6.8. Case #2. Appearance of the external genitalia at age 23:4, showing the effect of surgical feminization.

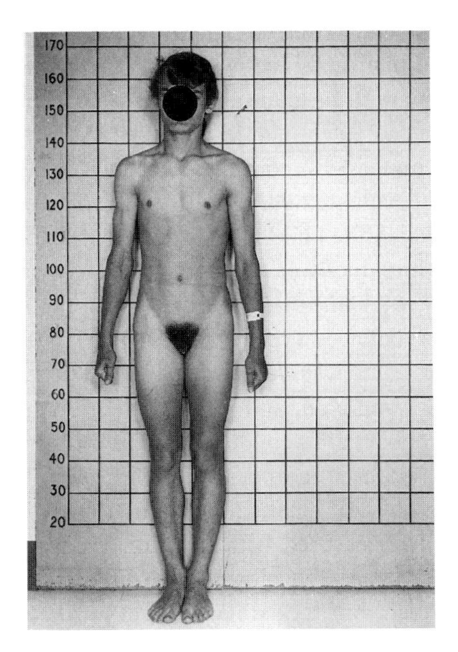

Fig. 6.5

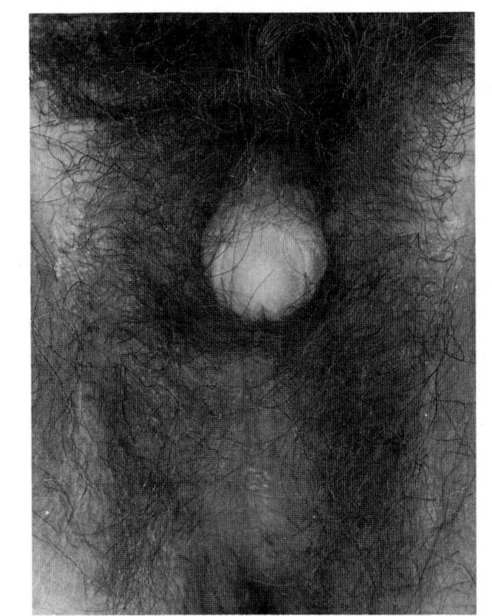

Fig. 6.6

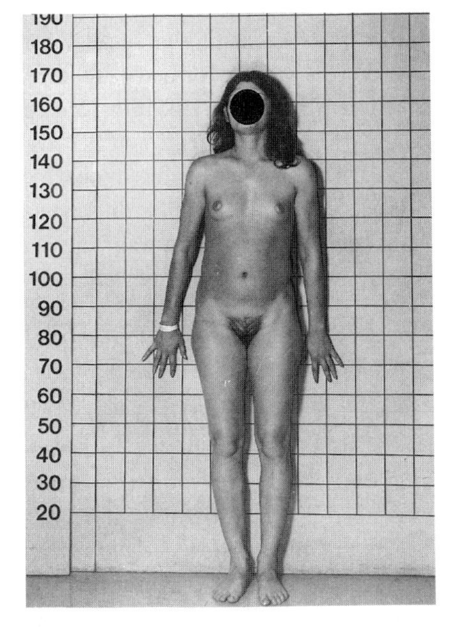

Fig. 6.7

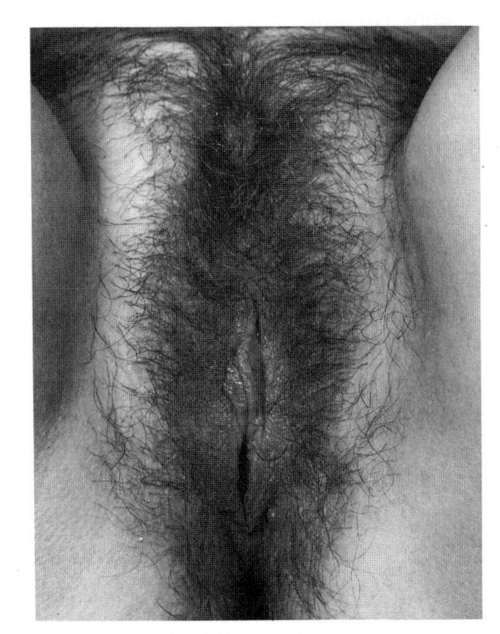

Fig. 6.8

abnormality of the first child at age 5, she had considered the younger one to be normally female from infancy through childhood, until around age 13. It was then, after observing hair beginning to grow on her upper lip, that she made a point of observing the child in the bathtub. She saw the clitoris protruding enough to make her suspect that the two girls had the same condition. The school doctor confirmed her suspicion, and concurred with her decision to have the child evaluated at Johns Hopkins. There had been no controversy regarding the sex to which the child rightly belonged, and the criterion of her being a girl was not brought into question.

At age 13:8 years, the girl measured 164 cm (64.5 in.) in height and 50 kg (110 lb) in weight. The body habitus was lean and masculine in type (Fig. 6.5). The voice was husky and contralto, exactly like her mother's, and was said always to have been so. There was mild acne on the cheeks, forehead, and chin, and a moderate growth of hair on the upper lip and chin. Axillary and pubic hair were developed at Tanner stage 4. The labia majora were wrinkled and scrotalized in appearance with partial fusion around a urogenital sinus into which the urethral meatus opened in approximately the female position. The clitoridean phallus measured 4 cm in length and was hooded with a redundant foreskin and chordee instead of either labia minora or urethral tube (Fig. 6.6). The glans was well formed, but lacked its urethral meatus. Though the vaginal opening was not readily seen in the urogenital sinus, by probe the vagina measured 3.5 cm deep. Gonads were palpable in the inguinal region too high to be measured. The chromosomal karyotype was 46,XY.

The outcome of the combined surgical, endocrine, and psychohormonal workup at Johns Hopkins was a unanimous decision, confirmed by the girl herself and her mother, in favor of surgical and hormonal feminization. Accordingly, surgery for gonadectomy, clitoral resection, and labial reconstruction was performed (Fig. 6.7). Postsurgically, long-term estrogen therapy was begun (Premarin, 1.25 mg twice a day). On this dosage, over the course of the ensuing two years, feminization was satisfactory (Fig. 6.8). The breasts developed adequately, the bodily appearance became more feminine, and facial acne disappeared. The second surgical admission, at age 15:9, was for vaginoplasty, which resulted in an adequately functioning vagina. Long-term follow-up was maintained once more in person at age 23, as well as by mail and long-distance telephone.

Gender-coded social biography

When the girl was first interviewed she remembered that, as a child, she had been interested in playing with paper dolls, and playing house with her younger brother and sister, with herself as the mother. She had also liked playing cowboys and Indians. She considered herself less tomboyish than her older sibling. She listed her favorite recreational activities as horseback riding, bicycle riding, swimming, snow-mobiling and fishing. She was on the girls' basketball team at school. She liked babysitting for neighbors and helping her mother with cooking. Though she was not an aggressive, fighting person, she could defend herself when attacked, and was perhaps less easily provoked than her older sibling was. She was stoical about the prognosis of amenorrhea and infertility. If she would be able to menstruate, she said 'it might probably be okay,' but added that 'I can't change it.' She was stoical about being able to have a family not by pregnancy but by adoption instead. Her long-range plans included the possibility of marriage. She estimated 25 as good marrying age and 'about four kids' as the right family size. At this time, her conjecture was that she would probably 'stay home' to raise a family rather than pursue an outside career.

She liked school, which she began attending at the regular age of 6. At age 13, she had completed 7th grade. Mathematics was her favorite subject. School achievement was consistent with the IQ (Wechsler Intelligence Scale for Children, at age 13:8: Full IQ, 112; Verbal IQ, 106; and Performance IQ, 117). She associated with boys and girls of similar age and had a close girlfriend with whom she would go bicycle riding and horseback riding. They would do school work together.

Concerning clothing, she preferred shorts or slacks and a blouse. She preferred pants of the hip-hugger type, and leotard-style body suits. She would wear skirts to school. She was not interested in cosmetics and jewelry, but liked to wear perfume on special occasions. She wore her hair at shoulder length. After age 16, by which time she had developed breasts and a feminine body build under the influence of exogenous estrogen, her clothing size was the same as her mother's, so they borrowed each other's clothing. She got along 'pretty good' with her mother.

At the age of 18, she graduated from high school with a scholarship for college. She could not take advantage of the scholarship, as it did not cover the cost of living. For the next three years she lived in a guest room with a family. Because unemployment was high in the region, she took whatever jobs were available — as a factory worker, for example, and once doing cleanup work in a disco bar. Her only means of support at age 23 was as a babysitter for her younger brother's children, in exchange for room and board. She was planning on going back to school, and had thought of joining the army in order to get further training. Seven months later she sent a letter in which she wrote about having started an educational program to become a medical secretary. She subsequently changed her academic interest and became a computer major. She was also working as a clerk in a government office.

The most recent follow-up was done by phone. At age 29, the patient was working full time on a night shift job as a switchboard operator. She had quit school owing to lack of funds, but had not given up the idea of going back someday. She was living with her mother and her youngest sibling, a sister, and sometimes staying overnight with her man friend.

There was a 14 year age gap between her and her youngest sister. 'I practically raised her for about the first couple of years,' the patient said, when, after her father's death, the mother had had to go to work every day. 'She used to call me Mommy, once in a while. We're real close. She thinks the world of me.' As an adult, the patient had also helped to raise her nephew during his first year, as she kept house for her brother and his wife, both of whom worked. 'The first time that little guy said my name, that was really decent,' she said, by way of illustrating her attachment to him. She envisioned herself either 'marrying someone with five or six children of his own,' or adopting children. 'I'm not crazy about newborns, because I'm kind of scared of hurting them. ... I'd say age four to ten or eleven is my favorite age.'

Lovemap biography

When this girl's mother brought her to the clinic at age 13 for the first time, she said that the girl 'had never said anything one way or the other' about wanting to get her female pubertal development, nor about wanting, maybe, to change over and be a boy. She did not know whether her daughter might be covering up her concern. 'I never thought much about it, either,' she said, 'until she started growing hair on her lip.'

Despite her initial reticence, the girl herself was able to specify that the beginning growth of hair on the upper lip and failure of breast growth and of menstruation were the signs that led her to believe there was something wrong with her development. Although she compared herself with her older sister, she had been too reserved to approach her sister and talk about their similarity. She had no certain knowledge as to how the size of her clitoris and her vaginal anatomy compared with those of other adolescent girls. There had been some sex education at school, but at home, 'We had practically none,' she said. She knew the basic biology of reproduction and about positioning in sexual intercourse but was uninformed concerning copulatory erotic sensation and feeling. The gaps in her knowledge were filled during the course of the evaluation interview.

The details of what could be accomplished, surgically and hormonally, for either feminization or masculinization were presented in straightforward language, without evasion or euphemism. The girl listened, rather than talked. Eventually she had a personalized understanding of her diagnosis, possible treatment, and prognosis. At the end of the first session, she summarized the most important thing about it as: 'Being here to talk,' and: 'About being normal.' She meant being normal as a girl. The vantage point from which she talked in all her interviews was that of being a girl who needed hormonal and surgical normalization as a female, for which she consistently gave her informed consent.

She disclosed a history of clitoral erection, but not masturbation. Erection was present sometimes upon waking. It occurred also from the friction of clothing, and sometimes from 'just thinking about boys,' she allowed diffidently, and sometimes from pictures of boys. Her pictorial stimuli had been limited to boys in the underwear and swimwear sections of the Sears, Roebuck catalogue. Boy-girl romances, in such magazines as *True Story*, and in various paperbacks, had also appealed to her. Whereas her older sibling at age 14 had read explicit or 'spicy' erotic novels, this girl claimed to have not been interested in them. She had not had any wet dreams, orgasm dreams, or love dreams. Apart from urine, she had no knowledge of fluid secreted or ejaculated from the urinary meatus, and she could not talk of orgasm as a first-hand, subjective experience.

When the two siblings were together admitted for vaginoplasty, this younger one, then aged 16, had no clear recall of her earlier report of having been erotically stimulated by the fashion designs of boys in the Sears, Roebuck catalogue. By contrast, she considered herself to be at the boy-watching stage — 'the good-looking ones,' she specified, but not if they were know-alls. From early teenage onward, she and her girlfriends would sometimes get together and talk about boys. If she discovered that a girl had romantic feelings for her, 'I'd probably go away,' she said. She was nonjudgmental about homosexuality. 'It's their life,' she said; but it wasn't for her. By age 18, as disclosed in a follow-up phone call, her attitude toward lesbianism remained unchanged, as did also her attitude toward being attracted to boys. However, she had not yet had a love affair, and had not yet attempted to have first sexual intercourse.

Between the ages of 18 and 23, follow-up was limited to sporadic correspondence and contact by long-distance telephone. Then, at age 23, the patient was admitted for a four-day, intensive follow-up evaluation, concurrently with the female-to-male reassignment evaluation of her older sibling. In the course of this evaluation, she reminisced on whether she had had doubts, as a child, about which sex she belonged to. 'I think I did,' she replied. 'I remember when I was little. I always thought that I'd rather be a boy because boys don't have to wear dresses, or they don't have to worry about having a period and stuff like that. ... In my mind it was really confusing when I was smaller, because in a lot of things I felt like a man or a boy, you know, and then I would feel like a girl. So it went. You know, during the moods. Like I used to play softball and baseball and football with the boys in school — tackle football and stuff like that. I was a lot closer with guys when I was in the younger grades, and even as I got older, you know, after the operation. But it wasn't like boyfriend-girlfriend, like other girls were getting along. It was more like buddy-buddy. And it seems to me like I can understand guys a little better than some women, although I don't know if I can.'

Recalling the appearance of her sex organs, she said: 'To me it looked like a small penis, you know, the beginnings of one. But like it — to me, what I remember like [it protruded] about maybe three quarters of an inch or an inch. From what I know of penises now, this is what it reminds me of. At the time it didn't. But as I look back now, it does.' She referred to what she had read and seen illustrated regarding prenatal sexual differentiation, as in the adrenogenital syndrome, and asked: 'So, when they did the operation, I had the beginnings of ovaries like that? Instead of testes?' The explanation that followed allowed her to understand that, in embryology, ovaries and testes look the same to begin with, and that her gonads had differentiated so that, under the microscope, they would look like testes. The conversation led on to her sibling's wrong decision at age 14 to be feminized. She answered a question as to whether her own feminizing decision at age 13 had been the right one, beginning with: 'I don't know. I had doubts about it then, you know, because they didn't really explain it to the point where they said we could go either way, you know.' Without specifying who they were, she added: 'It was, like, we're going to go this way. ... I did have little doubts, and stuff, but then I thought, well, the way — I've been pretty active and close to a lot of people, and stuff. And I couldn't, it'd be harder for me to say, well, my name isn't Leila any more, it's going to be Larry. I figured it'd be better if I just went ahead and did it this way, you know.' After the surgery, there were 'no regrets,' as such, but 'wondering if I did the right thing. But I knew I did. Because I couldn't have done it the other way.'

Considering what made her and Matthew, the older sibling, turn out differently, she said she hadn't figured out an explanation, 'Not unless, maybe, could he have more male hormone than me? Maybe personality, some lifestyles, because my father kind of brought him up more like you would a boy than you would a little girl. Taking him hunting, fishing; and you don't usually take your daughter down to your girlfriend's house. I've heard people take their sons, but not very often their daughters — unless it was the mother taking her daughter to her boyfriend's.' She herself got 'more of the feeling that I was his little girl, sort of like that. He'd tell me I was his little girl, and he called me Sis, or Sissy, and hold me more and hug me more than he would Matt. You know, where he'd pat Matt on the head, he'd give me a hug — that kind of difference.' He had no nickname for Matt. The mother, by contrast, was 'standing on the outside, with Toby, the only boy. Even now she dotes on him,' her third child.

Extending the consideration of masculinity in Matthew and herself, she talked free-associatively: 'I always have, like — I wonder if, I don't know, it's just, like the masculinity that could have been in me, I wonder if it would have come out if I'd let it. ... I think I've done away with most of it, you know. If I felt like, you know, in my brain that a certain thing was wrong, then I wouldn't do it. But if it was okay, like to play sports, it's not necessarily all masculinity, you know, and to be good at them isn't wrong. So I go ahead. And if I play, I do my best. In high school I was much better than most girls, you know, and a lot of guys too. So, I was always kind of afraid that people would connect Matt's problem with me having the same problem because, mostly, the girls in my group may talk about their menstrual cycle. Right? And I never did. So I worried if that made them wonder. I didn't want to lie and make up well, yes, I bleed pretty hard or, no, I don't, it don't bother me, or nothing like that. So I always wondered if, in their brains, they thought about the fact that I was good at sports and that I had a lot more muscles. You know, the framework, the muscles were a lot more dominant than in the other girls. The hair on the lip at the time — I always wondered if, wow, you know, what did they think of it, and stuff like that.'

Concerning Matthew's sex reassignment, she compared herself with him: 'I think

I'd be a little more afraid of ridicule. I don't think I could take the pressures of going through with something like that. And now I don't, I wouldn't want to — maybe because I've had relationships with men that I cared about more than Matt did.' She reported the effect of having taken estrogen, thus: 'It brought out, like, the female in me. It was quite an experience to go — well, I was 14, 13 or 14, and my best friend, when she was 9 years old she must have worn a size 36C cup, and here I was flatter than a board. So, to start getting breast development, that was really something, you know, psychologically. And at the time, it was right after the operation, so I couldn't really play football, baseball, and do a lot as much as I did. That cut down on that kind of thing. The sexual thing never came along until a lot later. I was eighteen, the first time I had sex. It was not too far before that I realized, you know, that I would want to.'

The first experience of sexual intercourse was with Dave. It 'was terrible. It was a disaster. I felt like, if this is it, well I don't know if I want to really waste my effort. ... No tenderness, no affection, nothing. No foreplay. Just right down to the nitty gritty and that was it. Slam bam, that was over, and that was no good at all.' A large part of the problem may have been that her first few experiences of sexual intercourse were uncouth, and happened when her partner was high or intoxicated. He was not in love with her, nor she with him, but she was in love with someone else.

'It's with Bill,' she said, 'and it's hard to explain love. Like I thought, I fooled myself into believing I was in love with Dale, Dave's younger brother. And I was almost going to marry him, but that wasn't love. It's hard to explain love. Now, with Bill, I felt that I fell in love the first time we met. Or, I knew that the possibility was there. That could have been because the chemistry was there. He's a pretty good-looking guy and everything, so that could be there. Love at first sight, that's the way I felt. But then I just forgot about him because — that was the summer before my eighteenth birthday, and he was engaged to be married that Christmas. So I just put him out of my head.

'This was before I had any kind of relationship with other guys. It was before I moved to live with the Bartletts. That's a family I lived with, and Dave Bartlett was the first guy I had sexual intercourse with. And his younger brother, Dale, was the one I got engaged to. The next year, at Christmas, I found out that Billy hadn't got married. ... The feeling kept coming back every time I'd meet him, but by then I was engaged to Dale.'

Like his brother Dave, Dale was not romantic and not an accomplished lover. After he went to Korea, 'I wrote and told him that I didn't want to get married. Then me and Billy started having our relationship. It lasted for about six months. The only time I ever saw him was when he wanted to get laid. ... We never went out anywhere. The only thing our relationship had was sex. Which wasn't bad, because I enjoyed it. But I thought there was a lot more to a relationship than that. Then, when I found out that he was going out with his old fiancée again, that's when I broke up.'

For two years they were apart: 'Then he came by again on a Friday night about two o'clock in the morning. He said, I'm all yours until nine o'clock in the morning. I thought, well, big deal. But I was a little horny, so I let it pass and said okay. I did care about him. So why not?' She was at this time 23, five years older than when she had first fallen in love with Billy. Her love was destined to remain unrequited. Billy had enmeshed himself in a messy relationship, complicated by pregnancy and abortion, with the girlfriend to whom he had once been engaged, but did not marry. 'We used to talk about her a lot, but that's mainly his trouble,' she said, with philosophical resignation.

During the follow-up evaluation at age 23, her doubt about whether she had ever had an orgasm or not was an issue that concerned her. What she had felt was 'just a, well, like a draining in my vagina. And just a feeling overcame me, you know, like relief, a feeling of relief or something.' Upon thinking about it, she allowed that there had been only a few times that she could have expected to have an orgasm, and that 'at other times it was just that it might not have been right to have sex at that time, and I didn't feel comfortable with my partner.' Another complication was that, for over two years, she had not been taking estrogen. She had gone to a local doctor who refused to renew her prescription for Premarin. It had upset her very much, she said, that he told her, as he had told her sibling also, that she had male chromosomes and should not take estrogen.

Information about the range of stimuli that the patient found erotically arousing was obtained in the course of the audiotaped interviews at age 23. She remembered having claimed at age 13 about boys in underwear and swimwear in the Sears, Roebuck catalogue, and explained: 'I kind of fibbed about it, too. Because I thought maybe, if I wasn't that sexually aroused by the men, that you might think something was wrong with me. So I told you I was. I just said what I thought you wanted me to say. But, actually, looking through a Sears book, didn't bother me at all. ... Maybe if they had the books, *Playgirl* or something, that would have been different.'

For sexual turn-ons, as an adult aged 23, she mentioned: 'Playgirl magazine, books, men. Just the whole masculine body, everything about it, not just the penis but the shoulders and the arms, the eyes, and the way their hair is cut, you know, the beard and moustache, and everything.' In the imaginary experiment of covering parts of a male centerfold, she covered everything except 'the shoulders and bottom torso. ... I'd like to keep the head on. I don't know. If there was just one thing left standing there, I guess it would be the penis, because then you could use it for a vibrator or something.'

Checking the other senses, hearing was dismissed, 'unless they're real close, and I can feel their breath;' and so also were smell and taste. Narratives were graded as erotically arousing 'a little, not a lot. I don't usually like to read sexy stories because they don't have a plot. It's mostly just sex. With a plot, then you get to know the people. They're better. I may get a little more horny than if I didn't read them, but not hot and bothered.'

The chief avenue of arousal was skin feelings: 'My ears, neck and shoulders, and my breasts and vagina.' There was no one part of the vulva that gave a clearly better turn-on than another: 'Not really, because most of my sexual experience has been straightforward intercourse. I've given head, but I've never had a guy go down on me.' The urinary meatus was not erotically noteworthy, and she had not been able to palpate, through the vaginal wall, any erotically sensitive tissue corresponding to the Grafenberg spot or the prostate.

Upon request, she tried to imagine the kinkiest kind of sex she could think of herself getting into, and replied: 'A lot of people have different words for kinky. Maybe a threesome, if that's what you consider kinky. ... Taking on a dog? I'd never do anything like that. Some people would think just the idea of a threesome would be really bad. So I don't know. I think I'd try just about anything, if I had the willingness of someone else. ... Maybe I'd accept an invitation to a sex party. I'm not saying I would stay, but I'd probably go.'

Responding to the theme of bisexuality and lesbianism, she said: 'Yeah, there was this one girl. She was a lesbian, and she kept patting me on the leg and telling me how cute I was, and stuff. They were driving me home, about a half an hour ride. ... It didn't feel right because her hands were so small and delicate, and it didn't seem right on my leg, because I'm used to a man doing that to me.'

In the event of a threesome, 'I think it should be shared between the three. Everybody should do everything to everyone. I mean don't be afraid to touch the other woman or man. ... If I could have a threesome, I probably could go down on her too. ... I think if I was going to do it in a threesome, I could do it with a man alone, or with a woman alone, too. I'm not sure. I've never been in that situation, but I think I could do it either way.'

By the time she was 29, she reported in a telephone interview that she was having sexual intercourse about twice a week, and having an orgasm 90% of the time. For a period of a year she had had a regular boyfriend, a man in his early forties who was separated from his wife. Her policy was that she would have to live with him for at least five years before deciding on marriage.

She was limited as to how much she could say without being overheard, but it was established that her sex life was fine, and that she liked to try new things. She liked oral sex and penovaginal sex. They did not have anal sex, and they did not do anything kinky. Once again, she was not taking estrogen, and had been off treatment for about a year, because of the expense. She worked nights for an answering service. Realization of her full potential, occupationally and otherwise, was thwarted by the impoverished economy of the region in which she lived.

EXPOSITION

These two siblings together with two others from another pedigree (both of whom have lived exclusively as girls and women) were the subjects of a comprehensive endocrine investigation into the fetal etiology of their condition (Meyer et al., 1978, already referenced in the Synopsis). In both pairs of siblings, it was found that the synthesis of testicular testosterone was normal, as judged by plasma testosterone and LH and FSH concentrations, as well as by incubations of testicular minces with labeled precursors. It was also found from a study of cultured skin fibroblasts that there was no deficiency of 5α-reductase, the enzyme that converts testosterone to dihydrotestosterone in peripheral target cells, and that the cells themselves had normal receptor affinity and capacity for dihydrotestosterone. There was no deficiency of 17-ketosteroid reductase. Histologically, the testicular specimens revealed a mosaic pattern in which some areas contained seminiferous tubules with active spermiogenesis, and others Sertoli cells only. Hormone-producing Leydig cells were present between the seminiferous tubules, more densely in some areas than others.

Early in prenatal life, the testes had functioned normally to produce antimullerian hormone and prevent differentiation of the mullerian ducts into female internal organs. Evidently, and for reasons unknown, they had not functioned normally and on time to produce the androgen needed for differentiation of the male external genitalia. Thus, the syndrome has been provisionally attributed to a defect in fetal testicular maturation which delayed the synthesis of fetal testosterone, and also impaired the differentiation of germinal elements within some of the seminiferous tubules. At the time of puberty, the testes produced testosterone to induce pubertal masculinization. The seminiferous tubules of the testes remained impaired, however. In consequence, so did fertility.

The two siblings of the present report were the oldest of a sibship of five. They were born 20 months apart. Two years later there was a normal brother, followed in another two years by a normal sister; then, after 10 years, another normal sister. There were no other confirmed cases of birth-defective sex organs in the extended family pedigree, but there was a married sister of the father who, according to family tradition, had never menstruated and was childless. She was assumed to have been otherwise normally female in development.

When hermaphroditism occurs more than once in the same pedigree, and especially from the same two parents, the chances of two different etiologies are statistically infinitesimal. In the present instance, both siblings were proved to be chromosomally sexed as 46,XY male. Their gonadal sex was male, though imperfectly so, because, when they were gonadectomized after the onset of nonfeminizing puberty, the testicles were shown under the microscope to be defectively formed, though hormonally competent to secrete testosterone.

In cases of birth defect of the sex organs attributable to an error in the genetic code presumed to be recessively transmitted, as in the present two cases, the genetic error is translated into an error in either the timing of the hormones of sexual differentiation, or of their synthesis, or of their cellular uptake and utilization. This error is, in turn, translated into an error of embryogenesis of the organs of procreation, and also, possibly, of the prenatal or neonatal differentiation of sexual dimorphism of those tracts and centers in the brain that will subsequently play a part in the regulation of reproductive behavior as masculine or feminine.

Even though the error may be diagnostically the same in each of the two siblings, there may have been minor differences in the time of onset of the hormonal error, or in its magnitude or degree of penetrance, such that each case is not a complete and perfect replica of the other. Thus one must not reject or disregard the hypothesis that the difference between the two siblings in adult outcome of G-I/R (gender-identity/role), socially and sexuoerotically, might have been derived from a minor difference in fetal differentiation of sexual dimorphism of the brain. If such a difference had, indeed, existed, then it may have made it easier for the younger sibling to differentiate a feminine G-I/R, postnatally, concordantly with the female sex of assignment, and eventually to fall in love as a female, erotosexually attracted toward men, and attractive to them. Conversely, it may have made it easier for the older sibling to evolve a progressive, subjectively experienced discordance with the female sex of assignment, culminating in sex reassignment in young adulthood in keeping with his erotosexual attraction toward females, and his experience of falling in love only with a female.

An alternative hypothesis to explain the difference between the two siblings in adulthood might also apply to a prenatal difference, though one related not to the hermaphroditic error, per se, but to the determinants of other traits and dispositions more peripherally related to femininity and masculinity in adulthood. In other words, individual differences in temperament and personality routinely found among brothers, and among sisters, may have made it easier for the younger sibling not to become sex reassigned, and for the older one to become so.

Individual differences in temperament and personality are not exclusively the product of prenatal determinants, but are subject to input from postnatal influences as diverse as nutrition, infection, toxicity, and social nurturance or neglect. These postnatal influences may contribute not only secondarily to femininity and masculinity, but also primarily. Social influences directly affect the social processes of identification and complementation (Money, 1972, reprinted in 1986), the two great principles by which the prenatal dispositions toward femininity or masculinity are augmented and stereotyped postnatally. Social influences within a family are never identical for any two siblings, even identical twins.

Retrospectively, in the present two siblings, it became confirmed that there had been a difference between them in the way that they and their father interacted. The older one became more like a son to the father than did the younger one. The degree of masculine identification and complementation established by the older sibling in this relationship with the father was not sufficiently powerful, at the time of surgery and hormonal treatment early in teenage, to override other considerations that prevented her from electing masculinization instead of feminization. She became a victim of the antisexualism and moral taboos of her rearing, insofar as she avoided self-incrimination and potential reprisals by not disclosing erotosexual attractions and genital practices that would have favored masculinization, or at least a postponement of feminizing surgery.

Ten years later, at the age of 24, the patient was confronted by the realization that living as a woman was a failure, just as conversely, in another similar case, mentioned in Money and Norman (1987), the realization was of failure in living as a man. The primary evidence, as well as being genital, was also romantic, erotic, and limerent (falling in love). As a woman, she could be sexuoerotically attached only to another woman — not to a woman with a lesbian orientation, but to a woman whose partner should have been, in her estimation, a man with a penis.

It was at this stage of personal development that the patient's oral and written biography metamorphosed. The data remained unchanged, but their meaning underwent reformulation: she had made a mistake about hormonal and surgical feminization, and had done so in deference to the covert demands and imposed dictates of her parents.

Retelling one's biography is simple in comparison with changing one's life from female to male. For this patient, the hurdles were not only genital, but also economic. Living as a male, academically he was an underachieved dropout, ill-prepared for employment as either a woman or a man in an economically depressed area. No hospital or clinical expenses could be incurred without prior approval through the bureaucracy of within-state medical assistance, for which reason counseling sessions at Johns Hopkins had to be without charge. Even if the plastic-surgical costs of constructing a penis had been approved, the outcome would not have been satisfactory for copulation. Thus the ideal of copulating as a man with a penis could not be achieved. Unable to achieve full status as a man, and unable to live as a woman, the patient became an itinerant in search of the impossible. He was medically indigent, in trouble with the law, unemployable, drinking too much, and playing Russian roulette with death on a motorcycle. This was not a standard case of two names, two wardrobes, and two personalities (Money, 1974), one of which had taken over completely and engineered a sex change. Rather, the male personality had taken over incompletely and without an all-conquering obsession to surmount all obstructions and obstacles.

With the knowledge of hindsight, one can make a case for the argument that the decision in favor of surgical feminization of the genitalia was made too precipitously in the case of the older of the two siblings. She needed far more talking time in which to become free enough to speak about that which was literally unspeakable in her life. She had assimilated too well the prohibitions of the sexual taboo in her upbringing. She was a victim of obedience to the taboo, which forbade her to disclose in full the ambiguity of her status and knowledge of herself as a sexual and erotic person at age 14. Her predicament illustrates all too well the limitations and constraints imposed on today's practice of sexual medicine, to the detriment of the victims of a birth defect of the sex organs.

To some extent, these limitations translate into bureaucratic and institutionalized pressures toward cost-saving and speedy action. In the present instance, there was no source of funding to ensure presurgical outpatient treatment and weekly supervision, away from home, for a Two-Year, Real-Life Test (Money and Ambinder, 1978), as is required in the case of transexuals before they embark on surgery of the sex organs for sex reassignment. The demands of this test are that the patient should become rehabilitated socially, economically, hormonally, and erotically in the proposed lifestyle, prior to the final, irrevocable step of genital surgery. Hormonal rehabilitation would have allowed this patient to experience the effects of hormonal feminization on the body and on its sexuoerotic functioning before the genitalia were surgically feminized. These effects are reversible upon discontinuance of treatment, except for breast growth, which is surgically reversible.

During the course of hormonal treatment, the patient would have been maintained in weekly sexological counseling therapy designed to promote self-discovery, especially with respect to romantic, sexual, and erotic attraction to either male or female partners, or both, in dreams, fantasies, ideation, and participant experience. In our society today, it hardly needs saying that the explicit nature of such therapy for a 14 year old carries a threat of moral veto, even by one's professional colleagues, if not of a criminal, malpractice charge on grounds of contributing to the delinquency of a minor. Society today prefers to sacrifice some of its teenagers to the dragon of sex, like Andromeda chained to the rocks as a sacrifice to the sea monster, rather than to have them rescued by a sexological Perseus. In some instances, the expert in sexology gets entrapped in a no-win situation, prevented by a child's parents and even, covertly, by colleagues, from serving the child's best interests.

To avoid the error of premature genital surgery, subsequently regretted, it will be necessary to legitimize and pay for the Two-Year, Real-Life Test, in at least some adolescent cases of hermaphroditism. If it had been applied to the younger of the two siblings in the present instance, it would not have penalized her. Because it was not applied to the older sibling, she, who should have become he, was irrevocably penalized.

These two siblings had the same diagnosis as did the second pair of siblings reported by Meyer et al. (1978). They also have been followed in the Psychohormonal Research Unit (unpublished). The older of this second pair would have been severely penalized by postponement of feminizing genital surgery for two years, as her masculinized genitalia and overall body shape and vocal pitch were for her a teenaged disaster. Her affected sibling was too young for a real-life test of feminizing hormonal treatment, and was not, on the basis of long-term follow-up, harmed by surgical feminization in childhood.

All things considered, the case of the matched pair of hermaphroditic siblings herein reported points to the significance, often overlooked, of developmental ambiguity or androgyny of the G-I/R; of the failure of our society to have institutionalized tolerance of ambiguity; and of the institutionalized professional and economic pressures on health-care providers to resolve ambiguity too rapidly. The two cases also point to the power of postnatal developmental factors in affecting masculine and feminine differentiation of the social and sexuoerotic components of the G-I/R. Whereas the prenatal differentiation of the one sibling replicated, so far as can be ascertained, that of the other, the postnatal differentiation did not.

The G-I/R discordancy in these two cases illustrates the primacy and importance of the sex of the partner with whom one falls in love and toward whom one experiences sexuoerotic attraction in defining one's own G-I/R not only as masculine or feminine, but also as heterosexual, bisexual, or homosexual. Indeed, it was the sex of the partner with whom each sibling was able to fall in love that was the ultimate, defining characteristic of the masculinity or the femininity of each one's G-I/R. The younger one had no history of doubting her own femininity. Her family, friends, and workmates accepted her as a female, and so did her male lovers. They defined her as heterosexual, and so did she herself.

By contrast, the older sibling defined herself as homosexual when she copulated with a man, even though she had a vagina with which to receive a penis in copulatory intromission. She was not sexuoerotically attracted to males, nor able to fall in love with a male. Conversely, her attraction to females and, eventually, her love affair with a woman finally led her to the conviction that for her to continue to live as a woman, even as a butch-dyke, impersonating a man, was to live a lie. She could not accept herself homosexually as a lesbian, no matter whether other people did or not. Lesbians found her too much like a man, and men found her too strange to be accepted as one of themselves. Her solution was to redeclare herself as a man, and to undergo the hermaphroditic equivalent of female-to-male transexual sex reassignment.

Since each of these two siblings has the same diagnosis, namely male hermaphroditism, whereas the sex of their lovers is different, there is a conundrum regarding their lovers: Which lover is homosexual? The man who has sex with the younger one, the woman? Or the woman who has sex with the older one, the man? The answer is neither. It is an answer that confronts us with the very basics of the concepts of homosexuality, and heterosexuality, and bisexuality, and the principles by which we define them.

BIBLIOGRAPHY

Meyer, W.J. III, Keenan, B.S., de Lacerda, L., Park, I.J., Jones, H.E. and Migeon, C.J. (1978) Familial male pseudohermaphroditism with normal Leydig cell function at puberty. *J. Clin. Endocrinol. Metab.*, 46:593–603.

Money, J. (1972) Identification and complementation in the differentiation of gender identity. *Dan. Med. J.*, 19:265–268.

Money, J. (1974) Two names, two wardrobes, two personalities. *J. Homosex.*, 1:65–70.

Money, J. (1986) *Venuses Penuses: Sexology, Sexosophy, and Exigency Theory*. Buffalo, Prometheus Books.

Money, J. and Ambinder, R. (1978) Two-year, real-life diagnostic test: Rehabilitation versus cure. In: *Controversy in Psychiatry* (J.P. Brady and H.K.H. Brodie, Eds.). Philadelphia, W.B. Saunders.

Money, J. and Norman, B.F. (1987) Gender identity and gender transposition: Longitudinal outcome study of 24 male hermaphrodites assigned as boys. *J. Sex Marital Ther.*, 13:75–92.

CHAPTER 7

Matched-pair comparison of two cases of congenital micropenis, discordant for sex of assignment, rearing, puberty, and transexualism

SYNOPSIS

Two infants born with a micropenis were neonatally concordant for chromosomal sex (46,XY) and gonadal sex (testicular), and discordant for assigned sex. In Case #1, the rearing was as a boy. His genital defect could not be corrected surgically. Pubertal masculinization was induced by hormonal treatment with testosterone. The physique became very well virilized, and the appearance masculine.

In Case #2, after three weeks of doubt that the child should have been declared a boy, the sex was reannounced and the rearing became that of a girl. The genital defect was corrected by surgical feminization in two stages. The surgery for stage one, feminization of the external genitals, was performed at age four months. Stage two, vaginoplasty, was performed at age 17 years. Pubertal feminization was induced by hormonal treatment with estrogen. The physique became very well feminized and the appearance as a woman very attractive.

Hypothetically, in both cases androgenization of the brain prenatally and neonatally may have been attenuated, so as not to preclude postnatal femininity in the differentiation of the gender-identity/role (G-I/R). In the case of the child reared as a boy, the G-I/R differentiated ambivalently. As early as age 8 or 9 he had begun contemplating the possibility of sex reassignment to live as a female. An agonizing period of vacillating between suicide and sex change was resolved at the end of his junior year in college in favor of living as a male. Heterosexuality failed, not only coitally, but also because sexuoerotic attraction was homosexual, exclusively. After surmounting moral reluctance to a gay lifestyle, he eventually established a permanent living relationship with another gay man.

In the case of the child reared as a girl, the G-I/R differentiated as feminine. Her romantic life in teenage was that of a girl, as was her experience of falling in love with the man whom she married. She became a mother by adoption.

Both cases, having been followed longitudinally since infancy, provide explicit evidence of the juvenile origins and precursors of the sexuoerotic component of adult G-I/R. The two cases have bearing on the definition of homosexuality, and on the theory of its genesis. They show also that the sexually dimorphic hormones of puberty do not automatically preordain a heterosexual status. Additionally, they show that a child's destiny in G-I/R differentiation is multiplexly determined within the family, as well as within the individual.

CASE #1: MALE SEX ASSIGNMENT

Diagnostic and clinical biography

There is no record in the clinical history of how the parents were told of their first-born baby's genital defect. Though from the beginning the baby was referred to as a boy, a urological consultation was obtained because of the extremely small size of the penis with minor, first degree hypospadias of its meatus at the corona of the glans, and bilateral cryptorchidism. The urologist assured the parents that one testicle could be palpated high in the inguinal canal, and that the baby was indeed a male. It was common practice in those days to use the gonadal criterion as the ultimate criterion of sexual status. It was too soon for the Barr body, discovered in 1948, to be used clinically as a test of genetic sex; and too soon for sex-chromosome karyotyping, as it was not until 1959 that human chromosomes could actually be visualized and counted. It was too soon also for H-Y antigen (Ohno, 1978; Silvers and Wachtel, 1977), discovered in 1976, to be considered as a criterion of sex, and likewise the gene for TDF (testis determining factor) (Page et al., 1987).

When the baby was six months old, the urologist reported that both testes could be palpated, and he optimistically declared that the child would develop normally. His optimism was not convincing, however. The pediatrician referred the baby at age 11 months to Lawson Wilkins, well known as a pediatric specialist in intersexuality, at the Johns Hopkins Hospital. There it was decided on the basis of the physical examination that exploratory surgery should be performed in order to rule out the possibility of covert internal intersexuality. The external genitals comprised a normally formed, empty scrotum; undescended testes, the right unpalpable and the left questionably high in the inguinal canal; and a penis that appeared to be 'a small fold of preputial tissue about 1 cm in length' (Fig. 7.1). Urination was from the tip of this micropenis, through a glans that was not quite perfectly fused.

Surgically it was established that the internal reproductive structures were masculine, with no evidence of intersexuality. The testes were bilaterally palpated with difficulty near the internal inguinal rings where they were left in the expectation that they would descend spontaneously.

Postsurgically, the following statement was entered in the baby's medical record: 'These findings clearly indicate that the patient is a male with an exceedingly small penis; with the absence of all female genital organs it would have been impossible to have converted him into a female even if this had been desired.' Today, this criterion against feminization no longer holds. Even at that earlier time, in related cases of gonadally male intersexuality or hermaphroditism, the appearance and surgical correctibility of external genitals constituted in rare cases the criterion whereby female assigned sex was decided upon. The possibility of feminization in the present case was mooted specifically by the psychologist and psychiatrist, by reason of the fact that the penis was too small ever to be made adequately functional for coitus in adulthood. This possibility would have necessitated a sex reannouncement, and was not acceptable. There was no precedent for sex reannouncement in a case of microphallus. Moreover, the parents were strongly committed to getting help for a son.

The parents were first referred in person to the Psychohormonal Research Unit on the occasion of the first follow-up visit when the baby was age 2. The referral was made in response to the parental quest for information regarding the psychologic as well as the somatic prognosis in their son's case. The recommendations were ameliorative: being honest rather than evasive with the boy, defining his problem explicitly in an age-suitable way, and giving him guidelines about how to prevent and/or deal with potential peer-group problems (see below, Biography of knowledge of condition).

Two years later there was another follow-up visit. At that time, a buccal smear

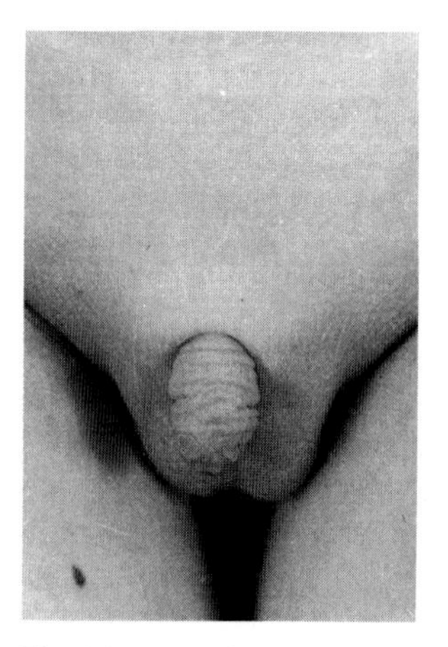

Fig. 7.1

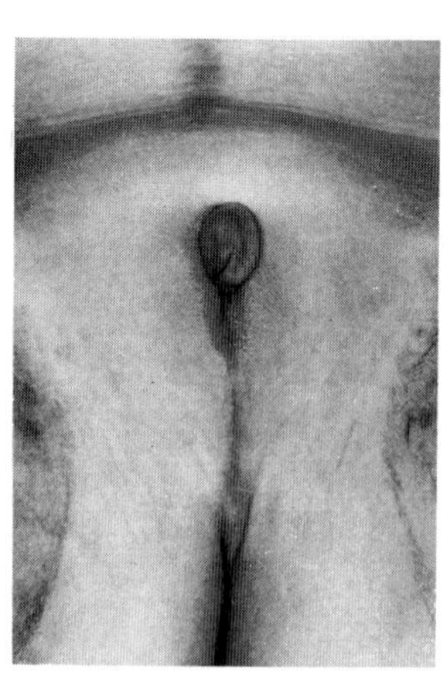

Fig. 7.2

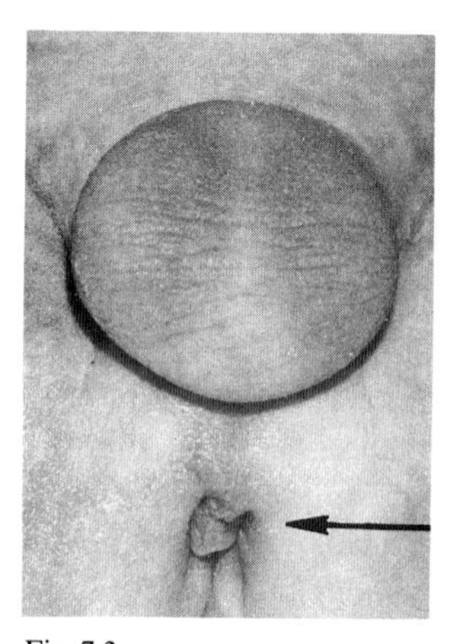

Fig. 7.3

Fig. 7.1. Genital appearance at age 8 years of the child assigned and habilitated as a boy.

Fig. 7.2. Genital appearance at age 3 years of the child assigned and reared as a girl.

Fig. 7.3. Neonatal appearance of complete agenesis of the penis. The arrow points to the urinary meatus at the edge of the anus.

was taken and it was found to be chromatin negative, as expected in a gonadal male. The chromosomal karyotype, 46,XY, was not actually verified until 1977, in preparation for this present report.

At age 4, the boy was subject to attacks of dermatitis, a symptom he and his father shared. Attacks had been more prolonged earlier in infancy. With advancing age they became shorter and less severe, though persistent sporadically through adolescence and into adulthood. The same applied to asthma attacks. Cortisone controlled the skin outbreaks, and antihistamine the asthma.

During the years of childhood, follow-up visits were scheduled one or more years apart. Some were at the boy's own request, for he took up the self-demand option of appointments with his 'talking doctor'. When he was 10 years old, he disclosed the existence of a private dilemma regarding the idea of being a girl (see below, Lovemap biography). He opted in favor of retaining male status, with the help of testosterone cream (0.2% testosterone propionate in stearin-lanolin cream) applied locally to the penis in an attempt to increase its size immediately, without waiting for the normal age of puberty. The cream did, as desired, bring about a premature induction of genitally localized pubertal development: in 19 months the penis grew from 1 cm to 4 cm in stretched length. There was a concomitant increase in morale. In adolescence, a final increment of growth increased the stretched length of the penis to its adult dimensions, 6 cm long and 2 cm in diameter.

The onset of adolescent development was at age 13:7, when it was decided to induce puberty with exogenous androgen, testosterone enanthate, 200 mg every three to four weeks, on which dosage androgenization was subsequently maintained. This decision was made on the basis of the lack of signs, upon physical examination, of the onset of puberty, together with the laboratory finding of elevated gonadotropins. The exact fate of the patient's own undescended testes will be definitively ascertained when he is psychologically ready for further gonadotropin testing and possible surgical exploration. For cosmetic appearance, the scrotum now contains silicone prosthetic testes, surgically implanted at age 14. At age 18, as a convenience while away at college, he substituted the oral for the injected form of testosterone (methyltestosterone, 20 or 30 mg daily). Compliancy was self-regulated, so as to strike a balance between being tired too often and 'horny' too often.

The hormonal masculinizing effect of testosterone therapy was completely effective with respect to general physique and appearance in adulthood. One encountered a tall, hirsute young adult male, whom both men and women found handsome and attractive, according to the criteria used in everyday life.

Gender-coded social biography

The biography of that part of social development which does not pertain directly to sexuality and eroticism begins with the fact that this child was the oldest sibling and only son to carry on the family name and its Jewish tradition. There were two younger sisters. The father worked in a medically related profession. The mother, also a professional, had her office at home. Both parents accompanied their son on visits to Johns Hopkins until he was 8 years old. Subsequently, he was accompanied by only his father. It was recorded at the time that the father appeared to take a more laissez-faire approach than did the mother to the problem of having a son with a micropenis. He was not laissez-faire, however, about his son, but was, in fact, very attentive to his welfare.

When the son entered first grade, the father wrote a progress report. There were complaints from the school that the boy was a poor listener and did not follow instructions well. He said he didn't like his teacher. She was a strict disciplinarian. Once after a distressing day at school, he was threatened with detention, whereupon he cried and vomited. That night he awoke from a nightmare and talked about his difficulties at school, some with other children, which he was unable to resolve. None of the difficulties could be directly traced to the genital condition. Offered a change of teacher, he elected against it.

'In ordinary, outside-the-class behavior,' the father wrote, 'he has been quiet, completely nonaggressive (except occasionally toward his sister), and perhaps oversensitive and insecure. He will not join a group of children who are not well known to him. Under friendly circumstances, he gets along well with children of either sex. He avoids the rough and tumble play of boys, but enjoys other types of play. The teacher reported that at school he does not get along well with the boys, but frequently plays alone, or with girls.'

At the age of 8, for the first time, the boy initiated a request for another appointment. Information about social development obtained during the visit confirmed prior evidence of extraverted sociability in general, but limited participation in boys' sports: 'I don't like to get beat up,' he said. 'My friend Sidney knows a lot about wrestling, and I do too – I can get him in some holds. ... Sometimes, it hurts. And I don't like to fight.'

The father, whose own boyhood social development resembled the son's, made a point of spending time with his son, teaching wrestling, for example, which the boy enjoyed.

Of most of the boys in his class, he said: 'Well, they're kind of rough, you know. And sometimes they don't want me to play kickball with them, and baseball. ... They say I don't play good, and they want the best man on their side. So I just swing on the swings. Some other kids do, too.'

After school, he was always responsive when his friends wanted him to join their play, but he was reliant on their initiative and leadership rather than his own.

His recreational interests included swimming, gymnastics, bicycle riding, and stamp-collecting. His collection expanded into a small dealership in adulthood. In high school, as a boarder, he took up tennis and soccer which, he said, helped him to make friends and not to feel like an outsider. He also took up writing and produced both light verse and short stories (see below, Lovemap biography). In college, he became an active organizer and leader in minority rights.

The IQ, taken at age 10 on the Wechsler Intelligence Scale for Children, was Verbal IQ, 128; Performance IQ, 108; Full Scale IQ, 120. Two and a half years later, upon repeat testing, the corresponding IQs were 119, 107, 115.

Up through his 11th year, the boy's school grades were A or B. Then the first sign of academic underachievement began to appear. There was much contention between mother and son regarding neglect of homework assignments. Ambitious for her family in the culturally Jewish tradition, the mother spurred her son on to academic excellence. One might have inferred a subliminal hope that a big intellect could offset a small penis. As time went on, mother and son became locked in an intensifying adversarial relationship in which continued academic underachievement became the son's chief weapon. The struggle escalated in junior high school. Academic performance, formerly average or better, dropped below average. Disputes with the mother increased, until father, son and mother eventually reached a consensus whereby the son completed the last two years of high school away from home at a private boarding school, where his grades improved.

From high school, he went to college and then graduated with a master's degree. He was accepted into a Ph.D. program in human sexuality in another institution, but did not complete the first year, partly because the chairman of his department was insensitive to the effects of his personal handicap, which he dared not reveal. Discontinuance of the doctoral program, and transfer to a different institution and training, might also be construed, hypothetically, as a manifestation of achievement panic, a theme that has long characterized his academic progress (see below, Biography of behavioral difficulties).

To help finance his education, during the summers of his student years he worked as a counselor in diverse institutions serving the handicapped or emotionally disturbed.

He has always been masculine in appearance. When he came to the hospital, it was obvious to all those with whom he had contact that his clothing, hairstyle, and jewelry fitted the style of the times as appropriately masculine. So too did his body language. As a teenager he grew a moustache.

Biography of knowledge of condition

An excerpt from the report of the first interview with parents, written when the boy was exactly 2 years old, is as follows: 'I limited my conversation with these parents to matters pertaining to the stable psychological adjustment which I have observed possible in children with severe handicaps. Much depended on a frank and straightforward way of talking with the boy as time goes on – not trying to spare him with evasions, euphemisms, and false hopes, which to the child becomes pretending that he doesn't have a disability. ... They asked about how to handle this first child's problems with other children they hope to have. Again I advised straightforwardness. I gave specific recommendations about preparing the way at school, so that the boy can avoid being seen by other children, advised in advance to guard against exposing himself to teasing. ... Both parents found the talk helpful and hopeful.'

The next visit took place in the week of the child's 4th birthday, and the following was recorded: 'The penis has a tiny glans, the rest seeming to be nothing more than a skinny tube lacking corpora. It is very small and reminded me of the protruding hood of a mildly enlarged clitoris except that the midline of the perineum below it was fused. ... In rearing their child, the parents had drawn on recommendations given them two years ago, but they were sometimes in a quandary as to whether to wait for the boy to ask questions, or to take the initiative themselves in telling him about himself. The mother has been inclined to do the latter, remarking, for example, in a confirmatory sort of way that the penis is a small one. The father was rather more inclined to view such affirmation as making too much of an issue of penis size. The boy himself had not actually made any size comparisons, but he had said that one day he will have a big penis. I suggested that the parents need not feel obliged to act as though they know all the answers, and they were relieved to have the recommendation that they could simply quote to the boy what his doctors had told them. As a matter of fact, I don't presume to know all the answers as to what plastic surgical reconstruction might be able to help out with, after puberty. ... At nursery school, he does not appear embarrassed by having always to sit to urinate, whereas other boys stand. ... He mentioned to his mother something about sitting to urinate, to which she replied that she has known other little boys who also sit.'

In kindergarten the next year, he still was obliged to sit, but without being self-conscious, as some other boys copied him. The reason for his hospital visit that year he gave as, 'Because I have a little penis, you know. ... That's just how I got borned,' On his finger, he marked off about an inch as the size it would get. He had seen his playmate's penis, but did not reveal his own. His father said, jocularly, that he had become quite a connoisseur of penises and their size, even to the point of asking a visiting mother when she would change her baby's diaper so he could see the penis. There was no evidence at this time of an awareness of cryptorchidism. He had seen the genitalia of his baby sister as well as the penis of his male playmate, and considered that he looked like the latter. Sometimes, he conceded, unconvincingly, he liked to play with his penis 'just for fun', and that the feeling was, 'Oh, very good.'

By the time of the follow-up interview at age 8:3, the boy and his best friend had compared penis sizes. He was convinced by his parents that, because the news might spread, he would not do the same again. He said his sister also had been told to stop making comparisons, as she had, one day, within her brother's hearing. He reaffirmed that his penis would grow, though not to full size, and in this context was told for the first time about the possibility of teenaged hormone treatment, if penis growth proved too slow. His parents reported that on an average of once a week he said or asked something about the size of his penis. He himself reported that he had not ever dreamed about his penis. It still remained necessary to sit to urinate, and would remain preferable to do so until the penis reached its maximum size in teenage.

He reported also the basic rudiments of sexual knowledge that he had obtained by age 8. He knew about the egg, and the baby growing 'in the mother's stomach. ... I know that there's water in the egg and the baby is still in there. ... And when the baby comes out, it can't see. ... It comes out from somewhere near the tookus. I don't understand how the baby can get through the real small hole. ... I forget what the father does, but I know. Oh, I've forgot.' Reminded about the sperms, he realized that he had confused the term with germs.

On the next visit, when the boy was 10, the father reported that his son had asked his parents some very straightforward questions about intercourse, which they answered equally straightforwardly, though in retrospect they realized they had omitted anything about the pleasure of it. Very pragmatically, the boy asked if he could see how the penis was put into the vagina, and accepted the parents' explanation of the rule of privacy. He applied his new knowledge of reproductive genital function to his own personal, micropenis predicament, and grappled with the possibility of changing sex (see Lovemap biography).

In subsequent years, he was kept as well informed of the facts of his case and the rationale of its management as was his father who accompanied him to the clinic. He was able to comprehend the information given to him, and to give his informed consent to both tests and treatment.

Lovemap biography

Inquiry into genital self-stimulation at age 5 produced replies that did not clearly differentiate adjustment of clothing from erotic good feeling, but it was clear that 'my mother doesn't like me to play with my penis.' Without evasiveness, the same lack of specificity persisted when the topic came up again at age 10, following treatment with testosterone ointment. The father had reported that his son had habitually masturbated a fair bit, but in an absent-minded sort of way, as when watching television without much concentration. The family policy was that masturbation should be private, not prevented. The treatment with testosterone ointment at age 10 did not produce any changes in localized eroticism according to what could be ascertained from both the father and the boy himself. Digital manipulation of the crotch area continued, but it seemed to be less an erotic act of masturbation than a pacifier, like rocking or thumb-sucking.

During childhood, there were no reports of sexual rehearsal play, nor of romantic rehearsals and juvenile love affairs. Many years later, however, at the age of 20, there was a retrospective recall: 'When I was nine, I had this crush on a guy in my third grade. ... just like the one I had at camp, this summer. It just struck me; I remembered back to the third grade. I said to myself: "It's just like the time I had this crush on this one guy in the third grade." ... It was more than hero worship, or anything like that. It was that, well, he just struck me as being really, like handsome. I just had him as a friend, right? Because he was so handsome. And it wasn't like having the next door neighbor as a friend. I mean, I had to have this guy as a friend. ... So I just assumed it was a crush, something like that.'

The only evidence of a potential duality of gender-identity/role prior to age 10 was the prevalence of pacific play over agonistic play. Then, at age 10, a highly differentiated duality was disclosed on the occasion of a follow-up visit scheduled at the boy's own written request. There is an entry in his psychohormonal chart which reads: 'At the outset of the interview, he said that he had asked for the appointment because he wanted to talk to me, but that he had nothing special to report or ask. Pursuing a cue given me by the father, I took up the topic of his having done well academically in fourth grade, just completed, but perhaps not so well socially with other boys. In his usual candid way, he said that he doesn't go off with the other boys to play, but usually joins in the girls' group. There was a certain amount of face-saving: the girls wanted to have some boys with them, and they played the same games as boys play, like softball and handball. He wasn't the only boy in the group. He mentioned specifically his friend, a diabetic, whom the other boys don't want to bother with, because he can't run fast enough. He himself can run fast enough, but for him the other boys are too rough, always fighting, and he doesn't like to fight. ... He gave no evidence of attraction toward girls' toys, clothes, or domestic activities.

'I decided to probe further into the significance of recreational alignment with girls by using the parable technique, namely, a narration of situational relevance used as a personalized projective test. The narrative was of another boy with a small penis who had told me that he sometimes wondered if God had really intended for him to be a girl – in fact, he once had a dream in which he actually turned into a girl. The boy's response to this narrative was that he himself had thought of changing to a girl. This was the point I switched on the tape recorder.' The conversation ran as follows, in small capitals for the interviewer, and in lower case for the boy.

I ASKED YOU IF YOU EVER HAD THE FEELING THAT YOU WOULD LIKE TO BE A GIRL. Yes, I would like to be a girl, but I wouldn't like to do the things that a girl does. I'd like to be born a girl. YOU HAD BETTER EXPLAIN THAT SOME MORE TO ME. I'd like to be doing the things that a boy does. ... Well, I would like to be born a girl – but I might want to do most of the things that a girl does.

WHY WOULD YOU LIKE TO BE BORN A GIRL? Oh, because you can have a baby. You could have a baby and take care of it. And that would be fun when it was a little child, you know. You would play with it sometimes when you are around with it. But, probably, getting it to sleep wouldn't be too much fun. Because sometimes it's harder to get babies to sleep. A FATHER CAN PLAY WITH A BABY AND PUT IT TO SLEEP AND EVERYTHING. Yes, I know that. ... I would rather be doing what a mother does. I've seen my mother do the things. I like the things like ironing clothes and washing – and it's fun. I've ironed clothes a lot for my mother. And I take them off the line when they are dry; and when they are coming out of the washing machine, you know.

And, it would just be fun for me to be a little girl.

WHEN DID YOU FIRST FIGURE OUT THAT YOU WOULD LIKE TO BE A GIRL? A long time ago. I wanted to be a girl ever since I wished to be a girl. HOW LONG IS A LONG TIME AGO? A year or so. I kept on wishing that I was a girl.

YOU SAID THAT YOU WOULD LIKE TO HAVE A BABY. HOW MUCH DO YOU KNOW ABOUT WHERE BABIES COME FROM? The sperm. That's where they come from. WHERE DOES THE SPERM COME FROM? The father. But I probably wouldn't have a sperm. So why be a boy if you can't have a sperm and your penis is small?

There then ensued an explanation of donor insemination and the sperm bank, and a lengthy conversation on how sperms get out of the penis. He was uncertain as to whether his own penis had ever become erect, and whether he had good feelings in it. Then the talk turned to pregnancy, and the way that the baby comes through the vagina, the same place where the penis goes.

Reflecting on this information, he decided: I would like to be a girl, and I would like to have a baby. I wouldn't like very much carrying it around, you know. I only would like to be a girl. I don't know anything else to say about it. HAVE YOU FOUND A BOYFRIEND YET? What do you mean? IF YOU ARE GIRL, AND ARE GOING TO HAVE A BABY, YOU HAVE A BOYFRIEND AND THEN A HUSBAND. Well, I'm not a girl. I haven't got a boyfriend. But if I was a girl, I could find a boyfriend. There's a lot of boys in college. You can go to college. You can find a lot of boyfriends. That's how my father found a wife. YOU ARE NOT IN LOVE THEN, YET? No; well I have two girls that I love.

DO YOU EVER HAVE ANY THOUGHTS IN YOUR MIND ABOUT BEING ABLE TO HAVE A BABY, BUT THE WAY A FATHER HAS ONE? Yeah. I thought that I could have a baby, you know, if I had a different penis. I don't know how the sperms could get up there, you know. I know that there was some fluid that sperms would swim in, and to go up. I couldn't do it. The sperms just went up, but I don't know how ... they came out, because you just can't make a sperm come out like that. ...

I WONDER HOW YOU WOULD FEEL IF YOU COULD GET SOME SIZE TO YOUR PENIS. Well, I wouldn't want to be a girl any more. Because, if I had a penis big enough to have sperm, I would like to have the sperm to put it up in the vagina.

Rather lengthily, the interviewer then outlined two hypothetical treatment options, masculine and feminine, neither totally satisfactory, and both without fertility, and asked if the decision between them would be difficult. 'No,' he replied, 'I want to be a boy. I would like to be the man. If you have to pay all that money just for an operation so you can have a vagina, and I would think to look like a girl, and can't get a baby out of you, what's the use of wasting all that money – if you have to go and get another baby?'

Supplied with information about the use of testosterone ointment to give an immediate boost in penis size, he had no hesitation in opting to try it, while leaving open the other option of reconsidering being a girl, should the going get too tough. In further conversation about the origin of his idea of changing sex, he said that he had read some media references to Christine Jorgensen, the transexual whose case had first astonished the world several years earlier.

The return appointment was in three months, by which time the application of testosterone has begun to take effect. Eventually he did concede that the growth of his penis had definitely confirmed his decision to keep on being a boy. Before that, he had circumvented reference to the last session's talk about being a girl. Then, directly questioned, he said he'd changed his mind because: 'I thought it would be more interesting to be a boy, because there's lots of things you can do, and girls can't be that many things. There's not that many girls can be airplane fliers, or train runners and engineers, or taxi cab drivers. ... I just don't like the things girls do. They are nurses, or stewardesses, or just plain American housewives. ... Look, girls' things just aren't that interesting, most of them. Maybe to girls they are. If I'm a boy now and would change, maybe I wouldn't find the girls' things interesting, and I might just want to change right back to a boy. I might want to be an engineer and, if I was a girl, I'd have to change right back to a boy, because there aren't that many girl engineers.'

Six months later, he fielded the question of feeling like wanting to be a girl with a dialogue as follows: 'Sometimes my penis, sometimes it feels like, you know, it has to be rubbed or something like that. HOW DOES THAT TIE IN WITH WANTING TO BE A GIRL? Well, I sometimes I pretend I'm a girl and sometimes before going to bed, if I get out of the bath. I feel like I want to be a girl. So I put on a towel, pretend it's a skirt, and I'm a girl. But before I go to sleep, it goes away. Most of the time I don't have that feeling.'

At this time he was aged 11 and still using testosterone cream. Further conversation established that this feeling of being a girl would more accurately be termed a masturbation fantasy. At this time masturbation produced good feeling, an uncertain degree of tumescence, and no ejaculation. The use of testosterone cream was discontinued at age 11. In follow-up interviews, inquiry on psychosexual imagery was open-ended and not addressed specifically to being female. In three follow-ups, over the next two and a half years, the topic was submerged by talk of heterosexual imagery and prospects. Developmentally the boy had remained prepubertal until at age 13:8 the first injection of testosterone enanthate to induce puberty was given.

Eight months after the first testosterone injection, prosthetic testes were surgically implanted. Thereafter follow-up visits were on an annual basis. The masculinizing effect of testosterone was as expected. It included an increased prevalence of erotic imagery, and of masturbation and masturbation fantasies which were talked about on the occasion of each visit. Erotic feeling in masturbation was localized in the usual place, the glans penis and area of the frenulum (there had been no circumcision).

The theme of being female remained submerged for four more years. The boy referred to heterosexual erotic imagery only. He took the initiative in dating, but the relationships were companionable rather than romantic or erotic. He knew that he was holding back lest he be embarrassed by the size of his penis. He was not prudish about nudity, as in the communal shower at school, but threatened by the potential failure of a penis less than three inches long in erection. He considered the possibility of professional surrogate sex therapy, but was too young to be accepted into a program. He was phobic to the idea and prospect of using a prosthetic penis or vibrator as auxilliary sex toys – intensely so.

The sex-change theme resurfaced as a studious interest when, at age 15, he purchased a copy of Harry Benjamin's *The Transsexual Phenomenon* (1966), newly published. Three years later, as a high school senior aged 18, he wrote a knowledgeable term paper on the subject. In this context, he answered a question about the possible self-application his term paper might have: 'Well, I don't know. Sometimes when I masturbate, I make believe that I switch, like I'm a woman and someone else is the man, or something like that. WHERE DOES THAT FANTASY COME FROM? I don't know. It's just that sometimes when I'm in a bad mood or something, or you know when I have a lot of problems with school, or something like that. I can't seem to masturbate, and you know to get the tension off any other way, so I just make believe that I'm a woman, or something like that, and somebody else is the man. And the boy in the fantasy doesn't have a face or anything. It's just somebody. AND WHAT'S HE DOING TO YOU? Well, he's fucking me. AND IN THE FANTASY YOU GIVE YOURSELF A VAGINA? Yeah. WHAT DO YOU DO WITH THE REST OF YOUR BODY IMAGE, IN THE FANTASY? Breasts, and I'm always naked. I always have a breast and a vagina. And we're always running around the bedroom.

'It's a toss up,' he said, to explain whether this or another masturbation fantasy would present itself. 'It just depends what kind of mood I'm in, you know what I mean? Like I'll start masturbating, like fucking some girl, but you know, sometimes if I'm in a really bad mood and had a lot of work or a lot of problems at school, I'd make believe in my fantasy that I was a woman. ... I used to have another, where I made – like I would masturbate, but I wouldn't be the one who was getting screwed or screwing. Two other people were screwing. That's another one. ... If I'm in it, I make believe I have a bigger penis, if I'm fucking some girl.'

He was about 10 years old, he said, recalling when he first started to think of being female and having a vagina: 'It was really funny. I used to masturbate when I was about ten years old, and I used to like – I'd take two pairs of socks and put them in a shirt and make believe that I was a woman, you know, and then stuff some shirt down my pants and pretend that I was pregnant, you know, and then I'd masturbate like that. BEFORE YOU COULD EJACULATE? Yeah. And then I heard about transexuals, and stuff like that and, I don't know, it started when I was pretty young, about ten years old or so. ... I remember one time wanting to change sexes when I was a little kid, real little, I don't know, seven, eight, or nine. I WONDER IF WE SHOULD HAVE LET YOU CHANGE? I don't know, because, like I don't know what I would have been like now – like I would have said I was a boy, why didn't I stay a boy? Or something like that. And I might have had feelings that I was still a boy, you know, because I would have found out sooner or later, you know, about being born male, or something. But then, maybe some people feel strongly that they are a girl, and maybe it would be good for them.

WHAT CAN YOU READ FOR THE FUTURE, ON THIS PARTICULAR ANGLE? I don't know. It might be tough, because my penis is very small, to have sexual intercourse. And it might be tough, if I marry, on my wife. ... Well, it might not, because if I find out that, you know, my wife enjoys sex with a small penis, or something like that, then I just might believe that I was a guy with a large penis, and leave it at that. I might find out that my penis is big enough to fit, you know, her vagina. Then I might not have transexual fantasies.'

No one had been told about the masturbation fantasies and their content prior to this present disclosure, but he had wanted someone to tell them to, 'because sometimes they're pretty weird.'

The imagery of being female did not include imagery of cross-dressing: 'I never really imagined walking out on the street dressed as a woman. I only thought of it if you're in private, or something. I never thought of going to a job dressed as a woman, or being a woman in body and mind, and so on.'

He had never been in transvestite costume, nor impersonated a woman's mannerisms in private or in public: 'But, I have tried to exaggerate a man's aspects,' he said, 'just to see what the women will do, what the girls will do, you know. Sometimes I get nice stares.'

Drawing the female comparison, he explained: 'I really have never practised it, wiggling my hips. That's not one of my hobbies. I wouldn't know exactly how to go about it, because I, you know, my sports and stuff like that, none of my sports are concerned with moving my hips, and I wouldn't know exactly what motion or muscles to use to do it, or what coordination it would take. So I might have to practice for a week or so, or, just do basic, you know, swinging the hips, you know. Also, I wouldn't want to wear those high heels. Goddamn, at Halloween, one time, I went out in high heels, and my feet were so sore after walking three blocks – and my friend went out with me too, and we both dressed up as girls. And after three blocks, he said: Boy, this is really bad, these high heels are really killing me. Kick me in the ass, I'm a schmuk. And I kicked him and I fell on the ground, and my shoe went flying up in the air. It was really funny. That was the last time I went out for Halloween.' He was then 14, and in eighth grade.

The imagery of being female with a male partner did not extend to encompass imagery of being male with a male partner, as is evident in the following: 'I WONDER IF YOU CAN IMAGINE YOURSELF CLOSELY HUGGING A BOYFRIEND AND KISSING HIM? I don't know. It's just that, if I did, I don't know what kind of a – I can't imagine myself now, you know, because I'm still a man, and, you know, I would think it was homosexual right now. And I can't really stand, you know, homosexuality that much because, you know, I don't mind homosexuals, but I'd hate to do it myself, you know, because it would strike me as perverted, you know. If other people do it, all right. But right now, if I did it, it would be man to man, whereas if I were doing it as a woman, he would be hugging and kissing me because I was a woman and not a man. ... I shy away from stuff like that, man to man. You know, it doesn't even appeal to me, doing something like that, as a man, you know.'

As an 18 year old, he had his first romantic friendship. The partner was a mannish girl, a fellow high-school senior, psychologically troubled and, because of it, an outcast among most of the students. 'She had never been kissed,' he said, 'when I kissed her in October. She told me, she groaned, she had to go to the psychologist because I kissed her. She wouldn't talk to me for two weeks because, you know, it must have been a real traumatic experience to be kissed by a boy.'

In summing up the 18 year old follow-up interview in which the foregoing psychosexual information had been revealed, the interviewer gave his opinion that the man's side of things was too strong to let the other side come out much further than in fantasy. Wisely, in view of what lay ahead, the gateway to further discussion was left unconditionally open.

In the follow-up at age 19, a year later, the psychosexual biography overtly remained essentially unchanged. As an appendage to the interview, there was an explosion of self-revelation, for the first time, of virulent attacks of suicidal depression (see Biography of behavioral difficulties). The attacks fluctuated in frequency over the ensuing years, loosely in parallel with fluctuations in the polarity of female versus male. Sleep dreams during this period of development combined explicit imagery on the themes of tunnel or bridge while in an automobile; accidents, death, incest, and homosexuality; and being female with a vagina.

Six months later penovaginal intercourse was first attempted. He wrote: 'I could not get my penis in at first, but I found that if I put a pillow under her buttocks, I could get in about $2^1/_2$ to 3 inches. I'm glad that she didn't comment about how small my penis is because, if she had, I might have almost cried. The only trouble I had was that, if I came out too far, I wouldn't be able to get it back in, and as you know, there isn't too much leeway with the size of my penis. She liked it, though, which I was glad about, and kept begging for more. In fact, now she's even boasting to a couple of her friends how many times we fucked that one night. And, sort of unfortunately, she's so gone on me that she's even talking of marriage.'

After this first experience, he mostly evaded invitations, some of them unequivocal and explicit, to coital encounters, though he did attempt penovaginal intercourse again with a couple of partners. The sum total of his experiences led him to write that he had no confidence in his heterosexual sex life.

A month before his 20th birthday, and now a sophomore in college, he was the young adult who volunteered to be interviewed in the psychoendocrine case presentation for senior medical students. Afterwards, he taped a summary interview in private, in which appeared a change toward acceptance of homosexuality. The change was manifest in masturbation imagery: 'I have four kinds. I can have myself in there, or I can have two different people in there with me looking at it – it's usually about a movie or something I've seen. I can have me as a lady. And me homosexual, that's the new one. And let's see, homosexual, I'm the receptor, of course. And heterosexual with me in it, I have a large penis. And the one with me as a lady, I don't know what my – like I can see part of my body. I don't know what my face looks like.'

Prevalence of the different types of imagery he estimated as: 'Maybe 20% homosexual, and around 20% me as the lady, and around 20% me doing the fucking, and around 40% I'm watching the thing.' The different types of imagery had different degrees of vividness. Watching others was clearest. The self with an enlarged penis came next. The self as a lady was 'like I'm seeing it through a fog.' The self as homosexual received no mention regarding vividness.

He had come to the conclusion that at some time in the future he would probably try a homosexual experience, and gave the following explanation: 'Well, when I couldn't fuck that second girl, I really felt bad about it. I felt like I'd failed, you know, and I decided – I was reading a book on homosexuality, Martin Hoffman, and it gave some vivid homosexual interludes, and I thought it really sounded good. I thought it might be a nice idea to try that.'

He tested relaxation of the anal sphincter digitally. He could not estimate when he might embark on an actual experience because: 'I'd ruin my whole reputation at college. At the moment, I'm making jokes about homosexualism. All the kids in my dorm take it as a joke, and they'd never believe that I'd do it.'

Reading the future with respect to sex change, he stated: 'I think it would work if I was about, if I was still ten years old; but I don't think it would work right now, because I'd still see myself with a moustache and I wouldn't really like – I couldn't picture myself right now walking on the street with high heels or anything. Maybe ten years from now, if I decided then, it might work. But right now, if I changed into a woman, it would be like I was living in a dream state, because I might not believe that I was actually a woman.'

There is in his psychoendocrine record a note as follows: 'It was on this occasion that I first presented the hypothesis that gender polarity of female and male had fused with a linguistic polarity of French and English. He had begun studying French in high school and, with remarkable rapidity, began writing French verse. In English he wrote verse and short stories. The French verse was lyrical and refined in imagery. The English writings were forceful and more vernacular in imagery. Insofar as one can classify the content and style of writing by the sex of the author, the French verse was feminine, and the English verse and prose masculine.' To this formulation his reaction was: 'That's interesting, because when I started speaking French and – like, I never told you this, but when I was in 8th and 9th grades I used to pretend, like to masturbate, I used to pretend I was pregnant, you know, and I used to stuff pants, I pretended I had breasts by putting in two pair of socks. And when I was like, when I started doing French, I thought well, if I didn't really make it as a guy, you know, like I figured if you're going to change sex, you might as well leave the state or country, or something like that, because by the time you're seventeen or eighteen you know a hell of a lot of people in the United States, but you don't know anybody in Europe. I don't know; I guess I probably thought that maybe by learning French

I could go to France or something like that, and nobody would recognize me.'

At this time, he had already finalized arrangements for his forthcoming year as a junior in college, namely, as an exchange student abroad, at a French-speaking university. When suicidal ruminations intruded on this plan, a rider became added: 'For a while, I was planning to kill myself when I came back,' he said. The unplanned, but logical alternative was to not return, but to remain overseas living as a female. Another logical alternative was to return home and become rehabilitated as a homosexual or bisexual man.

Obviously, the psychosexual agenda for the year abroad far outweighed the academic agenda. So it is not surprising that course grades were poor, except in conversational French. Psychosexually, there was much agonizing on whether or not to try a homosexual relationship. This issue remained, for the time being, unresolved. The sex-change issue was resolved insofar as the fictive French-language female did not materialize. Her place was usurped by an English-language male, severely depressed, grossly malnourished, physically debilitated, and dangerously ill when he returned to the United States to recover.

Even though the French-speaking female would not materialize, the past validity of being female was maintained. On a sentence completion test, for example, on the item that began: 'If I were young again,' the completion was, 'I would request a sex change.' Eighteen months after living abroad he wrote in a letter: 'I'm scared that my boyfriends (or girlfriends) are going to start judging their attraction to me by my small prick, rather than me. If you ever let parents who have a baby with a microphallus raise their kid as a male, you're a damn fool.' In another letter, a couple of months later, vehemently rejecting the idea of a prosthetic, strap-on penis, he wrote: 'I'd rather be gay the rest of my life than have to function heterosexually with the use of a dildo. If I had known all I know now when I was ten, I would have told you to cut my penis off at that time. I'm having fantasies about mutilating myself. That's why I need the Elavil [antidepressant medication] back. Luckily for me, it's just fantasies.'

His sex life during this senior year, back in the United States, continued to be heterosexual in expression, but bisexual in fantasy. 'Willard and I have been having some excellent talks about sexuality and homosexuality,' he wrote. 'In talking with him I discovered that I really don't consider myself gay, or bisexual, or anything. I sort of feel neuter, stuck in the middle.'

Though Willard himself was a member of the gay movement, for months the two maintained only a comradeship without sex. 'I was afraid that I would discover that, if I went to bed with a guy, I wouldn't like it, and would have been bluffing myself the whole time in my fantasies. I decided that I had to conclude whether I liked it or not, and not pussyfoot around any more.' This was the matter-of-fact way in which he reported 'about Will and I romping' – the first overtly genital homosexual interaction of his life, shortly before his 22nd birthday. He enjoyed the experience, despite being beset by misgivings about the size of his penis and being rejected because of it.

A year later, a letter contained a succinct statement of his ambivalence about being bisexual: 'I've been having head problems every now and then, again. Fortunately, it's not as troublesome as two and three years ago. Maybe it's because I've noticed an interest in myself in chicks (romantic), but no real urge to get them into bed. I'm afraid of a possibility that could happen in a few years: that I would marry a girl because of a romantic fantasy and because I don't want to feel abnormal in the eyes of my family, and then would have homosexual relations on the side. Here's something you'll probably find interesting. When I get a trick in bed, I feel satisfied if he fucks me in the ass. I get pleasure out of being touched and felt. I tried letting guys give me blowjobs, and I haven't been able to come. I sometimes can come being jerked-off by my partner, at times with great effort.'

The positional problems of anal intercourse, with himself as the insertor, made him feel intolerably inadequate, and realistically so when the partner's primary turn-on was from deep anal penetration.

Whereas the problem of being able to establish a lasting erotic pairbond as a heterosexual lover had been primarily a deficit in his input into the relationship, homosexually it was the other way round. Partners toward whom he felt a strong and lasting attraction, and who at first reciprocated, did not continue the relationship erotically as long as he was prepared to do. It was not possible for him to ascertain whether they were intrinsically fly-by-night lovers, or whether they rejected him because of the inadequacy of his penis size.

In his middle twenties, he identified himself personally and politically as gay. He was active in the gay liberation movement, and found morale-building support from working with like-minded colleagues.

His experience of masturbation imagery no longer included imagery of the self as a woman with a man. It was bisexual, though predominantly homosexual between two males. Explicitly erotic movies and graphics of male-male homosexual content were a better erotic turn-on for him than were heterosexual ones, though the latter were effective too. He was able to use explicit pictures as cues for masturbation imagery, should spontaneous imagery not present itself. Touch, however, according to his own evaluation, prevailed over vision as the major channel of sexual turn-on: 'All this guy had to do was touch me on the knee, right here, and I just, sort of, you know –' he finished the sentence with his hands, 'and I put my arm around him.' Smell, taste and sound were rated as erotically noncontributory. Narratives, however, could be arousing.

'I don't know,' he said, trying to foresee his sexual future. 'I guess I'll try to stay with male friends, like if I can latch on to an affair with a guy. But, if I'm still floating around for a few more years like I have been in the gay thing, I'll try to find a woman, then, who maybe doesn't care if I sort of trip outside.' His life turned out differently. He settled into a homosexual, interracial living arrangement with a gay professional colleague and, in the era of AIDS, did no tripping on the outside.

Biography of behavioral difficulties

Prior to age 8, the boy had had an episode of school-related difficulty early in first grade. It was transient and did not recur. Before and after age 8, his gender-dimorphic behavioral development did not conform to the idealized stereotype of American boyhood in assertiveness, dominance rivalry, fighting, and competitiveness in sports. However, that was more a manifestation of lifestyle than of behavioral difficulty. The father as a boy had been not too different.

The first mention of something that might qualify as a behavioral difficulty came at age 8. The mother described her son's habit of sitting and swaying or rocking, singing to himself a song of his own making, bouncing his head against the upholstery or pillow, and if lying or crouching face down, swinging his feet, or hitting his forehead on the pillow. The latter he sometimes did in his sleep. Asked about rocking, he told his interviewer: 'That's what I always do. I don't know why I like to do it, but I just like to do it. ... The couch we have is soft, and I like to bounce my head against it, like this. Now, uh, I like to sing to myself. I make up songs ... like: There's a pretty little bird sitting in a tree. ... No. I never sing about my penis. ... I'll tell you something else that I do. See, I like to kick my feet against the bed, you know.' He demonstrated it on the floor.

With much diffidence, when a graduate student in his twenties, he again mentioned this rocking routine, which had persisted. The associated head banging might hurt at first, if the contact surface was hard; but pain gave way to pain agnosia, as the rhythmic monotony of rocking and banging became soothing and hypnotic. He resorted to rocking as a relaxation whenever time was available. It became more insistent in times of stress. He did not want his student housemates to know about this habit, so he disguised it by wearing headphones, and swayed as if in rhythm with the music. As he described it, the motion induced a dissociative, trance-like state, 'spacing-out,' which provided a temporary respite from nightmarish attacks of suicidal and self-mutilatory fantasies, crying spells, and intense depression.

These nightmarish attacks were of adolescent onset. At one time in late adolescence there was a recurrent eidetic hypnogogic image of a gun: 'This vision in my head, you know, of an unhandled gun, rising up and firing in the air, and boom, blowing it to bits. ... I wrote that one poem about a head being blown to bits.'

This imagistic experience was so traumatizing that the only defense against it was a compelling fascination with gun shops, and with thoughts of purchasing a gun and using it to complete the nightmare once and for all. The purchase was never made.

Another diffident disclosure, withheld until adulthood, pertained to obsessive-compulsive enactments that had persisted since middle childhood, namely, compulsive checking that things were locked or in place; and compulsive, self-injurious picking or chewing. The compulsive severity of these symptoms, like rocking, fluctuated in synchrony with the crises and stress of family life, academic life, and love life.

The most severe crises evoked morbidly severe and suicidal attacks of depression, which usurped inordinate amounts of his time and energy. An attack could be alleviated by treatment with Elavil, but therapeutic talking, many times by long-distance telephone, on self-demand in times of crisis, was also imperative.

On some occasions, a depressive attack was temporally related specifically to sexual failure and penis size. More often, the relationship to micropenis was indirect, the apparent precipitating event being academic underachievement, secondary to work blockage, tertiary to achievement panic, insofar as academic success would lead ultimately to the rites and responsibilities of adulthood. The latter meant marrying, procreating, and carrying on the family name. The patient himself endorsed this formulation, witness the following quotation from a letter prior to graduation: 'You are right to a certain point about the self-sabotage. I never told you this, but the exam I had way back in November, I sort of fucked that one up on purpose. ... I don't think my penis thing will ever end, because I will always have a constant reminder, as I have in the past. I hypothesize that it won't matter that much to me, if I finally get my mind into the framework of a feminine homosexual. The only trouble with that is that my body is too masculine. But you are right about the never wanting to be a mature man being synonymous with not making the grade degree-wise, because of my small prick. I always wondered what it was that was making me rather powerless to do anything about my schoolwork. My parents and former teachers always said I was lazy and sloughed things off. ...'

Obstacles notwithstanding, amazingly they have been surmounted. It is reasonable to suppose that progress will continue, despite the enormity of the task ahead.

APPENDIX: Coping with microphallus, an autobiographical report

Introduction

I became incensed the other day as a result of a seminar I attended on intersexual conditions, including the condition I have. I became furious because the seminar leader lectured on all aspects of intersexual disorders save the problems that an individual afflicted with an intersexual disorder would encounter. I decided that it was time to write an article and tell my story.

Psychosocial history

I was born as an individual diagnosed as having micropenis. It was discovered that my gonadal constitution was decidedly male, and the doctors assigned to my case decided that I should be brought up as male. No apparent thought was devoted at this time to the idea that, in the future, I might not be able to function sexually or psychosocially in an adequate fashion as a male as a result of my disorder.

I grew up through childhood knowing that I was male, but aware that I was quite different from other males because of a small penis. I had seen other boys' penises at a local YMCA where I went to swim. My penile deficiency I assumed would be physiologically rectified as I grew older. This was a mistaken assumption.

As a child, I joined girls' play groups because at times I felt I should have been a girl/should be a girl. At other times, I felt I was neither boy nor girl and that it would be easier in the long run to associate with girls. I was a quiet, unaggressive child and indulged in rather passive activities and hobbies.

When I was an adolescent, I was forced to take gym classes because it was a required part of junior high and high school curriculum. Prior to my entering junior high and high school, I had been told by my hospital counselor that I could be excused from gym because of my genital status, as he understood that I might be afraid of changing my clothes in front of my classmates. My reasoning for taking gym was that I assumed that I would be alienated more by my classmates because of being exempted from gym classes, than because of my abnormal genitalia. As it turned out, I was alienated and ostracized because of my genitalia.

As a college undergraduate, I lived in dormitories with public showers. The first year in college, a guy who was living on the same floor with me in the dormitory noted my abnormally small penis and knew that it was an abnormality that one could maliciously tease about. He used his knowledge to brand me verbally in public as being abnormal in mind and body, because of my physiological defect.

My family and family life

My parents are both high level professional people. I have two sisters and am the oldest of the three of us. I get along with my youngest sister the best. I have a fair relationship with my father and poor relationship with my mother.

My family does not talk about my genital condition. My parents, older sister, and one set of my grandparents know about my sexual condition, but have never asked me any questions about the status of my condition aside from: 'When's the next visit to the hospital, have I had a liver function analysis recently (I take oral testosterone), or does the counseling that I get, help any?'

I have had problems relating to my mother for a long time. When I was a child through about age 14, I used to bang my head and rock my body (awake and asleep) when in bed. My mother used to beat me when I was younger, if she heard me banging my head at night. Since my head banging occurred sometimes when I was asleep, I was frightened that I would start banging my head in my sleep and awake to the tune of a beating. My mother would inflict very strange punishments on me for other things which she considered wrong-doings. For example, I used to have a penchant for saying 'no' to my mother, if she asked me to do something which seemed especially disagreeable. If I said 'no', my mother, more often than not, would make me write 'I won't say no' 100–200 times or risk the loss of dinner or allowance. My father was never the disciplinarian, always my mother.

I had many arguments with my parents about my academic performance. When I was in grammar school, I was an excellent student. In junior high and high school, my grades slipped very badly. In fact, my grades got progressively worse each year of high school. Consequently, my parents stripped me of my privileges and allowance for many weeks during each school year. It was very difficult living in my home during my junior high and high school years because of my academic problems. It got so tense in my house when I was attending high school, between my parents and myself, that I went to a boarding school for the remaining two years of school. My home situation got considerably better by my absence from home.

Medical

Each year from the time I was born, I have made a trip for a physical exam to the hospital where my diagnosis was made. I have been disturbed, at times, by the attitudes of the doctors who have examined me. I have been lucky, though, in that I have had counseling help in dealing with my endocrinological problem from psychoendocrinologists at this same hospital. I felt at times that my head was going to split from thinking about problems I had associated with my condition, and I was reassured that I could call someone at the hospital for help. It was extremely helpful knowing that in times of stress I could fall back on counseling support, and that the support would be offered in a compassionate, helpful way and not in a way overladen with pity.

When I was 14, prosthetic testicles were implanted in my empty scrotal sac. This was done for cosmetic reasons.

When I was 15, testosterone therapy was begun. Fortunately, I was one of those intersexual persons who did not have the androgen-insensitivity syndrome, so my body did masculinize with testosterone therapy.

Each year when I went to the hospital for my annual medical check-up, the student doctors and new fellows seemed to ask questions and make statements which were designed to annoy and make uneasy patients like myself. One particular question I distinctly remember being asked many years in succession: 'Was I satisfied with the size of my penis?' This seemed to be a very inappropriate question, and I vaguely remember others of a similar ilk. On one occasion I agreed to allow the doctors to take a piece of my skin using a high speed drill. They wanted the skin for a cell study. The doctor who was drilling the piece of skin (taken from my wrist) said that they would like to take a piece of skin from my penis, but he wouldn't do that. I was disturbed by this statement, and others like it, as they appeared rather crude to be used in conversing with intersexual patients on medical issues.

Sexual history

I have been masturbating for a long time. I have no qualms about masturbating. My masturbation fantasies are quite varied and very creative.

My first experience with sexual intercourse happened when I was 19. It was a real nightmare. I had to get very inebriated to attempt it. I was forced into telling the girl I was sterile, because my penis is too small to accommodate a rubber to fake not being sterile. The girl didn't believe that I was sterile and she was not on birth control pills, but we had intercourse anyway. I had intercourse with her acting like there was nothing abnormal about having an abnormally small penis. The first attempt at intercourse was with me lying on top of her. I couldn't get my penis into that girl's vagina, and future women's vaginas, lying on top of my partners; my penis is too small. All further attempts at sexual intercourse with women had to be done with me lying on my back, because my penis was not big enough to have intercourse in the missionary position.

I discovered as I engaged in further sexual relations with more women that I had no real sexual feelings for women. At this point, I decided to engage in homosexual experiences. My first homosexual experience was not a nightmare, as my partner and I engaged in mutual masturbation. I found that I had more sexual feelings for men than for women. Homosexually I am at a disadvantage, because I can't be the insertor in anal intercourse. When I have sex with partners who want me to be the insertor, I tell them that I don't like to be the insertor in anal intercourse. I prefer to tell partners that I don't like to be the insertor, over stating that physiologically I can't be the insertor because of having a micropenis. When my partners and I do have sex, we usually engage in mutual masturbation or fellatio. Sometimes my partners and I have anal intercourse, with my partners always being the insertors. Sexually, I have found it easier for me to be homosexual than heterosexual because of my condition, and a lot more enjoyable as my sexual inclinations go.

Conclusion

Here, I have attempted to give some insights into the problems encountered by intersexuals with visible sexual defects. I have also attempted to point out that (many?) individuals in the medical profession who come into contact with intersexual individuals may not be properly trained in how to talk with them. From my sexual experiences, it is apparent to me as an individual possessing an intersexual condition, that many intersexual individuals may have problems in carrying out sexual interests and activities.

CASE #2: FEMALE SEX ASSIGNMENT

Diagnostic and clinical biography

At birth, the mother of this baby was told she had a boy, and no mention was made of the genitals. When she got her first look at the baby, she counted the fingers and toes to see if they were all there, but it didn't occur to her to look at the genitalia. When the obstetrician told the father that he had a son, he added that the penis was not completely formed. This same information was then given to the mother.

Three weeks later, on their own initiative, the parents consulted a pediatrician who was the first to tell them that the baby might be female. Subsequently the mother learned that this possibility had been rumored at the hospital where the baby was born, but not within the parents' hearing. The pediatrician made an immediate referral to the Pediatric Endocrine Clinic at Johns Hopkins for a diagnostic evaluation and determination of the sex of rearing. At that time, the infant was three weeks old. Physical examination of the external genitalia revealed a pencil-thin phallic structure, a micropenis, that measured 1 cm in stretched length. Urination was from the tip of this micropenis. No gonads were palpable. A buccal smear was taken and found to be chromatin negative. The following was entered in the baby's medical record: 'It is immediately obvious upon looking at this baby that it will never have any structure resembling a normal penis; it has no apparent erectile tissue. In view of the absence of the possibility of ever providing an adequate penis, it seems only logical to raise this child as a girl.'

The decision as to the sex of rearing rested not on the diagnosis, but the prognosis, namely, that what could be accomplished surgically and hormonally to habilitate the child as a girl would be more effective than as a boy (Money et al., 1981). In accordance with the policy that parents cannot make a sex reannouncement and rear their child to be psychosexually healthy without fully comprehending what they are consenting to, they spent an entire afternoon in consultation in the psychohormonal unit, asking for and receiving information as outlined in *Sex Errors of the Body* (Money, 1968; see also Money, 1975; Money et al., 1969).

A prediction of an auspicious outcome was, correctly as it has turned out, recorded at the time. 'My impression was that it is hard to imagine people more organized and efficiently operative in the face of an emotionally impressive problem. Neither of them was shaken-up, though the wife showed a little more of this than the husband. They were not flippantly nonchalant, nor blandly indifferent. They were able to recognize and discuss the problems confronting them for what they are, in about the same way as the pediatric endocrinologist and I were able to discuss them together. ... Toward the end of the afternoon, I explained my policy in a case like this one of giving parents all the relevant information, so that they could follow the line of reasoning and agree or disagree. I was in favor of rearing a child like theirs as a girl. They didn't need to make an immediate decision. Perhaps they would prefer to sleep on it. Actually, the decision-making had been going on throughout the interview. To begin with, both parents began by referring to their baby as he. Then the father, who talked more, on the whole, than did his wife, began using the pronoun, it. Later, he used she a few times, with some appropriate proviso – she, if that's how we will decide. The mother did not disclose so much in pronoun usage, but when I spoke about making a decision, she quietly said that she thought she knew what her decision was – girl. ... Back in the endocrine clinic, the father a couple of times still used he. Then, smiling, he corrected himself to she. He was asking about procedure and the written statement necessary for changing the birth certificate.'

Because the surgical consultant was out of the country, the admission for exploratory laparotomy and cosmetic genital correction was postponed until the baby was four months old. Laparotomy established that the internal reproductive structures were of wolffian origin, with no uterus, and no evidence of intersexuality. The undescended gonads, which were removed, proved microscopically to be immature and testicular. The cosmetic feminizing procedure consisted of moving the urethral meatus posteriorly into a more feminine position, while preserving the glans and penile skin as a clitoris. The labioscrotal skin remained fused in the midline of the perineum. The appearance was that of labia majora that had fused, since the skin was taut and not pendulous. Thus, throughout childhood, the external genitalia looked feminine on superficial inspection, though not when closely examined (Fig. 7.2). Vaginoplasty and separation of the labia was postponed until mid-teenage, when the body was full grown.

At age 17, the patient elected to undergo vaginoplasty rather than wait longer. When the midline incision was made for the opening of the new vagina below the urethral meatus, for the first time the existence of a cryptic, blind vaginal pouch was revealed. It measured an estimated 1.5 cm wide by 3 cm long. It had not been found, though looked for, at the time of surgery in infancy. There was a defect in the urethrovaginal septum which made the urethra rather short, but did not interfere with the adequate surgical construction of the introitus of the vagina. The canal of the vagina was lengthened inward by splitting a cavity between the bladder and the rectum, and lining it with a split-thickness skin graft taken from the buttocks, below the bikini line. The completed vagina measured 10 cm in length. The grafted lining was held snugly in place by a vaginal form until it had healed. For several weeks thereafter, it was necessary to wear a form, molded from clear plastic, in order to keep the vaginal cavity dilated. Later it was sufficient to insert the form only at night. Eventually, sexual intercourse alone was sufficient to maintain vaginal patency. The cosmetic result was excellent.

During childhood, the father always accompanied his daughter and her mother on follow-up visits which were two or three years apart. Such family solidarity is almost always an auspicious sign, so far as the child's health and well-being are concerned, and this was no exception.

Growth and development during childhood were unremarkable. In stature the girl was relatively short in comparison with the average for girls her age, but not in comparison with family norm, both parents being short. Relatively, she grew somewhat taller than her sisters. When she was 7, the mother associated her being 'built bigger than the other girls' with being 'rougher' and more tomboyish. The father mentioned that, in comparison with the shoulders of her older brother, hers were broad and square. This was the final occasion when the parents made family comparisons of body build.

At 12 years of age the onset of pubertal feminization was induced with oral estrogen (Premarin, 2.5 mg daily, later reduced to 1.25 mg). It was at this time that chromosomal karyotyping (46,XY) was done. Feminization of the physique was ultimately complete, and there were no vestiges of a masculine body build. The breasts grew to normal female size, and the overall appearance was exceptionally attractive. Low-dosage estrogen treatment will continue throughout adulthood, indefinitely.

After the neonatal sex reannouncement was made, the parents explained the biomedical rationale of the change in lay terms to those who needed an explanation, including the two older children. So effectively did they manage the entire episode, on the basis of the guidance given them, that no subsequent gossip came to their attention. As the child grew up so obviously a little girl, some people who knew the family saga could not remember which child had been reannounced – a great reassurance to the parents.

Gender-coded social biography

This girl is the third in a sibship of one brother, the oldest, and three sisters. The father has a small farm and agricultural contracting business. The mother is a housewife.

All four children grew up to engage in the outdoor activities and recreation available to rural children. The father considered this daughter more tomboyish than the other two, because of her interest in horses and horseback riding, and in playing football, basketball, baseball and volleyball with other children, boys included. The mother did not quite go along with her husband's evaluation. She counterbalanced the more traditionally feminine aspects of her daughter's life against her outdoor recreational interests.

As a young teenager, the girl considered herself somewhat tomboyish, partly because she had liked to play 'Indian football' with her brother and friends, though by age 14 she had quit, at her mother's request, because 'all the neighborhood boys come out, and then I'm the only girl that ever plays.' She continued playing 'flag football,' however, on a girls' team, at school.

At 16, queried again about tomboyism, she said: 'Well, I consider myself, I'm a girl, but I still like to do a lot of things that boys do. I build stuff ... a guinea pig cage and a mouse cage, two guinea pig cages, not really good, kind of lopsided, but it holds them.'

From childhood on, she had an abiding interest, which she did not classify as tomboyish, in horses. Finally, at age 15, she bought her own mare, worked to pay for its upkeep, and planned to breed it. She liked to read about horses and watch them in westerns on TV. 'I'm crazy about them; I guess I'm a nut.'

Her self-styled tomboyism at no time in her life expressed itself as envy of masculine status or imagery of attaining it. As young as age 7, presented with a question about rather being a boy or a girl, she replied: 'A girl, because my mommy is a girl.

I like girls. I like all girls, also colored girls.' As she grew up, cooking and sewing were among her nonboyish activities. So also was doll play. Retrospectively, as an adolescent, she said: 'Oh yeah, I used to play with dolls – had a Barbie doll, and used to play dolls. We played dolls all the time, me and my sister and my neighbor. ... I don't play no more, but I still have my dolls. I'm saving them to give them to my kids when I get married. I got a real old one, old as my grandmother's sister, and when I was in the hospital, she gave it to me.'

At age 7, queried about what she wanted to be when grown up, she replied: 'A mother. I want to have two children, a boy and a girl.' By age 13, she took up babysitting and among her small cousins was the favorite babysitter, because she participated in their play activities. Soon she became employed as babysitter for a neighbor's adopted baby girl. By age 16, she was the sitter for two families, each with two adopted children.

At this same age, her experience with parenting became full-time at home, for her mother, with her own two oldest children now married, decided to accept babies in need of temporary foster care. Queried about his daughter's reaction to these babies, the father said: 'Well, I think she just likes small children, and she applies herself. She's had two of them that are twenty months old, and one three years old. She's taught them to swim. ... She loves kids, so – is there anything else to add?' The mother's response to the same query, with an emphasis on the word, knows, was: 'She said that she knows that she will be able to take one like this and love it, the same as having one of her own.' This outcome was a bonus, not an explicit goal of fostering, the mother explained, adding: 'See, I babysat for this little neighbor boy for three years, and I decided that instead of just having one eight hours a day, I'd have one twenty-four.'

In the cumulative psychoendocrine history, there are no entries pertaining to temper tantrums, quarreling, fighting, hating, or being angry or enraged. The girl herself said that there were occasional disputes between her and her siblings, but they were trivial and short-lived. She disliked fighting and avoided being provoked into it. 'I like to try and get along with everybody,' she said.

She has always liked all the members of her family. She said she takes after her father in being friendly and sociable, and she felt a special affinity with him. All six family members coexist cooperatively and amiably. They are able to find constructive resolutions to such issues as the son's following too closely in the father's footsteps in neglecting academic grades and overemphasizing sports; and the senior daughter's getting pregnant in order to get married.

Friendly herself, the girl had no shortage of friendships as she grew up. Her friends were of both sexes. She classified four or five as close friends, and one as a best friend. She arrived at the dating and romantic, boyfriend stage in synchrony with her agemates. She did not consider herself a leader, or the dominant or organizing one in a group. In high school she was on the Spirit Committee, a morale squad for school teams, and was a member of the Girls' Athletic Association. Her activities were school rather than community or church related. Her church affiliation, protestant, was nominal, and attendance was on special occasions.

Throughout childhood and adolescence, her dress, hairstyle, ornamentation, and use of cosmetics, while changing to suit the fashion of the day, were femininely defined. Young men turned their eyes in her direction, attracted by her good looks and fresh bloom of youth.

At school, from first grade through high school graduation, she worked consistently at the level of average. Her IQ, taken at age 17 on the Wechsler Adult Intelligence Scale, was Verbal IQ, 112; Performance IQ, 109; Full Scale IQ, 112.

As a 12 year old, she had a daydream of being married and living on a farm with lots of animals. Later she considered being a veterinarian, then a nurse, and then a medical lab technician. Upon graduation from high school, with the career/marriage option unresolved, she took an interim job as a sales clerk, prior to marriage and adoption of a son.

Biography of knowledge of condition

On the recommendation of their psychoendocrine advisors, the parents of this child avoided a policy of cover-up with respect to their child's nudity at home and her genital difference from her brother and sisters. In consequence, her knowledge of her medical condition was something that grew in detail gradually, and without traumatic impact. From early life, she knew that she was sexually 'unfinished,' and would one day 'get finished, down there' at the hospital. Sometimes this explanation needed to be refurbished, in a different context. Thus, when the girl was 9, the mother recounted: 'On the way down to the hospital (a long trip by car), she asked why does she come down here. I told her that, when she was born, she wasn't quite finished. This is what Dr. Money told me to tell her. And that, later on, when she got to be a bigger girl, they would finish her. I told her that that's why we come down here. She will be later on.'

In this same incremental way of building up information, the child always knew that motherhood could be by adoption, if not by pregnancy. Eventually, she knew also that some girls don't menstruate, in which case motherhood is by adoption, and that girls like her never wait too long for a pregnancy, but apply for adoption when they need a baby. She was 14 when she replied to a direct question concerning her feelings and thoughts about not getting menstrual periods. 'I don't know,' she answered, 'I guess you get along without them. My neighbor does. She can't have kids either. She's adopted one.'

By this same age, she had been instructed in the wisdom of not disseminating medical self-information indiscriminately, and not confiding too much, too soon to a boyfriend, no matter how serious her attraction to him.

Long before the teenaged boyfriend stage, when at age 9 one interviewer asked her to recapitulate her knowledge of why she had come to the hospital, the reply was:

'Because I was born different than other girls.' She conjectured 11 as the age when she would want surgery to be genitally 'finished,' and added: 'That's too long a time,' when 14 was suggested as a more likely age.

During this same visit preparatory information concerning estrogen and pubertal development was first given, and the girl knew that she would help decide when the treatment would begin, so that she could keep in step with her girlfriends, but not at the expense of commencing too soon, and thereby being too short in stature.

In keeping with standard policy, this girl like all others talked first with her counselor alone, and with a guarantee of confidentiality, except insofar as she consented to the subsequent sharing of information in a final joint interview with the parents. Through the joint interview, the three of them became aware of what diagnostic and prognostic information they shared and could take for granted in talking at home. This policy applied also to general sex education which, in progressive installments, began with standard information about the baby nest, the swimming race of the sperms, and the penis pumping out the sperms within the vagina. This information was, of course, integral to the child's understanding of her own special case and, as a matter of policy, could not be omitted. The parents endorsed this policy. However, they were ill at ease as participants, because of that aspect of the sexual taboo that decrees avoidance of sexual conversation between the generations. For them, their daughter's sex education became the province of psychoendocrinology, school, and friends at school. The school contribution included, early in teenage, talk about girls who became pregnant.

The girl herself reciprocated her parents' reticence. By the time she was 12, like many girls her age, she had become very diffident when interviews were directed to sex information, though she was never too diffident to collaborate in decisions about her own hormonal and surgical therapy. Nonetheless, her body was a reproach to her, and in the early postpubertal years she avoided nudity in the shower room of the school gymnasium because her breasts were slow to grow large enough, and she was apprehensive about not being genitally normal. By age 17, immediately prior to surgery for vaginoplasty, she confided to her younger sister, apropos of her genital defect and her boyfriend, that she would be cheating him. Her parents noticed that she avoided him, whereas, postsurgically, she 'went after him,' allowing the mother to conjecture that they might marry, as eventually they did. Both parents were sure that surgery had lifted a burden and made her much happier.

After surgery, the girl was able to talk openly again with her doctors, topics of sexuality included, whereas for the preceding three years hospital visits, with their physical examinations, medical photography, and psychoendocrine interviews had been dreaded.

'Now she has had her surgery,' the mother reported two months later, 'she's been kind of inquisitive, and she wanted to know why we had her operated on when she was little, because this other girl (her hospital roommate who also had a vaginoplasty) hadn't been operated on in infancy. ... She started to fire questions at me –

why we had her operated on when she was four months old; and how we discovered [what was wrong] at twenty-two days old, and this girl didn't know nothing till she was fourteen. And, so then her father decided we'd better tell her.'

'We sat down one night,' the father said, 'and talked to her and told her that the first time we came to visit Johns Hopkins Hospital that she was named a boy but, through the discovery of the hospital, they recommended that we rename her a girl. ... She said she thought she may have been something like this.' The mother added: 'That she was maybe meant to be a boy.' In the father's judgment, 'It didn't seem to upset her, I mean we just sat and talked about it, and it didn't upset her any, I don't think.'

The girl manifested her own reaction of frankness engendered by this discussion when six months later she followed her mother's advice and asked her psychohormonal counselor for more advice on 'when to tell whoever it is that I'm going to marry about this whole ordeal.' By ordeal she meant: 'Going through about being born without ovaries and not menstruating and – I'm getting good at this – having to have the operation on the vagina and all that. I'm really getting good at this, man I got to pat myself on the back. And I'm not even blushing, am I?' She did, in fact, need the help of prompting as she tried to piece together and recite aloud the story of what she actually knows about her condition, but she said with confidence: 'I know as much as there is to know. I've been told about it since I was two years old, three, maybe four. ... I feel that I know everything there is to know.' She opted against additional technical details, saying: 'I wouldn't understand them, anyhow,' and settled for obtaining them whenever she might want them – if she ever pursued her ambition to be a lab technician, for example.

There were, as of this visit, only two items of diagnostic information, both of which she was prepared for, that would need to be explained, namely, her chromosomal status as an XY woman having a sex chromosome that in her case is better conceptualized not as a Y, but as an X with broken arms; and her gonadal status at birth (testicular under the microscope, but imperfect and unable to do their proper testicular work). With this information, she was eventually equipped to overhear or read, without traumatic effect, anything contained in her medical record, or in popular medical articles, books, or television programs.

Lovemap biography

When at the age of 17 the interview topic was how long she had been 'turned on by guys,' the girl replied: 'Oh, since I was, I don't know – hey, when I was little, I used to sit on my dad's friend's lap.' Twelve years earlier, when she was 5, a story involving this friend, actually the father's adult cousin, had been entered in the record as an example of the coquetry typical of girls at the kindergarten age. Admiring her cuteness, the parents encouraged their daughter to repeat for her psychoendocrinologist the sayings she used in playfully teasing this man, whom she called her boyfriend. 'Why don't you come and kiss me, sometime, baby?' was one of them. The other, spoken in a husky, sexy voice, probably in imitation of Charles Boyer, the actor to whom it was attributed, was the famous line: 'Come with me to the Kasbah, baby.'

She contested her parents' verdict that the romance was mostly her own idea, and said it was Jack's too. She told of running so that he might chase her 'and then he thinks of catching me up.'

The parents said she talked often about boyfriends and getting married. For example, another relative had just finished college and married. She said: 'I was going to marry Jerry, but I can't now, because he's gotten married.' One of her kindergarten boyfriends remained a regular friend through adolescence.

At age 7 she answered another interviewer's query about a boyfriend. 'His name is Randy,' she said. 'He is in the same grade. He is chasing me. I hid from him, but I like him.' The next follow-up visit was at age 9, by which time there was another boyfriend, but he didn't like her any more. There was a three-year interval before the next visit. Then, at age 12, talking about babysitting and imagining how many children she would like to raise up, the girl allowed that she did not yet have a husband picked out, then coyly added: 'I have a boyfriend, though; he is my neighbor.' Sometimes they walked home from school together. On each of the next three annual visits, she was more offhand about boyfriends. She had phone calls, cards, and visits from boys, but no steady relationship. The family rule forbade the daughters to go out on dates until age 16. She was still only 15. In addition she was concerned about the implication for a boyfriend of not yet having had a vaginoplasty. At 16, she actually cooled off her boyfriend relationship until surgery the next year was finished and successful.

Postsurgically, she was diffident, shy and ambivalent about reconciling the moral decision of when to begin her sex life with the postsurgical prognosis that intercourse would not be injurious, and that regular penile penetration would substitute for insertion of the vaginal form. It was already a foregone conclusion, however, that her sex life would begin independently of being married. Her condition was that: 'I just couldn't go out and jump into bed with anybody; I'd really have to love somebody before I could, you know. ... He told me to call if I wanted to go to bed with him, but I know it's just to screw me and run away, you know, back to his girlfriend. That's the way he is. So I don't know. ... Guys are all alike. ... I'm not like that. I just can't help it. I couldn't be.'

Reflecting on what it was about this particular man that 'turned her on' the most, she said: 'He ain't very cute, but I think he is, you know. ... Something about him. He's got sexy eyes. ... I don't know, I think if you like somebody enough, anything could turn you on. ... I made out with a couple of guys I don't like; I mean they really just don't, you know, they don't do nothing for me.'

On the occasion of the follow-up visit 18 months postsurgically, the girl was able to report that her sex life had begun. She and her boyfriend had had intercourse about 20 times. She wanted her mother to know, so that 'she wouldn't have to worry that I had any hang-ups on sex, or anything like that.' Yet, she had reservations: 'I know, if you tell my mom, she'll tell my dad, and I don't know if that'll be cool or not.' She elected to let her mother know, but to wait for an opportune moment when she would tell her herself. The mother's policy toward premarital sex was that 'you can't stop it these days,' and she was not too much bothered by it. Nor was the father.

Prior to having had intercourse, the girl had not been able to give information on the location of erotic sensation, through lack of experience. Even the manipulation of the postsurgical vaginal form did not lead to erotic stimulation. Throughout childhood and adolescence, the history had been consistently negative for masturbation. 'Actually, I never picked up the habit, that I know of,' she said in late teenage. 'Somebody said that everybody masturbates whether they realize it or not, is that true?'

Searching for words to describe the feelings of intercourse, she said: 'Oh, it's pleasant, and I'm getting better at it. It's nice. ... I get to feeling like – I don't know, it's just different. ... In my stomach, it just feels funny ... tingly.' About orgasm she stated: 'I never had one, but I'm not worried about it because my sister told me that she never had one until after she'd been married and had her first kid. So, I just never really worry about it. It'll come in time, I hope.' Though the area of greatest genital sensitivity was identified as the outer part of the vagina, responses to queries on this topic indicated an insufficiency of experience for greater precision. There had been no experience of oral sex.

There has been also no accompanying erotic fantasy imagery: 'I never fantasized in my life. I thought about it [what intercourse would be like] but I guess I was just like anybody else, curious, you know, wanting to know.' There was no history of dreams with sexual or related content, nor of orgasm occurring while asleep.

There was also no history of homosexual imagery, perceptual or fictive. About two girls or two boys who have a romantic feeling for one another, she said: 'Well, I've heard a lot about it. ... I'd never do it, I'll tell you that much. ... But the Lord made all kinds of people, I guess.' She was nearly 15 at the time, and her response to the hypothetical proposition of having a girl make a friendly, romantic approach was: 'I'd go home.' Three years later she summed up her policy toward homosexuality: 'If that's their thing, that's their thing. I can't see nothing wrong with it, really I can't,' except that it wasn't her style. 'I'd start swimming,' was her reply, when confronted with the hypothetical proposition of being in a lifeboat and having to get turned on to women, or else losing her place.

Once embarked upon her sex life, she reviewed the sensory channels of erotic arousal that were most effective for her, beginning with: 'Okay, touch and sight. I mean not me looking at Chuck, but just, I mean, the way that he gives me these looks sometimes.' Tactual arousal was diffuse rather than localized. Smell pertained to perfume: 'I love the smell of musk oil for men. I think that just turns me on. And he wears it.' Tastes and sounds rated negatively, except insofar as the ears listening to conversation could bring two people close, whereas the eyes served at a distance. Romantic and sexy stories or shows were at best mildly erotically stimulating. The only explicit erotic materials she had encountered were the nude male photographs in *Playgirl* magazine: 'They got some good-looking ones in there. ... I mean not just – I mean their faces are cute, too.'

Biography of behavioral difficulties

Early in childhood, in comparison with her siblings, it seemed possible that the child's physique and energetic behavior might have been misconstrued by the parents as being incompatible with femininity, but nothing came of it.

As a 5 year old, the girl got into trouble when she wiped muddy hands on neighborhood laundry and, with a boy playmate, picked the tops off two young dogwood trees. In all the subsequent history, there were no further reports of disobedient or unacceptable behavior at home, at school, or in the neighborhood. As a teenager, the girl could recall having had only one dream during childhood that could qualify as a nightmare.

At around age 16 she began having attacks of stomach cramps that might have been ascribed to stress related to impending surgical vaginoplasty. A year after the vaginoplasty, however, another operation disclosed a twisted bowel and adhesions secondary to the exploratory laparotomy done in infancy. Both were corrected, and there was complete relief from abdominal pain.

All told, therefore, this biography is remarkable for a total absence of behavioral difficulties, and a clean bill of health, psychologically.

EXPOSITION

A micropenis is one in which the corpora of the penis are agenetic or vestigial to a variable degree, without other defects of embryogenesis of the external genitalia. The urinary meatus is in the correct position. The testes may be of normal size, and descended, though they may also be small with one or both undescended. The internal genitalia are differentiated as male, though not always perfectly so. The karyotype is 46,XY. However, the differential diagnosis includes not only a karyotypic and gonadal male, but also a 46,XX karyotypic and gonadal female, as well as various other chromosomal karyotypes found in hermaphroditic birth defects of the sex organs.

In complete agenesis of the penis (Fig. 7.3), the urethra may open on the edge of the anus, with the penis entirely absent, despite a scrotum and testes that appear normal. It is when the agenesis of the penis is extensive, though not total, that the condition is referred to as micropenis. The corpora of the penis are primarily affected. They may be all but missing in their entirety. The glans of the penis is not missing, nor is the foreskin and the skin surrounding the penis. The urethra is normally positioned. The entire organ is extremely small. At birth it may be no larger than a clitoris, or it may be up to 2 cm in stretched length, with a corresponding reduction in diameter. By convention, a penis is called a micropenis when its measurements are two or more standard deviations below the norm. In absolute terms, this means the stretched length of the penis is no greater than 2 cm at birth (Money et al., 1984).

When the urethral meatus is not in the usual male position at the tip of a micropenis, but at the penoscrotal junction, then the condition is designated as major, or third degree hypospadias, and the sex of the organ is ambiguous – either a clitoridean penis or penile clitoris. The criterion for defining the sex of an ambiguous micropenis is, by convention, according to the sex of the gonads. The sex of the gonads does not, however, necessarily dictate the sex of assignment and rearing.

Micropenis may be also associated with an epispadiac defect. Epispadias is the birth defect in which there is a failure of midline fusion from the navel to the tip of the penis with an associated unfused bladder. When the epispadiac micropenis is repaired as a penis (even though female assignment might be preferable), the single term, micropenis, is acceptable.

In addition to hypospadias or epispadias, micropenis may occur in association with other anomalies, or without them. In the former instance, the total number of possible syndromes in which micropenis is one of the anomalies has not yet been decided (Kogan and Williams, 1977; Lee et al., 1980a,b; Danish et al., 1980; Money and Norman, 1988).

The etiology of the micropenis defect in each of the two cases in this chapter is unknown. Thus, it is not possible to specify the extent to which hormonal masculinization of sexual pathways in the hypothalamus or elsewhere in the brain may or may not have failed to take place in fetal or neonatal life.

At the time when the two patients were born, today's technique of extremely precise radioimmunoassay of circulating sex hormones had not been discovered. Nor had it then been discovered that there is in infant males a surge of testosterone secretion from birth until the second or third month of life, after which the surge recedes until by the seventh month the plasma level of testosterone is down to the very low level of prepubertal boyhood, where it remains until the onset of puberty (Forest et al., 1973). There is no corresponding surge of hormonal activity in the infant female.

The function of the androgenic surge in male infants is not known. An organizing effect on the brain is not unthinkable, for its neural fibers at this age are incompletely myelinated and still differentiating. Should there be such an early postnatal androgenic brain effect, then the chances are that in both cases herein reported, the infants were equally exposed, for the one reared as a girl was not gonadectomized until age four months.

Thus, the two cases are as well matched for prenatal and neonatal hormonal history as they are for 46,XY chromosomal status. They are also well matched with respect to degree of imperfect external genital differentiation: the boy had a mildly incomplete urethral fusion, affecting only the glans penis, and not in need of surgical correction. In the case of the girl, surgery in adulthood revealed what in infancy had been too small to identify correctly, namely, a hidden and incompletely formed invagination, or blind pouch. In infancy it had been identified as the urethra itself, because the urethra opened into it, deep inside. Externally, the outlet for this blind pouch served also as the urinary outlet at the tip of the skinny micropenis.

The very nature of the defect in penile morphogenesis and descent of the gonads for which these two cases are well matched signifies impairment of masculine anatomical differentiation, prenatally. Thus one might feasibly hypothesize that, as compared with an ordinary 46,XY male infant, a masculine differentiation of G-I/R postnatally might also be subject to impairment. If so, then one has a partial explanation for how easily the girl differentiated a feminine, albeit tomboyish G-I/R, and fell in love with a boyfriend; and, conversely, how difficult it was for the boy to differentiate a masculine G-I/R, veering toward and then circumventing transexualism, and finally establishing a G-I/R weighted toward erotic homosexual bonding with a male.

A prenatal hypothesis by itself alone is not, however, sufficient to explain the divergent outcome of G-I/R in the two cases. Postnatal determinants also must be hypothesized in order to account for the divergence, namely, the cumulative experience, assimilation, and application of the social stereotypes of male and female, and the haptic and visual body image of being male or female.

In the case of the girl, there was no intrafamilial evidence of inconsistency or ambiguity in rearing the child as a daughter and sister, and no evidence of community inconsistency or ambiguity in defining the child as a girl. Nor was there any evidence of intrafamilial psychopathology that might have hindered the child's assimilation of a feminine G-I/R. Her genital appearance in childhood supported a feminine visual body image, despite the necessary postponement of vaginoplasty until teenage. Genital haptic awareness in childhood, at least insofar as was ascertainable, also did not contradict her feminine haptic body image. The outcome in adulthood was a univalent G-I/R as a woman.

In the case of the boy, there also was no evidence of intrafamilial or community inconsistency or ambiguity in his rearing as a male, other than age-mate teasing because of the size of his penis. By contrast, there was evidence of an intrafamilial, insidious power struggle in which the boy identified his mother as chief protagonist. It is possible that the disruption of this protracted dispute not only generated symptoms of psychopathology in the boy, but also spread to include ambiguity in the differentiation of G-I/R. It is somewhat more likely, however, that the chief source of male/female ambiguity was the inadequate genital body image, generated by the actual visual-haptic image of the genitals, and the social consequences thereof. With the knowledge of hindsight, one suspects that the body image might have been more pervasively masculine had a prosthetic, strap-on penis been provided from early childhood, so that it could have become incorporated into the body image (Money and Mazur, 1977).

Emergence of the sex-change premise was highly logical, especially in view of the juvenile age, 8 or 9 years, when it first began to formulate itself. In this respect, it differed from the dogmatic and inflexible insistence of the typical transexual's sex-change premise. The greater logical flexibility is consistent with the ultimate resolution of the transexual premise into the alternative premise of bivalency of G-I/R, with more emphasis on the homosexual than the heterosexual. This bivalency showed itself almost exclusively in that component of G-I/R which pertains to erotic pairbonding. It was not evident in the vocational and recreational components of G-I/R, except insofar as some activities and projects pertained to the gay community. In everyday activities, the body language and manner met the cultural specifications of masculinity.

A special feature of both cases is that, because of the longitudinal study design, it is possible to identify juvenile precursors of the sexuoerotic component of adult G-I/R which confirm that childhood is indeed the period when the basic patterns are laid down. The girl's record at age 5 contains a report of typical girlish coquetry and flirtation with a male, her father's cousin. As a teenager, she herself spontaneously recalled this early precursor of heterosexual pairbonding. The parallel phenomenon for the boy was a crush on another boy in third grade. Though not recorded at the time, this experience of pairbonding had, since its origin at age 9, persisted in memory until young adulthood.

The two cases have obvious bearing on the very definition of homosexuality. In both cases, if only the chromosomal and neonatal gonadal criteria are used, then the sex is defined as male; the neonatal morphologic sex of the genitals is arguable, though by convention defined as male rather than intersex. If the devil's advocate used only these criteria, and in the case of the girl neglected her hormonal and morphologic pubertal sex, which is defined as female, as is the sex of her surgically feminized genitalia, then he might dispute the evidence of common sense, namely that, when she has intercourse with her husband, she is heterosexual. Common sense would, of course, be right in declaring that the definitive criterion of heterosexuality is complementary erotic matching of morphology, and in particular, the genital morphology. By contrast, for homosexuality the criterion is parallel matching. By this criterion, his micropenis notwithstanding, the boy has a heterosexual history as well as a homosexual one, for he has had both the imagery and the erotic actuality of a partner with a vagina, as well as of one with a penis.

The two cases have an obvious bearing also on the postnatal contributions to the genesis of either a homosexual or heterosexual G-I/R. The girl's heterosexuality had unmistakable origins in her sex assignment and rearing as a girl, was playfully rehearsed as early as age 5, and developed monodically throughout the juvenile and adolescent years. In the boy by contrast, it was established at least by age 9, if not earlier, that psychosexual differentiation would not be monodic but duodic. He had had the experience of a homosexual crush, had formulated the concept of changing his sex, and had had erotic imagery of himself as a female as well as a male. At age

10, he might well have been rehabilitated as a sex-reassigned female, but instead underwent masculinizing hormonal treatment that temporarily enhanced his juvenile body with an early pubertal-sized micropenis. The duodic psychosexual status persisted, first in masturbation imagery, and later in erotic partnerships, eventually to yield to monodic homosexuality in adulthood.

In both cases, the hormones of puberty were exogenously administered and maintained. In the girl, estrogen was concordant with, and sustained her feminine sexuoerotic status. In the boy, testosterone sustained the strength and frequency of sexuoeroticism, but did not affect its heterosexual-homosexual balance.

Both individuals as adults rated tactual (haptic) stimulation above visual as an erotic turn-on. As an adult, the man had extensive exposure to visual erotica and was erotically aroused by it. The woman had little exposure and minimal turn-on. She had always had a negligible visual erotic fantasy life, whereas her counterpart had one of great magnitude. The relationship of these discrepancies in visual erotic imagery to postpubertal hormonal differences cannot be ascertained on the basis of the two cases alone. A likely hypothesis is that the two people have always been extremely disparate as visual imagizers, and that in the man the prevalence of erotic visual imagery fluctuates in parallel with compliancy in taking testosterone, and hence with fluctuations in plasma testosterone level between deficient and normal.

Heuristically, both cases demonstrate the principle that data recorded and preserved in a prospective developmental study surpass data retrieved in a retrospective study. This principle is particularly well demonstrated in the boy's case by comparing his own biographical statement, retrieved from memory, with that retrieved and illustrated with verbatim quotes from his cumulative record. Only the cumulative record gives detailed insight into the ontogenetic evolution of his G-I/R into its homosexual instead of transexual outcome.

Both cases demonstrate the powerful effect of parental and family influences, directly and indirectly, in the postnatal differentiation of G-I/R, and also in the genesis of symptoms of psychopathology or its absence. The boy was, in effect, an exile in his own home, unable to disclose the unspeakable monster in his life to anyone except his hospital-based counselor, whereas the girl had no such lonely trauma with which to cope.

BIBLIOGRAPHY

Benjamin, H. (1966) *The Transsexual Phenomenon*. New York, Julian Press.

Danish, R.K., Lee, P.A., Mazur, T., Amrhein, J.A. and Migeon, C.J. (1980) Micropenis. II. Hypogonadotropic hypogonadism. *Johns Hopkins Med. J.*, 146:177–184.

Forest, M.G., Cathiard, A.M. and Bertrand, J.A. (1973) Evidence of testicular activity in early infancy. *J. Clin. Endocrinol. Metab.*, 37:148–151.

Jost, A. (1974) Mechanisms of normal and abnormal sex differentiation in the fetus. In: *Birth Defects and Fetal Development: Endocrine and Metabolic Factors* (K.S. Moghissi, Ed.). Springfield, Charles C Thomas.

Kogan, S.J. and Williams, D.I. (1977) The micropenis syndrome: Clinical observations and expectations for growth. *J. Urol.,* 118:311–313.

Lee, P.A., Danish, R.K., Mazur, T., Amrhein, J.A. and Migeon, C.J. (1980a) Micropenis. III. Primary hypogonadism, partial androgen insensitivity syndrome and idiopathic disorders. *Johns Hopkins Med. J.,* 147:175–181.

Lee, P.A., Mazur, T., Danish, R., Amrhein, J., Blizzard, R.M., Money, J. and Migeon, C.J. (1980b) Micropenis. I. Criteria, etiologies and classification. *Johns Hopkins Med. J.,* 146:156–163.

Money, J. (1968) *Sex Errors of the Body: Dilemmas, Education and Counseling.* Baltimore, Johns Hopkins University Press.

Money, J. (1970) Matched pairs of hermaphrodites; Behavioral biology of sexual differentiation from chromosomes to gender identity. *Engineering and Science* (California Institute of Technology) *Special issue: Biological Bases of Human Behavior,* 33:34–39.

Money, J. (1975) Psychologic counseling: Hermaphroditism. In: *Endocrine and Genetic Diseases of Childhood and Adolescence,* 2nd edn. (L.I. Gardner, Ed.). Philadelphia, W.B. Saunders, pp. 609–618.

Money, J. and Mazur, T. (1977) Microphallus: The successful use of a prosthetic phallus in a 9-year-old boy. *J. Sex Marital Ther.,* 3:187–196.

Money, J. and Norman, B.F. (1988) Pedagogical handicap associated with micropenis and other CHARGE syndrome anomalies of embryogenesis: Four 46,XY cases reared as girls. *Am. J. Psychother.,* 42: 354–379.

Money, J. and Primrose, C. (1969) Sexual dimorphism in the psychology of male transexuals. In: *Transsexualism and Sex Reassignment* (R. Green and J. Money, Eds.). Baltimore, Johns Hopkins University Press, Ch. 6.

Money, J., Lehne, G.K. and Pierre-Jerome, F. (1984) Micropenis: Adult follow-up and comparison of size against new norms. *J. Sex Marital Ther.,* 10:105–116.

Money, J., Mazur, T., Abrams, C. and Norman, B.F. (1981) Micropenis, family mental health, and neonatal management: A report on 14 patients reared as girls. *J. Prev. Psychiatry,* 1:17–27.

Money, J., Potter, R. and Stoll, C.S. (1969) Sex reannouncement in hereditary sex deformity: Psychology and sociology of habilitation. *Soc. Sci. Med.,* 3:207–216.

Ohno, S. (1978) The role of H-Y antigen in primary sex determination. *J. Am. Med. Ass.,* 239:217–220.

Page, D.C., Mosher, R., Simpson, E.M., Fisher, E.M.C., Mardon, G., Pollack, J., McGillivray, B., de la Chapelle, A. and Brown, L.G. (1987) The sex-determining region of the human Y chromosome encodes a finger protein. *Cell,* 51:1091–1104.

Silvers, W.K. and Wachtel, S.S. (1977) H-Y antigen: Behavior and function. *Science,* 195:956–960.

CHAPTER 8

Two cases of 45,X/46,XY hermaphroditism discordant for sex assignment, preventive counseling and outcome after puberty: 20 and 24 year follow-up

SYNOPSIS

The two people in this matched pair are concordant for chromosomal status, namely 45,X/46,XY, a relatively rare form of mosaicism (Richards and Stewart, 1978). They are concordant also for prenatal gonadal dysgenesis, and incomplete differentiation of the genital organs in prenatal life so that, at birth, their sex was in doubt. By contrast, they are discordant for postnatal life history. Both were neonatally assigned as boys, but only one was continuously reared as a boy. The other had a sex reassignment imposed at age 4:6 and thenceforth was reared as a girl. To induce puberty, she eventually was treated with female hormone, whereas the boy received male hormone to induce his puberty. The boy and his parents were provided with preventive counseling beginning neonatally, and he grew up free of behavioral pathology. The girl and her parents received no preventive counseling. At age 13 and in crisis, she was referred for therapeutic counseling. She and her parents were behaviorally gridlocked into a three-way system of pathological family relationships.

The boy's gender-identity/role (G-I/R) differentiated as masculine. Sexuoerotically, in adulthood, his orientation was that of a heterosexual male. The girl's G-I/R differentiated as predominantly feminine. She lost from memory the biography of her first four and a half years of living as a boy until, at age 17, recall of the lost years began to re-emerge in dreams. Romantic attachments were with both men and women, without being explicitly genital until she was a woman in her mid to late twenties. Penovaginal intercourse was impossible until after surgical vaginoplasty at age 30. After the surgery, she married. Her sexuoerotic imagery and ideation as a heterosexual wife included paraphilic bondage and discipline of light to moderate degree. The erotization of bondage and discipline might be interpreted as the polar reverse of the traumatization of bondage and discipline in childhood. As a child she

TABLE 8.1
Diagnostic and descriptive data, case #1 (see Bergada et al., 1962, case #6)

Age at first psychohormonal visit	1:4
Age at last follow-up	24 years
Sex assignment	Uncertain for one week, then declared male
Genital anatomy	Asymmetrical intraabdominal gonads. Rudimentary left gonadal streak and mullerian structures (removed at age 16 months). Right dysgenetic gonad (removed at age 12:6). Hypospadiac phallus (repaired male)
Surgical history	Exploratory laparotomy, removal of mullerian organs and left gonadal streak, age 16 months. Release of chordee, age 2 years. Urethroplasty 1st stage, age 3, 2nd stage, age 4. Urethroplasty revision, two stages, age 9, 3rd stage, age 10, 4th stage, age 11. Right intraabdominal gonadectomy and cosmetic testes implanted age 12:6. Total of 9 surgical admissions. After age 12, 2–4 urethral dilations annually
Hormonal treatment	Oxandrolone 2.5 mg/day for height, age 9:7 to 10:10; 11:8 to 12:3; 14:11 to 16:6. Human growth hormone, age 16:10 for 4 months. Testosterone enanthate, 300 mg per 3 weeks, age 17:6 onward
Puberty	Induced hormonally with testosterone enanthate at age 17:6; masturbation and climax by age 18; no ejaculatory fluid; penis stretched (erectile) length, 8.5 cm ($3^1/_2$ in.)
Adult height and weight	Height 146.5 cm (4 ft. 10 in.). Weight 48 kg (106 lb)
Other clinical data	Long-term treatment with antimicrobial medications for recurrent urinary infections
Cytogenetic data	45,X/46,XY (XY in 10% of the cells). Sex-chromatin negative
IQ	Age 9, WISC Verbal IQ 115, Performance 117, Full 117

had been traumatized by the sex-change experience, by hypodermic needles to which she became severely phobic, and by multiple surgical procedures, and restraints applied to the pelvic and genital area. In a child's mind, such procedures may become a nosocomial equivalent of sexual child abuse.

For each of the two cases, pertinent clinical data are summarized in Tables 8.1 and 8.2, respectively.

CASE #1, 45,X/46,XY REARED MALE

Diagnostic and clinical biography

This patient (Bergada et al., 1962) was a baby 16 months old and actually undergoing exploratory laparotomy while his parents were engaged in having their first psychohormonal clinic interview. During the interview, a phone call from the operating

TABLE 8.2
Diagnostic and descriptive data: case #2 (see Jones and Zourlas, 1965, case #4)

Age at first psychohormonal visit	13:1
Age at last follow-up	32 years
Sex assignment	Declared a boy at delivery, then maybe a girl, until at age 3 weeks declared a boy. Reassigned female at 4:6 years
Genital anatomy	Asymmetrical intraabdominal gonads (removed at age 4:6). Rudimentary right testis and primitive left streak. Mullerian organs differentiated. Fallopian tubes (removed with gonads). Hypospadiac phallus and external urogenital sinus (repaired female)
Surgical history	Attempted hypospadiac repair, age 4. Exploratory laparotomy, age 4:6. Clitoridectomy and first-stage vaginoplasty, age 4:6. Second laparotomy, gonadectomy and salpingectomy, age 13. Second-stage vaginoplasty age 29
Hormonal treatment	Premarin, 1.25 mg/day, age 13:1 to 14:9; 15:3 to 17:7. Fluoxymesterone, 2.5 mg/day (for height), 17:7 to 18.4. Premarin, 1.25 mg/day, 18:4 to 20. Noncompliant, age 21 to 28. Sequential Premarin 1.25 mg/day (days 1 to 25) and Provera 10 mg/day (days 19 to 25), age 29 to 32
Puberty	Induced hormonally with Premarin beginning at age 13:3; breasts by age 16; menarche delayed to age 30 (hormonal noncompliance); vagina depth: 8.5 cm ($3^1/_2$ in.)
Adult height and weight	Height 133 cm (4 ft $4^1/_2$ in.). Weight fluctuated from 65 kg (145 lb) to 50 kg (110 lb)
Other clinical data	Congenital abnormalities: double kidney, asymptomatic; hypertelorism; poorly developed mandible; short neck with low-lying cervical hairline; hypoplastic nails
Cytogenetic data	45,X/46,XY. In some cells, the Y was not missing but was a Y fragment (identified by banding). Sex-chromatin negative
IQ	Age 13, WISC Verbal IQ 110, Performance 94, Full 103. Age 17:6, WAIS Verbal IQ 122, Performance 115, Full 120

room carried the information that a dysgenetic left gonadal streak had been removed along with mullerian structures. The right gonad could not be brought down and so was left in place. It had the appearance of an ovary, but upon biopsy proved to be testicular. In accordance with the policy of the times, the structure of this gonad was the criterion by which such doubts as there had been, concerning the fitness of the sex of rearing as a boy, were disposed of.

The ambiguity of genital appearance had been evident immediately upon delivery, and the decision was thereupon made to postpone the announcement of sex for a few days until the recently discovered test for the sex chromatin or Barr body (the technique of actual chromosome counting did not yet exist) could be performed at a major university center. When the result came back negative, that is typical for a male, the sex was divined as that of a male. Only the grandparents had known of the uncertainty. Other people had always known of the child only as a boy, and thenceforth there was no break in the consistency of his status as one.

As was their wont, the parents were phlegmatic and matter-of-fact in receiving the operating-room news that eliminated the upheaval of a sex reassignment. They asked for information about terminology and how to explain their child's condition to their 5 year old daughter, and other close relatives and friends. They learned for the first time the simplified concept of incompletely differentiated or 'unfinished' sex organs, and the technical term, hypospadias, for an unfinished penis. This was the starting point, and these were the concepts and the words that eventually they would use to explain the boy's condition to him as he grew older, without retracing all the details of the history that they themselves had had to live through. They did not evade talking to him as the occasion might require, but they found it more congenial to assign medical self-knowledge and sex education as topics for the boy to discuss chiefly with his psychohormonal counselors. His hospital visits were scheduled annually, except when surgical appointments were more closely spaced and when, in teenage, the hormonal induction of puberty required more frequent follow-ups.

After the exploratory operation at age 16 months and before the age of 11 years, he had seven more surgical admissions for repair of hypospadias (Fig. 8.1) and for subsequent revisions secondary to postoperative complications, namely fistulas that opened in the wall of the urethra, and strictures that blocked the urethra and prevented urinary flow. Urethral stricture remained a persistent and recurrent problem well into adulthood, when it abated. It required frequent dilations of the urethra by a hometown physician, sometimes as often as once every three months. This problem of urinary blockage was exacerbated by persistent recurrent urinary infections requiring regular antimicrobial treatment. The stretched length measurement of the penis, which is equivalent to the erected length, was 8.5 cm ($3^1/_2$ in.) at age 18.

The boy was always kept informed of what would happen to him, and why, at each clinic visit or hospital admission. Despite the excessive amount of doctoring which was his lot, he maintained a degree of phlegmatic matter-of-factness through it all, resembling that of his parents. When he was 9, his parents reported that he had little to say about coming to the hospital, and showed no associated behavioral disturbances, either awake or, in dreams, asleep. As they grew older, his older sister and younger brother had a casual knowledge of the reason for his hospital follow-up, but he did not go into detail with them. Such knowledge as they had, they kept confidential.

Fig. 8.1. Case #1. Appearance of the genitalia at age 11:8, after treatment with the weak anabolic steroid, oxandrolone.

Fig. 8.2. Case #1. Appearance of the body habitus at age 11:8, after treatment with oxandrolone to promote growth in height.

Fig. 8.3. Case #2. Appearance of the genitalia at age 13:1, after feminizing surgery at age 4:6.

Fig. 8.4. Case #2. Appearance of the body habitus at age 13:1, prior to pubertal hormonal treatment.

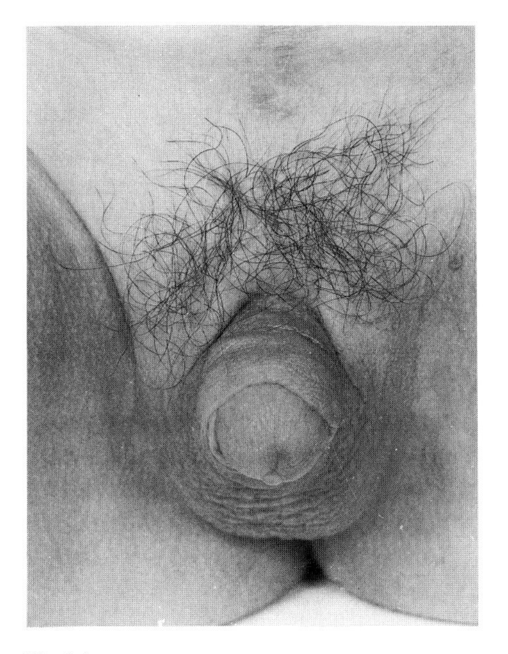

Fig. 8.1

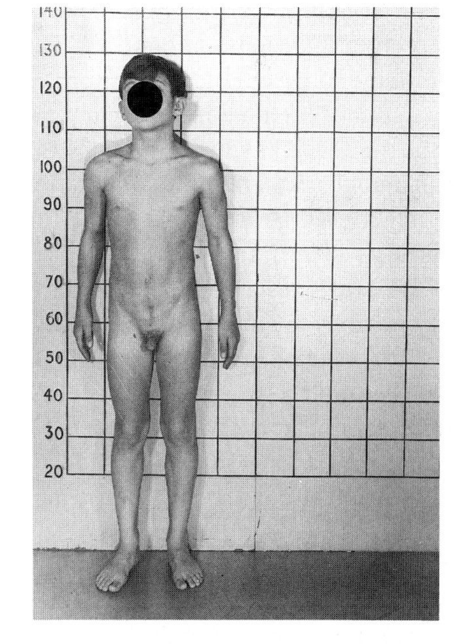

Fig. 8.2

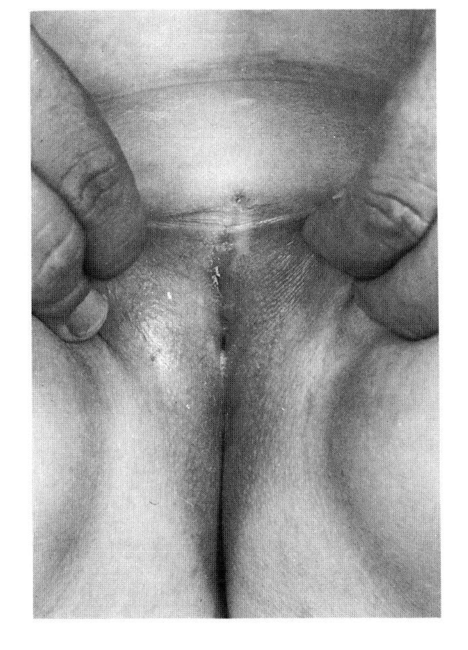

Fig. 8.3

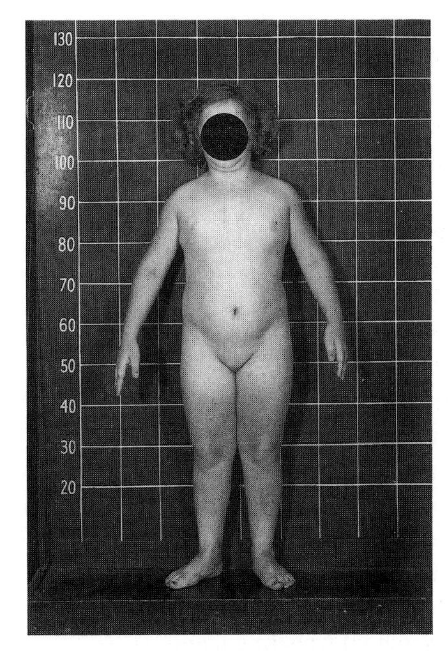

Fig. 8.4

From early years, during his counseling visits in the psychohormonal unit, the boy was given information about prognosis as well as treatment so that he had a basis on which to develop long-term projections of his future. He was 9 when the future prospects of his gonadal status first came under discussion. He learned of the cosmetic possibility of having artificial testes implanted in the scrotum. Since his own gonad could not be brought down and thermoregulated, it would not be capable of fertility. It might even be advisable in the future, he was told, to consider removal of this gonad as an insurance against the risk of malignancy in later adulthood. As a method of achieving parenthood, he could eventually consider resorting to the sperm bank, and not only to adoption or marrying an 'instant family.'

The foregoing information was put to good use three years later when it helped the boy to agree to yet another operation, his ninth, for gonadectomy and cosmetic testicular implantation, at age 12:6.

The only additional clinical information that he needed about himself pertained to hormonal replacement therapy, namely its relationship to pubertal sexualization, on the one hand, and to growth in height, on the other. So as to ensure the maximum opportunity for statural growth, there was a program of treatment, initiated at age 9, with a weakly virilizing anabolic steroid, oxandrolone. This treatment induced the growth of sparse pubic hair, but no other pubertal signs (Fig. 8.2). When a sufficient supply of human growth hormone (somatotropin) eventually became available, this also was tried at age 16:6, for four months. There was no increase in growth rate. Therefore, at age 17:6 virilizing treatment with testosterone enanthate (300 mg/month) was finally begun (see Table 8.1). Within six months the somatic maturation of adolescence and epiphyseal closure ('cementing over of the growing tips of the bones') was virtually complete. The final height was 147 cm (4 ft. 10 in.) and the weight commensurate. Hormonal virilization by age 19 included shaving on alternate days. The age appearance was that of male adulthood, but still youthful.

Gender-coded social biography

At age 10 this boy made an accurate guess of his ultimate height: 'Probably around five feet or something. I probably won't be taller than my dad.' His dad's height was 5 ft. 8 in., and his mother's 5 ft. 1 in. His own height in adulthood proved to be 4 ft. 10 in. At age 11 he completed one item on his sentence completion test thus: My greatest weakness is – 'my last name [Frost] and my height' (127 cm, 50 in., height age 7:6). Another item on the same test was: I could be perfectly happy if – 'I don't know that one; I'm happy now.' Here are exemplified the mixture of realism and contentment that throughout his life combined with compensatory striving to achieve. His height problem, he said at age 15, instead of defeating him, 'makes me want to do it more' – and the examples he gave were playing basketball and trying out for the school wrestling team.

At age 14, in the 8th grade, he was one of two student counselors for his home room. In the 9th grade, he was elected from among 350 students as class president. In the 11th grade when he was 17, he said: 'I try to make friends with everybody since I, let's put it this way, I have to be friendly in order to maintain my position in the class [he was junior class president]. I probably will be running for office – senior class officer – again next year [he was elected] and I will have to keep my social life above average.'

Throughout his life, he had joined his age-mates in play. At age 9, he liked to play cowboys, baseball and basketball, to ride his bicycle, to improve his muscles, climb, wrestle, and be the strongest, according to his mother's observations. The next year he joined a 'Bitty' basketball team and a 'Midget' baseball team of small boys his own age. At this same age he studied music (piano, and later cornet and trumpet) in the school concert band and orchestra, and won the title of best boy actor in the local little theater. His drama interest continued through college.

At age 12 he went to a Boy Scout camp before the cosmetic implantation of prosthetic testes, and after a sparse amount of pubic hair had appeared. He used the public shower room without special incident. Eight years later, he became a summer-camp counselor.

From grade school through high school and college, he was diligent in the sport of wrestling, despite the difficulty of finding competition to suit his diminutive size, and the difficulty of convincing strangers of his age. 'We were recently at the sectional wrestling tournament,' he said when he was in 10th grade, 'and some of the people there couldn't believe that I was sixteen. They had to have it proven to them, so I showed them my driver's license, and that kind of broke them up.' In his first year of college, he found that 'there were only three of us in my weight class in the freshman tournament, and I ended up second; I won one and lost one.' The following fall, there was no team for his weight class. Undaunted, he waited to see whether there would be a lower weight class in the spring.

Academically, throughout his school career, the boy always had passing grades. His IQ had been tested at age 9 on the Wechsler Intelligence Scale for Children, at which time the results were Verbal IQ, 115; Performance IQ, 117; Full IQ, 117. After high school, he entered the same local college from which his parents had graduated and from which his sister and he would graduate. He majored in communication performance and continued graduate studies.

Lovemap biography

In this case, because of the regularity of the boy's hospital visits and admissions throughout childhood, it was possible to ensure that his sexual knowledge was synchronized with what his parents knew, and what he, himself, needed to know. Thus information was given in synchrony with surgical correction of his penis, the prognosis of cosmetic implantation of artificial testes, and the probability of androgenic replacement therapy. It was possible also to coordinate sex education with the early appearance of sparse sexual hair in response to hormonal therapy, beginning at age 9:6 years, with the weakly virilizing steroid oxandrolone, given primarily to stimulate growth in height. In teenage, it became necessary to coordinate sex education with the prolonged withholding, until age 17:6, of the induction of full puberty with testosterone enanthate, while waiting for further growth in height.

The curriculum in the schools he attended did not include sex education. When he was 13:6, his church (Protestant) offered a sex education program, less complete than what he had already received, in some instances jointly with his parents, in psychohormonal counseling.

At age 9, he had summarized his sexual knowledge as follows: 'I remember about the baby story, when first a little dot, a little egg formed. And then after a while it starts to go through channels. And these little squirms, sperms, go after the egg, and there's only one winner out of two or three hundred million. And then the baby starts forming, and getting bigger. Then the mother has to go to the hospital to get it taken out. And then she takes care of it.' He remembered also the exodus of sperms from the testes through the penis, and the sperm bank as a source of sperms in case of insufficiency. At age 10 his sexual knowledge was enlarged to include information about sexual intercourse, and enlarged still more the next year with printed photographic illustrations of coital positioning. In the psychohormonal interviews at age 14 and again at age 16, sex education was reviewed and extended to include scientific terminology and concepts. It was also put in the context of the boy's preparatory planning for success in his own sex life, since his own unique condition in conjunction with his religious standards might make age-mate talk in the locker room particularly unavailable to him. At 16, he did in fact say of a sex picture book at school: 'I just saw a little bit of it, and then I walked away. It was disgusting, that's all.' His sexual knowledge became more personally relevant after his own sexuoerotic functioning was activated more strongly in synchrony with the onset of hormonally induced puberty at age 17:6. Thereafter he became more sophisticated and wordly wise about sexual knowledge, in the manner of adulthood.

There were no family reports of an early history of infantile play at being the little-man escort or being flirtatious, and no reports or self-disclosures of juvenile sexual rehearsal play.

When he was 9, he was presented with a narrated projective test, a story of a boy with a history of penis operations, who had a dream of changing to be a girl. His response was: 'Well, I don't think I'm going to have any dreams like that, because I'd rather be a boy. ... Boys don't have to wear dresses, and girls fiddle with their hair too much.' At age 10, when the same test was repeated, he said he 'wouldn't know' what to say to a boy if he asked for advice about such a dream.

His parents considered that he had shown no signs at age 10 of approaching the girlfriend stage. In response to information on sexual intercourse, he himself allowed that he knew that some teenagers have sex and some become pregnant while not married. For himself his prediction of first intercourse was: 'After I get married. If I find a girl that I like, I won't, you know – I'll take her on dates, but I won't do any other stuff. ... I don't know about kissing.'

A year later, at age 11, he took for granted the ultimate eventuality of marriage: 'Well, when I marry, if I need more sperms than what I have, I just go to the sperm bank.' Then, in the context of reviewing the discussion on medical self-disclosure to a future girlfriend or wife, he said: 'Well, I won't tell her about my problem too soon. I'll just wait until I'm sure I'm going to get married. ... Then I'll tell her I may need to go the sperm bank.'

Between ages 11 and 12, a girlfriend interest dawned. Initially it was a case of visual attraction at a distance – he saw the girl at a concert, but did not get a chance to meet and talk with her until the summer, when they both enrolled in the Children's Theater. The relationship was a talking one only, followed by an occasional letter. By age 14, he still did not rate himself as having arrived at the girlfriend stage, though there was one girl with whom he walked to class, and he had thought about kissing her. Friendship prevailed over romance during the ensuing years while he remained hormonally preadolescent. As a junior in high school he did not see his age-mates as overtaking him: 'There are very few of the people in our junior class that are actually going with each other. ... Everybody is just kind of at the same level, you know. The feeling is mutual, you know. I mean, if there is a school function, like the senior class play that you go to see with a girl, then you ask her.' After the hormonal induction of puberty, and during the first three college years, the pattern of dating remained essentially unchanged: 'On the weekends, like things on campus where you could go to relax, I'll ask somebody, some girl, to go with me.' It was during the final year of college that he developed a complete girlfriend attachment.

The strength of the bond was such that he bypassed two more prestigious graduate school acceptances in order to attend a smaller institution where his girlfriend was also accepted. 'I have to think about it for a minute,' was his response to the question of whether he would say he and his girlfriend had fallen in love. 'I think that uh, we have – I'm waiting to see how the next year goes because we haven't spent a lot of time together. ... We were in separate schools. We'd see each other only on weekends, maybe every other weekend. Now it'll be a different story. We'll both be in the same place. I'm waiting to see how the relationship goes.' Eventually they engaged in heavy petting but, respecting their religious precepts, eschewed penovaginal intromission until marriage – and neither was ready to marry while still a student.

During the juvenile and prepubertal earlier teenaged years, this boy's history was, in effect, negative for erections, masturbation and, indeed, any form of sexuoerotic rehearsal with playmates. By age 17, there was no history of sexual dreams, and no subjective experience of feeling an urge for genital stimulations. Erections were 'probably once a week'. A year later, in an interview after four months on treatment with testosterone enanthate, things had changed. Erections had increased in frequency and included those related to masturbation approximately once or twice a week. 'It's just changed since I've been on the testosterone,' he explained, 'and there are different sensations this time.' There was no ejaculatory fluid released in association with masturbation, nor in nocturnal emissions.

The absence of ejaculatory fluid remained a permanent feature of genitoerotic function over the ensuing years. The rate of masturbation remained constant at once or twice a week. The words he used at age 20 to describe orgasm were: 'There's a tingling sensation, you know. I can't describe it any better than anybody else. ... You know, it begins slow and then uh, reaches a climax and then you know it's over.' To this description he added, two years later: '... like tingle, or being just overwhelmed with, with a feeling of pleasure for that period of time ... not quite so much a spasm, a good word is like it's a release. When I think of spasm, I think of epilepsy. Mildly spasmodic might be a good word. ... It pretty well takes hold of the whole body. ... It lasts somewhere between eight to twenty seconds. ... It varies, I'd probably say, before I can do it again, I don't know, a couple of hours.'

No prominence was given to fantasy imagery in association with masturbation. 'I don't have anything in particular,' he said. 'I just fantasize as I see fit ... just about having sex ... with women.' Erotic imagery in dreams was equally without prominence. 'My dreams are very infrequent,' he reported. 'The amount of dreams I remember are very slim. I can't even remember the last time I had a dream, as a matter of fact.' Pressed to give one example, he responded, pondering at first: 'Umm. Umm. All right, I saw a girl that I didn't know or that I'd never met. I knew she went to school, but I had never met her. And, umm, I was running from something, when she and a bunch of other girls just attacked me, I guess, and raped me, and that was it. I woke up. ... No, I haven't had wet dreams. It stands to reason, I guess.'

The topic of homosexuality, which came up in the context of an inquiry regarding gay liberation on campus, led to a query as to whether the patient could imagine any exigency which might enable him to find out whether he was bisexual or not. His answer was: 'I can't think of anything, you know; there's just no desire, sexually or medically, or any other way, for males.' He said that he believed that, because of his size, his parents had brought him up 'sort of stereotypic male,' and that 'it helped my confidence.' He had difficulty with the 'Empire State Building Test,' namely the hypothetical choice between fellatio and being gunned over the parapet of a skyscraper by a crazed sex maniac. With an ironical smile, he explained: 'Umm, I don't know. I'd have to be in that situation to answer that. I would be – I just don't know.'

He listened to a question about the special senses and erotosexual turn-on, and pinpointed, as the major sense, 'touch, I think,' more than what he might see. The other senses were of minor significance. He had found himself with an erection in response to sexy pictures. Comparing the difference between men and women in sensory turn-on, his verdict was: 'I don't know. I think men are turned on faster.'

At age 24, he had not yet surmounted the diffidence that held him back from sexual intercourse, and in particular the diffidence due to penis size, which he estimated at 10 cm (4 in.) when erect, and of cosmetic testes that had some yield, but that he correctly identified as feeling pretty hard. Nonetheless, he had no diffidence regarding the ultimate expectancy of being a husband and parent.

CASE #2: 45,X/46,XY REARED FEMALE

Diagnostic and clinical biography

At the time when the girl (Jones and Zourlas, 1965) was first seen for psychohormonal evaluation at the age of 13, the family policy had been to protect her from exposure to overt information concerning her clinical history, including the history of sex reassignment at age 4:6 years. In family communication with the patient, the sex reassignment from boy to girl existed as a non-event.

The mother's retrospective account of the neonatal and subsequent history was as follows: 'The first doctor that delivered her said we had a boy. But he wouldn't give us any details after we felt as though something was wrong. When he did say something, it was that nothing could be done about it. ... So we tried a pediatrician. He felt as though it was a girl. When she was three weeks old, we went to Dr. Glasscroft, the urologist. He felt differently – that it was a boy. So we changed back again. ... He checked on her each year until she was four. Then he did the first operation, trying to enlarge the clitoris that actually looked like a penis. And then the second operation, when they did an abdominal, they discovered that she had all the female intestines. So then they changed her back into a little girl.' She was four and a half years old. Before age 5, the penoclitoris (or clitoropenis) was removed and first-stage feminization performed (Fig. 8.3).

There were three older siblings, the youngest of them 12 at the time. The mother protected them from neighborhood inquisitiveness and teasing. 'I kept her [the patient] inside at home, and drew the drapes. ... We kept her to ourselves, and we changed ourselves. We were there only two weeks. We moved and went straight to the farm to live. ... I got her fancy nightgowns and a few dresses. She just thought she was playing dress-up – just switched over gradually and changed to frills and baby dolls. I let her hair grow. She just looked like a boy that needed a haircut. ... As soon as I got to the farm, I stopped right away and put a permanent into her hair. It didn't take much to change it over; and she went right on from there and accepted the whole thing and never questioned it anymore. ... One time, when she was little, she said, 'Some people used to call me Jerry.' And I said, 'Well, that's just short for Geraldine.' And she never said any more. ... You see, we had kept it Jerry, in case. ... Once you've been switched back and forth, you don't know for sure, until the last minute. So we weren't taking any chances. ... My mother is the one who started calling her Eve (from her second name, Evan). ... All the time when she was little, she had acted and had looked like a little girl. She was a beautiful little boy. Really, I mean, people would actually say that she looked like a little girl. She just had the features. Of course they didn't know how much apprehension they were giving us, but ... she was very dainty with everything when she was small. ... [As a boy] she had always liked to play with dolls, so always in the back of my mind I had had doubts. Now [as a girl] since she's gotten older, she's gotten more tomboyish [despite her tiny size]. ... It's not in the strongarm sense, like boys play. She just likes to play those boyish games. ... I was a typical tomboy myself. ...

'Naturally you always worry and wonder if you made the right choice. You go through so much, there's always doubts in your mind. But I feel as though, through her general appearance and the way she acts, I feel as though it's right. The only doubt I have is that the organs are actually there. Are we right? And, of course, that's not my department. I had to take the doctor's word for it. ... I mean, you don't make a deformity out of somebody. I mean that's not our choice. So we just told Dr. Glasscroft to do exactly what he would do if it was his child. ... He is a very gentle kind of person. He takes a great deal of time to explain what he's doing, and why. ...

'The way she asks me things, sometimes, it makes me wonder if she has heard things and wants to ask me something, but it just won't come out, and she doesn't know how to put it into words. And that does worry me. ... Sometimes she'll go into a fit of tantrums, and she'll say, 'You're not telling me the truth. What is it? What's the matter with me, Mommy?' And then I sit down, and explain to her again how she wasn't, her stomach just wasn't properly developed. ... They enlarged the vagina ... and had taken this tissue from the stomach, which I didn't understand too much about, but that was to prevent her having pain when she had her menstrual period. ... What worries me is wondering if she remembers enough that something happened, and it's just completely confused in her mind. ...

'She remembers things from the old neighborhood, before we moved to the farm. The little girl that used to live next door to us in town was killed on a bicycle a year ago. And Eve said, 'I remember Patty. We used to play together in the yard. That was before the operation. ...' Of course, naturally, we try to wind up such conversation and get on to something else.'

All of the foregoing refers to the period up until the child was 13, at which time she first had her own private doctor-patient encounter, alone. It was with her psychoendocrine 'talking doctor,' exactly two weeks after surgery for removal of dysgenetic gonads, in preparation for the initiation of estrogen replacement therapy. She knew she had had three prior operations, her only understanding of which was 'that when I was born, well, my organs weren't exactly straight the way they should be. My intestines weren't straight, in their proper position.' To illustrate the recent operation, she drew a picture of intestines first closely coiled, and then correctively uncoiled like an S. 'I don't know anything about the other operations,' she said, 'but I'm pretty sure all four were right on the same spot.'

Postsurgically, estrogenic hormonal treatment was begun with Premarin, 1.25 mg a day. The hormone was taken from age 13:1 to 17:7, except for six months prior to age 15:3. It induced growth of the breasts which, at age 17:6, measured 10×12.5 cm (4×5 in.) in diameter and were described as well developed. The onset of menstruation was late, as the hormone was not taken cyclically until much later in life (see Table 8.2).

The mother was embittered that the outcome of the referral to Johns Hopkins had been another operation, the child's fourth. Her concern had been her daughter's diminutive stature at age 13 (Fig. 8.4), the height being only 120.5 cm (47 in.). She had expected only hormonal treatment to promote growth and maturation, not surgery to remove the gonads so as to prevent the future possibility of pubertal virilization. 'Guinea pig surgery,' she called it, and returned her daughter to the much-trusted care of Dr. Glasscroft, her private urologist. It was not until after his death, four years later, that the daughter was again seen in the psychohormonal clinic, in need of psychotherapy.

At the age of 18, a trial period of treatment with fluoxymesterone, a growth-promoting anabolic steroid, was instituted, with a view to increasing height. Though not potent as a virilizing agent, this hormone is, technically, an androgen. Its identification as a 'male hormone' made it suspect to both mother and daughter. The upshot was noncompliancy, which left the patient devoid of replacement therapy with any type of sex hormone for several years (Table 8.2).

In her late twenties, the patient moved to another state, where she established contact with the local medical center. There, at age 30, she underwent surgery to release constriction of the vagina. Postsurgically, the depth of the vagina was 8.5 cm, and penile intromission was possible for the first time. Cyclic hormonal replacement therapy was begun with estrogen (Premarin 1.25 mg/day for days 1 to 24 each month) and progestin (Provera, 10 mg/day for days 19 to 25). Under this treatment, menstruation had its onset at age 30. The periods are regular, but scanty. The adult height was 133 cm (4 ft. 4½ in.). The weight had fluctuated from a high of 65 kg (145 lb) to a low of 50 kg (110 lb).

Gender-coded social biography

Developmentally, everything about this girl's social life was dominated by the vast discrepancy between her chronological age and her height age, the latter being extremely deficient. When she was 13, she had the appearance of being 7 or 8; and when 18, of being maybe a prepubertal 10 or 11. In accordance with the principle of 'the tyranny of the eyes,' other people, at home, school, and hospital, invariably juvenilized her, the inevitable reaction being that she reacted as being socially too juvenile relative to chronological age. The mother epitomized the daughter's resultant dilemma of being socially isolated at age 18, by saying: 'It's like watching someone that life is passing by.'

Academic achievement suffered. Each year, she said, she made a new effort, but the pressure built up. She could not concentrate, and then she began to fail again. At age 17:6, the IQs on the Wechsler Adult Intelligence Scale were Verbal IQ, 122; Performance IQ, 115; Full IQ, 120. The corresponding scores on the Wechsler Intelligence Scale for Children had been suboptimal: Verbal IQ, 110; Performance IQ, 94; Full IQ, 103.

Problems at home made everything worse. They increased as soon as she had retrieved the lost early years of her biography (see Lovemap biography) and embarked on achieving more adolescent autonomy. The mother accused her of being obstinate and disrespectful, and of coming between the mother and her husband. The mother had, however, long had a plan to live separately from her husband as soon as the patient, their youngest child, graduated from high school. She implemented this plan, on schedule, immediately after her daughter's graduation party, two months before her 18th birthday.

Three months later the girl was sick with spells of sleepiness, joint aches, and hot flashes. Scheduled for an intravenous pyelogram, she melodramatically fled the radiology clinic when, after two attempts, the intravenous needle failed to penetrate a vein. Her phobic panic and inability to return to a clinic or doctor, despite her illness, did not bring the parents together to take care of her. They separated, and neither of them truly welcomed her. Eventually she gravitated toward her mother. They shared an interest in drawing and painting. She worked in a food business for her mother until, in her mid twenties, she went to business college for management training, which led to computer training, and also to meeting the program analyst with whom she would live and then marry.

Lovemap biography

When her daughter was 13 and having her first psychohormonal interview, her mother did not know what, if anything, the girl knew about sexual intercourse or any other matters of sex. Mostly she relied on her older daughter and daughter-in-law to deal with their sister's quest for sexual information – her interest in menstruation, for example. At the same time, the mother had an image of trying to answer her daughter's sexual questions 'as close as I can. But she's been so small. If she was at the stage where they want boys, and all like that, then I think that's more or less the time to bring up the subject.'

The girl herself said that she knew about pregnancy, but did not know how a baby is born, and had not tried to guess. She knew that a male and female get together before a woman gets pregnant. However, having not seen the male genitals, even on a baby, she did not know how they get together until, in the course of the interview, it was explained.

In responding to a query about having reached the boyfriend stage, she spoke as a juvenile: 'Yes, sir, I have one. I have one at school. There isn't too much to tell about him. He has blond hair and, um, he has kind of dark complexion, and he's real cute.' The boy knew he was her boyfriend, because 'one of the girls told him. I don't think I could have kept it from him too long. I have his name written all over – He saw his name written all over my binder. Of course, I found out he has my name written all over his binder, anyway.'

When, after an interval of nearly five years, she returned to the psychohormonal clinic, her sexual knowledge was adequate, except that she did not know how it applied to herself. Since she had not been able to approach her mother for information, it was supplied in the course of counseling sessions. The topics covered included the function of hormonal substitution therapy, hormonal regulation of menstrual cyclicity, and in the absence of some unforeseen obstetrical breakthrough, parenting by fostering or adoption. By itself alone, this information was not enough to ensure compliancy which was jeopardized by the intensity of the patient's phobic panic when confronted with intravenous hypodermic needles. The phobia did not let up, even though she was able to stick her own finger with a lance for a blood sample.

At 17, she still had no information or recall concerning her early years, before the sex reassignment at age 4:6. 'I don't remember too much when I was about four, but, umm, then I remember everything from that time on.' Recently, however, she had been having dreams backdating to age 3 or 4 – playing in a big back yard, getting a souvenir Peter-Pan hat at a movie, being given china plaques of elephants to put on the wall near her bed, and being in a crib, with her parents coming in and giving her a toy car.

Her search for her own lost biography manifested itself not only in dreams, but also in poetic expression. One of the poems written at this time was given to the counselor two sessions later:

A Lonely Search

I walk a long and lonely path
with not a soul to guide me,
I travel alone, without a friend
and my life is dull and empty.

I wander forever, a searching soul
watching always for what I seek
a wanderer without a goal
searching every hour of every week

My life is but a constant search
only my soul knows what I search for
While it sits on a lofty perch
and I search forever more.

Only when I find my quest
can my restless soul be at peace
But till that time my soul cannot rest
and my wandering cannot cease.

The lost biography had actually been found in a dream, dating from around age 7, recalled but not reported until this present age of 17: 'I had gone to see Dr. Glasscroft. He was examining my arms, and he made some remark about "They were definitely feminine." And that night I had a real wild dream about that, umm, when I was born, I was born a boy; and for some reason I'd started turning into a girl. ... I woke up in the middle of the night. I was shaking. My sister [aged 16] was sleeping in the same room. She sat with me for a few minutes, and I went back to sleep. But, umm, I thought quite a bit about it the next day, and I kept trying to piece things together, things that would point to, and – different things that were said to me, the fact that I had so much hair on my arms, and everything. And I tried to find facts to discredit the idea. ... I thought it was impossible for that to happen – that a boy could change into a girl, or a girl change into a boy.' She did not discuss the topic with anyone: 'I just decided that it couldn't have happened, and that it wasn't worth bothering with, or even thinking about.'

A month later, when the girl had her next counseling session, she was told the story of the embryology of ambiguous differentiation of the genitalia illustrated with diagrams. Then the biographic truth of her dream was confirmed, and explained on the basis of having been born 'genitally unfinished.' The latter term had been first introduced in the previous session. 'I started then to feel,' she said, 'that maybe the dream, maybe I had been raised as one sex, and then they found out I was more of the other sex, and had assigned me the other sex. ... It wasn't very upsetting, except that I wanted to know just what had happened.' Now she wanted to inform her parents of what she knew, 'to ease the burden of what they've been living with for quite a few years.'

That night, she dreamed again the dreams above-mentioned, the content of which dated back to when she had lived as a boy, aged 3 or 4. Previously, she said, 'I saw myself – I saw what was happening [in the dream] through the eyes of the person. ... The night after our last session, the dream was the same, only this time I saw the person – I saw myself, and I realized that it was myself, and I was dressed as a boy. It was almost like looking at a photograph. The next day, after I came home from school, my mother showed me quite a few photographs that had been taken at the time. One of these was the scene that I had seen the night before – the same toys, the same clothes. Everything was exactly the same. ... I was kind of shocked, and then I grabbed the picture, and told my mother that this was the dream that I had had. ... She told me I had just turned three when the photo was taken.'

While the search for the lost biography had been going on, the girl had formulated an explanation for her condition on the basis of high-school biology: 'I feel that it is due either to genetics or to my glands.' As if to pursue more information, she had developed an ambition to become a histologist, though in fact she eventually trained for computer operation. In counseling, she obtained an adequate working knowledge of her clinical history. Before she could be adequately prepared to learn about the chromosomal and gonadal data, however, extreme noncompliancy in keeping both endocrine and psychoendocrine appointments, secondary to morbidly intense phobia of venipuncture needles, had become a major issue and remained so for years.

With the priority issue of the lost biography resolved, the issue of adolescent autonomy rapidly took front place. By age 18, the adversities of being only 130 cm (4 ft. 3 in.) in height, even though the breasts were adequately developed, had restricted further development of romantic life. She herself was interested in boys, she said, but 'most of them tease me, and the others won't date me because their friends would tease them if they did.' One exception occurred when, at a friend's house, she met and became friendly with two mentally retarded boys on leave from the local institution where they lived. Her mother put a stop to that. 'We already have enough of a problem with Eve's condition,' she said. The daughter's response, given in a joint interview with her mother at that time, while crying: 'I hate my mother sometimes. I don't know why. I just can't understand it. I love my mother and all, but sometimes I simply can't stand her.'

To escape from the chaos of feuding within the family, the girl learned to be a hunter with her older brother and went off with him on hunting trips. When she was 19, she met another hunter, Hank, aged 26. They got into kissing and hugging, but nothing genital. 'I have to fight him off a little,' she said, but she believed their relationship had not yet reached the stage of his considering sexual intercourse. The partnership was short-lived and broke up in dispute related to the girl's antagonisms within her family.

In the ensuing decade, there were eventually other romantic attachments. 'I dated quite a few people and had a few serious relationships, both with men and women. ... Before I was thirty, when I had the surgery [second-stage vaginoplasty to release constriction of the vaginal orifice present since first-stage surgery at age 4:6], there was no way I could have a normal sexual relationship with any man. Other than heavy petting, there wasn't really any sex life with the men. With the women, there was a perfectly normal relationship.' She had first become aware of a bisexual potential in high school, but had not talked about it nor acted on it for 10 years or more, until after her heterosexual life came into bloom.

The planning of second-stage vaginoplasty had been delayed by a couple of years, as the financial details could not be worked out. 'I'd been denied health insurance all along,' she said.

Her husband and she had met at work. They are a complete contrast in height, he being over 183 cm (6 ft.) tall and heavily built, and she being 133 cm (4 ft. $4^1/_2$ in.) tall. 'We started dating,' she said, 'and after about eight months, he moved in. We decided to live together for a while to make sure the past medical history was not going to hinder a normal marital life.' Three months later, they married. 'It's working out very well. I couldn't be happier.'

On the subject of orgasm, her estimate was that she achieved it 95% of the time. 'As far as I can relate it to anything else it seems to be perfectly normal. ... I really don't know how I would describe it other than very enjoyable.' It was more diffuse than focused, 'but very intense. ... When we first start, the breasts are more stimulating than the genitals.' (The clitoridean structure had been excised as part of the feminizing surgery at age 4:6.) As self-estimated, her sex drive was average. She could not rate any particular stimulus, or any one of the senses as being especially likely to induce a romantic and sexy mood. 'But I am a rather touchy-feely type person,' she said, 'hugging and such.'

'I do tend to fantasize a lot,' she volunteered, 'and my husband and I do get into some light bondage and things like that. We do experiment with this. ... He'll come into the room dressed in full leather, or something like that, leather cuffs and thongs, tying [either partner] to the posters on the bed. ... It's not necessary, but we find it makes it much more enjoyable. We have gotten together with friends and done bondage scenes together. We have tried various costumes as well as the leather – various uniforms, like a military uniform, pith helmet, or a motorcycle type person wearing leather. And we'll act out a scene with me as the villager, and things like that. We make it up as we go along. ... Our fantasies do mesh quite well. I feel a lot of satisfaction, a lot of closeness, knowing I've brought enjoyment to him, as well as being well-satisfied myself.

'We do somewhat [have an open marriage],' she said. 'We mainly have one basic agreement – he doesn't go to bed with any other women, and I don't go to bed with any other men. Other than that, it's free, because that is one part of our sexuality that the other can't fulfill.'

The couple were interested in the possibility of a family by adoption. 'Actually, we would run into quite a lot of difficulty, trying to adopt,' the wife said, 'considering our past histories of bisexuality. But we haven't tried yet. We've only been married a year.' She preferred the toddler age for adopting, since a child that age would tie her less to the home and interfere less with her outside, salaried career.

EXPOSITION

So far as can be ascertained, this is the first report of its kind in which the cytogenetic status of the subjects is 45,X/46,XY mosaic. Mosaicism originates in a very early embryonic error of chromosomal mitosis which produces two or more cell lines, each with a different complement of chromosomes. When the two cells lines have a different X and Y chromosomal complement, as in 45,X/46,XY mosaicism, the embryological risk of gonadal dysgenesis is increased (Simpson, 1980; Money, 1980). Dysgenesis itself increases the risk of fetal gonadal insufficiency, which in turn affects the fetal differentiation of the genital morphology.

The 45,X/46,XY embryo is at high risk for defective gonadal differentiation, in particular for the dysgenetic formation of a streak gonad of the type associated with 45,X Turner's syndrome. However, there may be sufficient testicular differentiation, at least unilaterally, to permit the secretion of testicular hormones prenatally. In consequence, the mullerian ducts may undergo regression, as expected in a male, under the influence of mullerian inhibiting substance from the testes, although it may be incomplete. The external genitalia, under the influence of testosterone, undergo masculinization, although it also may be incomplete. If so, the genitalia look ambiguous at birth with a phallic organ that could be a large penile clitoris, or a small clitoridean penis without a urethra. The urinary orifice or urogenital sinus is more or less in the female position. Typically, the dysgenetic gonads or gonadal streaks are undescended. There may be associated kidney, urinary tract, or other morphologic defects not under the control of fetal gonadal hormones. Postnatally, the statural growth rate is consistently retarded, and height is deficient relative to age. Adult height is typically under 152 cm (5 ft.). Dependent on the dysfunctional state of the gonads, puberty may need to be hormonally induced.

Genital reconstruction as male or female follows the decision regarding sex of rearing which may vary, more or less arbitrarily, according to the policy of the physician or clinic responsible for the baby. Reconstruction as a female is usually accomplished in two stages: cosmetic feminization of the external genitalia in infancy, and vaginoplasty, as needed, in mid adolescence. Reconstruction as a male is technically more complex and requires at a minimum two or three surgical admissions, but quite often as many as 10 or more, because of the formation of urethral fistulas and strictures.

Ideally, the parents should have medical-psychological counseling (Money, 1968, 1980, 1987) available to them from the time of birth onward, so that they can comprehend the nature of the problem with which they are dealing, and the decisions which they must assist in making, notably the decision regarding the sex of assignment and rearing. They need counseling also with respect to various other topics: what to tell the older siblings, other relatives and friends; the risk of future pregnancies for themselves and relatives; the religious or ethical constraints they may have on contraception; what to explain to the child in later infancy and childhood; how to integrate the child's medical self-knowledge and sex education; and what role to prepare to play during adolescence. Rehearsal of all these issues far ahead of their application is detraumatizing and is essential to the ability of the parents to do an optimal job of rearing.

The matched pair of cases in this chapter replicates what has been found in matched pairs drawn from other intersex syndromes, namely that concordance with respect to prenatal determinants and fetal sexual differentiation does not predictively guarantee subsequent concordance of gender-identity/role, irrespective of the postnatal influences of the sex in which the person is reared and surgically and hormonally habilitated. Rather, there is postnatal discordance of G-I/R, insofar as differentiation is male and female, respectively, in accordance with the sex of rearing and habilitation. There is, however, a proviso.

The proviso is that the comparison of these two particular cases illustrates the long-term, disjunctive effect on G-I/R of a relatively infrequent phenomenon, a forced sex reassignment in childhood – at the age of 4:6 years – in the child changed from boy to girl. It contrasts with the consistency effect in the child who was continuously assigned as a member of only one sex, male. This boy, whose habilitation was continuously masculine, differentiated a consistently masculine G-I/R, small stature and late puberty notwithstanding. The girl, whose name and status as a boy had been abruptly changed, and her phallus amputated without explanation or consent at age 4:6, differentiated a G-I/R that, though predominantly feminine, was subsequently marked by crises and discontinuity in the social, erotic, and genital development of femininity, especially in teenage and young adulthood. The crisis list includes underachieving academically; rediscovering that she had once been a boy; noncompliancy in taking female hormone; being socially isolated; seeking escape as a hunter with her brother, despite the handicap of diminutive size; suffering parent-child behavioral pathology; and being delayed in establishing a love life. In young adulthood she became bisexually able to have male and female partners. Eventually she became heterosexually pairbonded in a mildly paraphilic relationship of light bondage and discipline. Thus was manifested, in erotosexual imagery and practice, a theme familiar in her life, namely being a member of the cast of a sexual drama of someone else's contriving.

In each of the two cases, hormonal puberty of the body was successfully regulated therapeutically to be congruous with the sexual status as male or female at the time. Thus, neither case is informative with respect to the effect on G-I/R of hormonally incongruous development of the body, though on the basis of other cases it is known that a feminine G-I/R can transcend masculinization of the body, and a masculine G-I/R, feminization of the body at puberty. The hormonal changes of the body at puberty do not automatically correlate with, or dictate the sex of the partner toward whom the person will be attracted in either fantasy or actuality, witness the evidence of numerous cases of hermaphroditism and transexualism, and hundreds of thousands of cases of ordinary male and female homosexuality. The role of pubertal hormones in the male/female dimorphism of erotosexual attachment is not to cause or create this dimorphism but to activate the masculinity or femininity already differentiated and waiting to be expressed in either imagery or behavior.

Hormones, insofar as causality may be attributed to them at all, would exert their influence prenatally, according to the evidence of experimental animal studies and clinical human observations (reviewed in Money, 1988; Money and Ehrhardt, 1972; see also Crews, 1987). In the present two cases, one may infer, on the basis of the gonadal findings, that each fetus had a similar endocrine history of incomplete androgenization not only of the external genitalia, but also of the brain. This hormonal effect did not rigidly impose on the brain either a male or female stereotype. Though there is no proof one way or the other, it may have left the brain more dimorphically flexible than usual, and postnatally unrestricted in continuing its differentiation as either male or female under the impact of the sex of assignment (and, in one case, reassignment), rearing, and surgical genital reconstruction. What effect, if any, incomplete prenatal androgenization may have had on masculinizing the role of erotosexual fantasy and turn-on imagery in the erotosexual practice of adulthood is a question that can only be asked, not answered. The personal content of this imagery, whether masculine or feminine, is attributable to early life experience rather than hormones, as is indicated especially by the bondage and discipline imagery evident in the lovemap, in the case of the girl.

Bondage and discipline, whether of the light or heavy variety, is one of the 40-odd paraphilias (Money, 1984, 1986) that may enter into the formation of a lovemap. Its occurrence in this case has special significance for the theory of paraphilia insofar as information relevant to its formation was collected prospectively in a series of children being followed in psychohormonal study. It was not anticipated that paraphilia would be one of the outcomes, which it has been in half a dozen or more cases (see also Chapter 3, Case #2). In each case it is possible to recognize a conceptual connection between the theme of the paraphilia and the biography of sexuoerotic development. In retrospective studies of paraphilia in adulthood, such developmental information may be irretrievable, as well as difficult to authenticate.

The endocrine control of growth in height is still so imperfectly understood that, as in the present two cases, statural dwarfism frequently cannot be prevented. A big discrepancy between height age and chronological age puts a great strain on the maturation of all components of social age, for people in general unthinkingly obey the 'tyranny of their eyes' and juvenilize the one who is short. Probably on the basis of preventive counseling, the boy of the present pair manifested a continuously good developmental alignment between birthday age and social age. The girl, whose counseling was corrective, manifested developmental disalignment sporadically, until young adulthood. Then the two ages became more parallel than would, with such a history, typically have been expected.

In the case of the boy, counseling began with the parents when the child was newborn and continued progressively as he developed. It was designed to be preventive and health-predicting, and its overall outcome did, in fact, appear to be successful. In the case of the girl, counseling began late and was, of necessity, instituted as corrective, not preventive. Inevitably it was sporadic and crisis-directed instead of progressive and developmental. It may well have been ameliorative, but it was not effectively remedial. The behavioral disability and suffering of adolescence did eventually resolve, though slowly, over a period of years.

From the point of view of preventive psychiatry, it goes without saying that a planned program of preventive counseling is superior to ad hoc, crisis-directed corrective counseling, despite the imperative urgency and necessity of the latter. Preventive counseling is not improvisational, but is designed ahead of time. It is designed on the basis of a knowledge of the syndrome independently of any particular patient, but accommodative, in its application, to individual and family idiosyncracy. The demand on counselors is, 'Know thy syndrome.'

In an intersex syndrome, as exemplified in this report, the case-management program must incorporate the functions of specialists (whom children readily recognize as the talking doctor, the needle doctor, and the cutting doctor), and it must begin neonatally, sometimes literally in the delivery room, with the parents. There must be unanimity among the specialists regarding the contents of the program, and the contributory sexological roles of medical psychology, endocrinology, and surgery in implementing it. Without this unanimity, the management of the case all too readily degenerates into what might be termed 'doing one's best,' but all too often deserves also the appellation 'hospital child abuse,' in general, or 'hospital sexual abuse' in particular. Multiple hospital admissions easily become experienced as hospital abuse by children who are not entrusted with predictive information about what will happen to them, and who are not consulted regarding elective options. The annals of hermaphroditism and birth defects of the sex organs are, unhappily, replete with examples. The two cases here presented show that hospital abuse can be circumvented in hermaphroditism, even when multiple genital surgery is part of the history. The prognosis then can be very good.

BIBLIOGRAPHY

Bergada, C., Cleveland, W.W., Jones, H.W. Jr. and Wilkins, L. (1962) Gonadal histology in patients with male pseudohermaphroditism and atypical gonadal dysgenesis: Relation to theories of sex differentiation. *Acta Endocrinol.*, 40:493–520.

Crews, D. (Ed.) (1987) *Psychobiology of Reproductive Behavior: An Evolutionary Perspective*. Englewood Cliffs, Prentice Hall.

Jones, H.W. Jr. and Zourlas, P.A. (1965) Clinical, histological, and cytogenetic findings in male hermaphroditism. *Obstet. Gynecol.*, 25:597–606.

Money, J. (1968) *Sex Errors of the Body: Dilemmas, Education, Counseling*. Baltimore, Johns Hopkins University Press.

Money, J. (1980) Hermaphroditism and pseudohermaphroditism. In: *Gynecologic Endocrinology*, 3rd edn. (J.J. Gold and J.B. Josimovich, Eds.). Hagerstown, Harper and Row.

Money, J. (1984) Paraphilias: Phenomenology and classification. *Am. J. Psychother.*, 38:164–179.

Money, J. (1986) *Lovemaps: Clinical Concepts of Sexual/Erotic Health & Pathology, Paraphilia, and Gender Transposition in Childhood, Adolescence, & Maturity*. New York, Irvington.

Money, J. (1987) Psychological considerations in patients with ambisexual development. *Seminars in Reproductive Endocrinology: Disorders of Sexual Differentiation* (J. Rock, Ed.), 5:307–313.

Money, J. (1988) *Gay, Straight, and In-Between: The Sexology of Erotic Orientation*. New York, Oxford University Press.

Money, J. and Ehrhardt, A.A. (1972) *Man and Woman, Boy and Girl: The Differentiation and Dimorphism of Gender Identity from Conception to Maturity*. Baltimore, Johns Hopkins University Press, pp. 151–160.

Richards, B.W. and Stewart, A. (1978) XO/XY mosaicism and non-fluorescing Y chromosome in a male. *Hum. Genet.*, 45:331–338.

Simpson, J.L. (1980) Disorders of sex chromosomes and sexual differentiation. In: *Gynecologic Endocrinology*, 3rd edn. (J.J. Gold and J.B. Josimovich, Eds.). Hagerstown, Harper and Row.

CHAPTER 9

Concordance for psychosexual misidentity and elective mutism: sex reassignment in two cases of 46,XX congenital virilizing adrenal hyperplasia (CVAH)

SYNOPSIS

The concordance of these two cases originated even before conception, insofar as, in both cases, each parent was a carrier of a single unit of the genetic program responsible for a deficiency of the hormone-synthesizing enzyme, 21-hydroxylase, which in turn is responsible for the syndrome of congenital virilizing adrenal hyperplasia (CVAH), otherwise known as the adrenogenital syndrome (Migeon, 1989; White et al., 1987a,b). The sperm of the father and the ovum of the mother in each case bore this genetic program so that, when sperm and egg united, the two copies of the program also united. This reduplication ensured that the developing baby would not, like its parents, carry the program recessively and invisibly, but actively, in a full production of the CVAH syndrome.

In addition to bearing the CVAH gene, both the sperm and the egg were, in each case, bearing an X chromosome so that, when they united, the chromosome count was 46,XX. Thus the two cases were concordant for female chromosomal sex. In embryonic life, the onset of sexual differentiation was feminizing. Bilaterally, the bipotential gonadal tissue differentiated as an ovary. With no mullerian inhibiting hormone (MIH), which is of testicular origin, the bilateral mullerian ducts developed into a uterus and tubes. By contrast, with no testicular androgen to stimulate them, the bilateral wolffian ducts atrophied, so that no masculine internal organs were differentiated.

It was at this stage of sexual differentiation, during the third month of embryogenesis, that the hormonal secretion of the glandular tissue of the adrenocortices had its onset. Anatomical feminization was interrupted, and both cases became henceforth concordant for the masculinizing effect of the CVAH syndrome, from the masculinizing secretion of the adrenocortical glands. Genetically programed to be deficient in 21-hydroxylase, the adrenocortices failed to synthesize their glucocorticoid hormone, cortisol, and in its place substituted another steroidal hormone, androstenedione, which, although similar in biochemical structure, differs in being bioactively able to induce masculinization. This masculinizing hormone is timed so as to be able to influence the sexual differentiation of the dual-purpose tissues which are the precursors of the homologous external sex organs, male and female. The consequence, in each of these two cases, was that instead of a female vulva with clitoris and divided labia minora and majora, the external genitalia were masculinized. Their masculinization was not, as it is in some CVAH cases, so complete that a clitoris and unfused vulva had become a penis and fused, though empty, scrotum. In these cases, the clitoris was hypertrophied so as to be the equivalent of a hypospadiac penis. This hypospadiac organ had a clitoridean hood instead of a full foreskin, and was held, bound down by divided labia minora that had failed to wrap around its shaft to form a covered urethra. The urethra was located at its base, in more or less a feminine position, opening into a urogenital sinus, or funnel, within which was also concealed the opening of a vagina. Below this sinus, there was midline fusion, so that the outer lips, the labia majora, formed a quasi scrotum or labioscrotum. It was empty, since there were no testicles to descend into it, and the ovaries were in their regular, intraabdominal, female position.

Prenatal concordance on the criterion of hormonal masculinization applied not only to the period of external genital differentiation in the second trimester of pregnancy, but also during the third trimester. It is during the third trimester, not before (Abramovitch et al., 1987), that hormonal masculinization of the brain takes place, with a possible extension into the early neonatal period. Thus, by inference, these two cases may be considered concordant for a degree of masculine dimorphism in the brain.

At birth, as a sequel to concordance for ambiguous appearance of the genitalia, there was a corresponding concordance for indecision regarding the baby's declared sex. In Case #1, the declaration was changed from girl to boy at age six weeks. In Case #2, it was changed from boy to girl at age two weeks. Concordance for ambiguity of genital appearance continued throughout childhood as there was no surgical intervention to correct the ambiguity until age 12 in Case #1, and age 11, in Case #2.

In the absence of early surgical correction of the genital ambiguity, there was continued concordance between the two cases insofar as there was incomplete commitment to the sex in which each child was ostensibly being reared. In Case #1, the rearing was ostensibly as a boy, and in Case #2 as a girl.

Despite the discordancy of the declared status of the boy and girl in society, their cases continued to be hormonally concordant during childhood, for the masculinizing influence of adrenocortical androgen was only sporadically suppressed by glucocorticoid replacement therapy, partly through failure of clinical follow-up, and partly through noncompliance.

During childhood, there was an additional manifestation of concordance, for in each case the child differentiated a gender-identity/role that was not fully consistent with the ostensible sex of rearing. The inconsistency in each case became, quite literally, an unspeakable issue. Its unspeakability qualified diagnostically as elective mutism. The outcome was concordance of the two cases for sex reassignment, in Case #1 to live as a girl, and in Case #2 as a boy.

CASE #1: SEX REASSIGNED AS GIRL

Diagnostic and clinical history

In the pediatric endocrine clinic and psychohormonal unit at Johns Hopkins, this patient was unknown until, at age 12:5, the case was transferred from another hospital where an impasse had been reached on the issue of sex reassignment from boy to girl. A few months earlier, in compliance with advice given at the other hospital, the mother had told the child that he could decide to remain a boy, or change to be a girl. 'I couldn't talk about it, then,' the patient said when looking back, 12 years later. 'I used to be ashamed of it. But I'm not anymore.' As a boy of 12, he had experienced a degree of elective mutism so selectively complete on matters of male and female that nothing could be verbalized, not even optional sex differences in haircut or clothing style.

Details of the birth and prior clinical history were given by the mother. She was aged 22 when the baby was born, and the father 30. The baby had two older siblings, sisters, and subsequently would have four more sisters and three brothers, none with the CVAH syndrome. The children's paternity was distributed among four fathers. The obstetrical clinic in which the pregnancy had been followed had no beds available, and so sent an obstetrical resident to deliver the baby at home. There were no complications. The birth weight was 3448 g (7 lb, $10^1/_2$ oz). The mother recalled having been told by the doctor who did the delivery that the protruding, inch-long genital organ would correct itself. The baby was given a girl's name.

There is no record of the baby's hospitalization at age six weeks, nor whether it was CVAH related, but at that time a doctor told the mother that the child was a boy, not a girl. She sought confirmation from the hospital responsible for the delivery, and thereafter changed the name from Deborah to Stanley, and raised the child as a boy. She returned to this same hospital when the baby, as a 2 year old, had grown too big for his age. He was admitted for surgery, an exploratory laparotomy, to examine the pelvic reproductive organs. The mother did not recall having been told what the findings were. She took the child home without medication and continued raising him as a boy. She took him back to the same doctor when he was 5, and recalls only that she was told not to worry, but to postpone starting school until after he had been operated on again. He was at the time showing the virilizing signs of pubertal precocity, namely pubic and axillary hair, and excessive height of 91 cm (3 ft.).

Though the proposed operation did not take place, the mother did keep the boy out of school until age 8. When he was 10, his mother returned with him to the pediatric clinic where he had been seen at age 5. He was admitted for three months. This time, the diagnosis of CVAH was for the first time correctly recognized, and he was begun on treatment with cortisone, 25 mg by mouth, twice a day. As the mother recalled it, when the medication ran out after six months or thereabouts, and she did not have the money to buy more, the doctor told her that the boy could get along without it. Evidently the duration and dosage of treatment had been sufficient to permit the ovaries to become pubertally activated, for by age 10:6, the child's breasts had begun to enlarge.

During the ensuing two years, there was no treatment for CVAH, only for minor intervening illness on two occasions. In addition, over the course of about a year, following the inpatient admission, there were outpatient sessions with a children's psychotherapist. The child's severe affliction with elective mutism was unremitting, so that the sessions were terminated. Soon thereafter, a house staff officer in the pediatrics clinic suggested that the mother transfer the case to the Johns Hopkins pediatric endocrine specialist, Lawson Wilkins, who had codiscovered the then 6 year old hormonal treatment of CVAH with cortisol.

At age 12:5, the height age was 13:9, the height 153 cm (5 ft. $^1/_4$ in.), and the weight 56 kg ($124^1/_2$ lb). The body build was described as stocky and muscular, with broad shoulders, small hips, and no excessive fat (Fig. 9.1).

The entry in the chart under the heading, sexual development, is as follows: 'Acne and seborrhea of the face. Thick growth of axillary hair. Pubic hair is curly and of very heavy growth. Phallus measures 9×2.5 cm ($3^1/_2 \times 1$ in.). There is a slight depigmented groove on its undersurface, and an opening of 0.5 cm at its base. There is labioscrotal fusion [Fig. 9.2]. No gonads are palpable. On rectal examination I can palpate two masses. I think they are uterus and cervix. The breasts are well developed with glandular tissue and measure $12 \times 12 \times 4$ cm. Areolae 4 cm, and hyperpigmented. Nipples stand out.' The buccal smear test proved to be chromatin positive, thus indicating a 46,XX chromosomal type.

The diagnosis was established as CVAH, and was confirmed when the lab results were returned, showing elevated levels of adrenocortical androgen as indicated by

Fig. 9.1. Case #1. Appearance after the onset of hormonal feminizing and prior to feminizing genital surgery.

Fig. 9.2. Case #1. Preoperative genital appearance.

Fig. 9.3. Case #1. Postoperative genital appearance.

Fig. 9.4. Case #1. The written message with which the obstacle of elective mutism was surmounted.

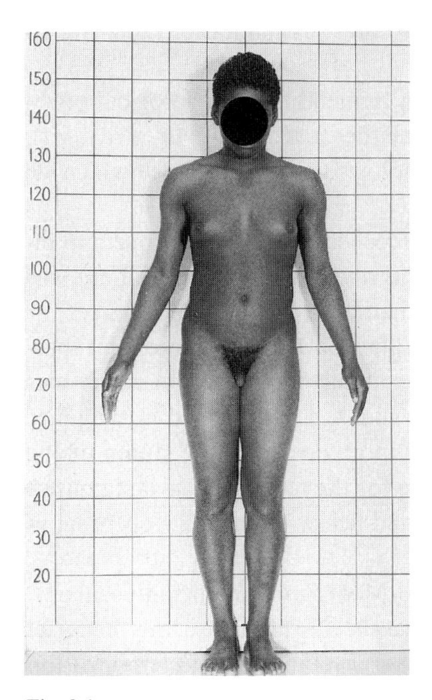

Fig. 9.1

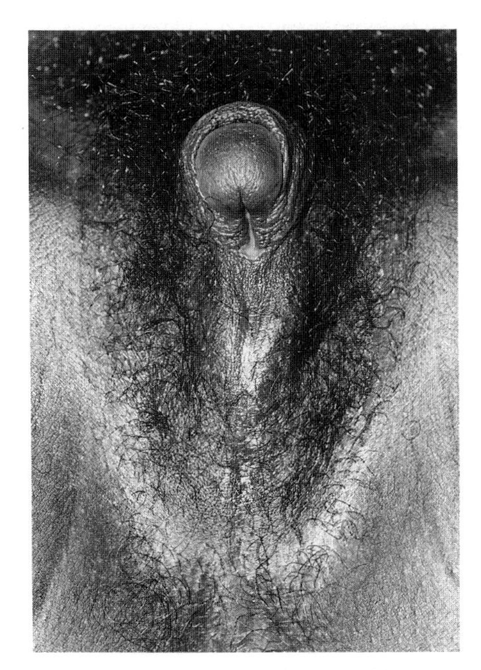

Fig. 9.2

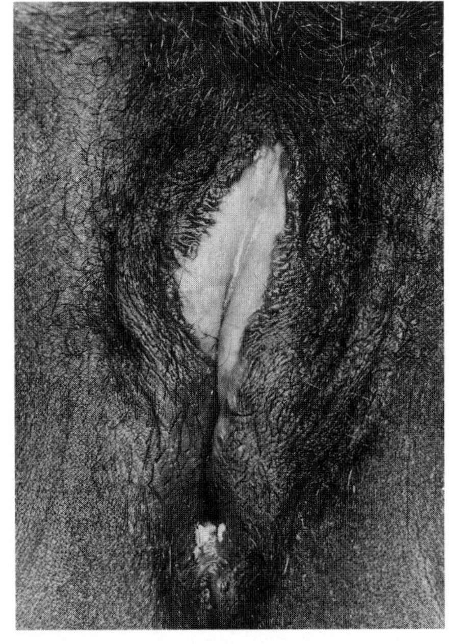

Fig. 9.3

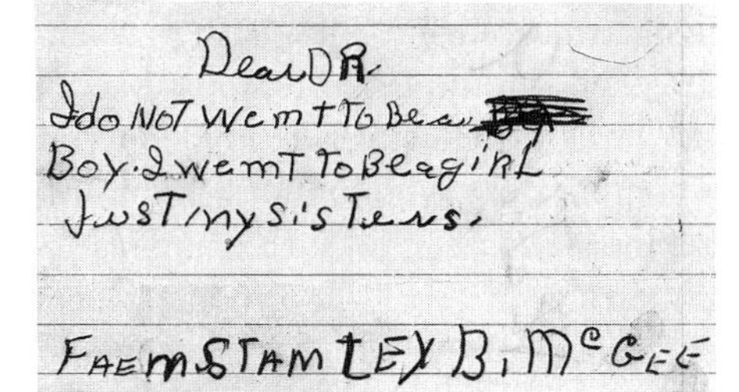

DearDR
Ido NOT wemt To Be a ~~Boy~~
Boy. I wemT To Be a giRL
JusT my sisTers.

FAEM STAMLEY B. McGEE

Fig. 9.4

the measurement of 17-ketosteroids (51.7 mg/24 h) and pregnanetriol (26.2 mg/24 h).

The examining pediatrician noted that there is 'a tremendous psychological problem – he has been told by his mother that he is neither a real boy nor a real girl, even though raised as a boy. It will require psychological probing to see which he should be.'

The details of the psychological evaluation are recorded below in the Lovemap biography. The evaluation was spaced over time. Thus it was not until the patient was 12:11 years old, six months after the referral examination, that the stage was set for an admission to regulate adrenocortical hormonal (glucocorticoid) replacement therapy, and to perform surgical feminization of the genitalia. The operative result was satisfactory (Fig. 9.3). For cosmetic appearance, a skin tab was formed in the place of the clitoris, though erotic sensation was located more deeply in the stump of the amputated organ. There had been no suitable tissue for the formation of labia minora. The vaginal opening was exteriorized and situated below the urinary opening, in a satisfactory position for sexual intercourse.

In the early period of hormonal treatment (Depo-Medrol, 60 mg intramuscularly, semimonthly), dosage regulation and regularity were successful in inducing the onset of menstruation and bringing about very good breast development and feminization of appearance. Then and subsequently, however, it has been consistently difficult for the patient to maintain compliancy in providing urine samples for the measurement of 17-ketosteroid levels against which to gauge the dosage level of replacement hormone. Consequently, blood levels of adrenocortical hormones have varied erratically, and there have been periods of weeks, months, and even years, when the patient has been off treatment completely. Without treatment, adrenocortical masculinization resumes. More through good luck than good management, the patient's body has not responded by becoming heavily virilized in voice or body and facial hair, as might have been the case. Overall, the maintenance of health as directly related to adrenocortical function has been problem free, except for pregnancy failure.

It is difficult, though not impossible, for CVAH women whose adrenocortical hormone levels are effectively regulated to become pregnant, providing they do not have the complicating symptom of being salt losers (Mulaikal et al., 1987). In the present case, there had been no such complication, and the patient very much wanted, after her marriage at age 17, to become pregnant like her older, next sister, who was her feminine model. At age 18:6, she came to the clinic in maternity clothes, having not had a period for three months. She had attacks of nausea, headache, and dizziness. Her skirts were not fitting well, as she had gained five pounds, with an increase in abdominal girth. When the verdict, 10 days later, was negative, she was as if bereft. In the ensuing years, she failed to become pregnant, possibly because of the ever recurrent problem of irregularity regarding treatment.

Over the years, there was a procession of general health problems of varying degrees of severity and disablement, some readily treatable, and some, of unascertainable etiology, chronically recurrent. The most disabling, recurrent symptoms have been severe headache, possibly of migraine type, and arthritic joint pain, especially in the neck, arms, and hands. The list includes dizzy spells or syncope, once with a fall down stairs; right-sided numbness and paresthesia; leg cramps; blurred vision; fatigue and sleeplessness; palpitations, sweating, shakiness, and breathlessness; labile hypertension; swelling of the right face and neck; polydipsia and nocturia with urinary tract infection; nausea, vomiting, and burning sensation in the stomach; vaginal discharge; and right inguinal hernia, surgically repaired. At age 25, the VDRL antigen test for syphilis was weakly positive, and a course of antibiotic treatment was successfully completed. Comprehensive diagnostic workups and specialty clinic referrals on different occasions failed to reveal a specific etiology for the symptoms, except for those that were infection derived, those that might have been sequelae of an automobile accident, and those that were surgically treatable, e.g., hernia. The origin of other symptoms as being those of a post-traumatic shock syndrome would not be ascertained until the patient was 30 years old. Then, nine years after the event, she was for the first time able to reveal the trauma of a brutal gang rape (see Gender-coded social biography).

The patient belongs in the first generation of CVAH patients with a long-term history of steroid-hormone treatment. The prevalence of symptoms similar to hers in the CVAH population at large is not yet known. Thus, some of her symptoms may be syndrome shared, whereas others are undoubtedly specific to the health stresses of her personal history.

Gender-coded social biography

At the time of the initial referral for sexological evaluation in the Psychohormonal Unit, the patient as a boy aged 12:5 alternated his surname between that of his natal father and that of his stepfather. He was one of a loyal and close-knit family of brothers and sisters, stepbrothers and stepsisters, all from the same mother. He could count on their allegiance, regardless of whether he lived as a boy or changed to live as a girl, regardless of the repercussions on themselves. He could count, in addition, on the unconditional acceptance of his mother who, unlike many other parents, did not become exasperated and impatient in the face of the void in communication imposed by the child's symptom of elective mutism.

During the years of boyhood, he mostly stayed to himself rather than socialize with outsiders. His mother's employment as a domestic left him to his own resources for long periods of time. She recalled that he would spend many hours playing solitaire. There were times when he could not be found anywhere in the house, and then would mysteriously reappear from the top floor. Presumably he had found a way out onto the flat roof, where he could be alone, but he did not tell where he had been. Outdoors, the layout of the city row houses left little play space for children, except on the sidewalks and streets. There, on some occasions, he would join with his brothers and sisters and other children to dance and sing, and to play dodgeball or basketball. He had had no great inclination for football, climbing, or other energy-expending activities that would, in a girl, be called tomboyish.

When he was 10, because he was so big, the school authorities transferred him from the school where at age 8 he had begun his education to an all boys' school. There he had the privilege of using the bathroom unaccompanied. That gave another boy, aged 11, the opportunity to follow him and 'make him take his pants down and use him from the back.'

It was this episode that precipitated his return to the hospital and the three-month admission. In the hospital he had a visiting teacher, and continued to do so at home for the ensuing year. The teacher's visits lasted only 20 minutes twice a week, so the mother initiated a change, and the child was enrolled in an opportunity class. By this time he was 12 years old. He had had breast enlargement for over a year. For the same period of time, he had known from his mother who had told him, on the advice of his psychotherapist, that he could make a decision about his sex. When she asked him about being a boy, 'he smiled,' his mother said, 'but when I asked him about being a little girl, he burst out and cried.'

As time went on, there were signs that he was testing out the possibility of being a girl. The teacher of the opportunity class had a conference with the mother and told her that the child would shy away from all the other children and go in a corner alone. She thought there were ways in which he acted more like a girl than a boy.

At home there were some occasions when he would play with dolls. On one occasion, his younger sisters surprised him in the bathroom trying on his older sister's clothes. The mother had never talked to these children about his condition – only to her oldest daughter who was then married and away from home.

At school, in the opportunity class, there was one occasion when the boy was victimized by another boy. The source of the trouble was the exchange of Christmas gifts. A girl chose to give her gift to the patient instead of the boy who had eyes for her, so he and his rough gang of friends out on the street attacked the patient several times with stones and bottles. One of them told the mother that they were picking on her son because 'he don't act like no normally boy is supposed to act, and he don't walk like no normally boy is supposed to walk.'

The awkwardness of being in transition from boy to girl was resolved with sex reassignment as a girl, and the patient took a girl's name, Deborah, originally given at the time of birth. To circumvent further victimization during the early period of rehabilitation as a girl, a plan was outlined whereby she would be discharged from the hospital to live in the country with a maternal cousin. There she would be able to enroll at the end of the summer in a school where she would have no history of having been a boy. This plan failed to materialize, for the patient herself elected against being separated from the teacher whom she liked in the opportunity class, in which she was catching up academically. Though the school system was not geared to receiving a child who had changed her sex, the initial perturbation gave way to accommodating to the girl's own resolve not to hide what she had accomplished.

The level of academic achievement was modest. There was a lingering persistence of elective mutism which interfered with communication and verbal learning. This interference showed as a discrepancy between Verbal IQ and Performance IQ. In serial testing, the IQs were as follows:

Chronological age	12:6	13:0	24:3	34:4
Wechsler Test	WISC	WISC	WAIS	WAIS-R
Verbal IQ	71	71	84	82
Performance IQ	85	86	94	98
Full IQ	75	76	87	87

Schooling continued until, at age 17, the girl withdrew from her 11th grade class to get married. Her husband was older than herself by seven years, and she had known him for five years. He had just returned from military duty. He wrongly estimated his readiness to become settled, and after about 18 months quit the relationship, except for rare visits, for example on her birthday.

After the marriage was discontinued, she returned to live at the house of her mother and stepfather. Over the ensuing years, she lived there for long periods of time, leaving on occasion to share an apartment with a sibling or friend.

Until she was 22, with some supplemental income from social service sources, she supported herself in various types of employment – nursing-home assistant, barroom waitress, lunchroom waitress, dry-cleaning clerk, diaper-laundry attendant, and babysitter. The employment she most consistently returned to was babysitting and the care of children. 'The only thing I ever found that I liked,' she said, 'was working in a hospital. And mostly, working around children.' Various health problems prevented long-term employment. In recent years she has received a disability pension from the Social Security Administration.

Not only the ability to earn a living, but the entire course of her life and health was catastrophically affected when, at the age of 22, she was lured from a neighborhood tavern by a woman who proved to be the accomplice of three sadistically assaultive men who brutally gang-raped her, vaginally and anally, at gunpoint. The post-traumatic shock syndrome that ensued has proved to be permanently incapacitating, with only slight improvement with the passage of time (see Diagnostic and clinical history). The event became subject to a return of elective mutism, and was totally unspeakable for nine years, at which time she disclosed it to her trusted psychohormonal counselor, a woman. It was not for another seven years, however, that she could speak about it fluently. Her account was as follows.

'I had my own apartment. I was living on Forest Hill in the 1600 block, and I was working. I had got me a job. I had just got off from work. I didn't go straight home. I went to a club where most of the people that I knew hung out. At the time when me and my boyfriend were together, that's where he had taught me how to shoot pool and everything. That's why I used to go there. There was a lady – there were quite a few people in there – but there was a lady sitting at the bar. She had started talking to me and asked me could she buy me a beer. I told her no because I didn't drink. She offered to buy me a soda, so I sat there and I talked with her. I drank my soda. I had told her that I had to leave because I had some place else to go, and I was going home to change my clothes. I'm trying to get it all coming back to me. I was living at home with my mother, but it was only because me and my boyfriend broke up, and I had went back to my mother. I left the bar that I was in, and I went home and changed clothes, because she had asked me to meet her back at the bar where she was. I told her I would, and I went home, took my bath, and changed from my uniform and stuff and put on some street clothes. I went back to the bar, and she was still sitting there. We talked for a few more minutes, and she asked me to go home with her while she changed her clothes.'

THIS WAS NOT SOMEBODY WHO WAS IN WITH YOUR FRIENDS, NOT ANYBODY THAT YOU KNEW? 'No, I had just seen her when I went into the bar that evening. We left the bar to go to her house. I don't know if we were on Clinton Avenue or if we were on Heather, because they're kind of close and the houses look the same just about. I don't know what block we were in. We were in the house not even a good half hour, I think, and she was taking a shower. I think she was because I heard the water running, and there was somebody at her door because they kept ringing the bell. I went to the bathroom door to tell her somebody was at the door. She asked me to open it, and I opened it. There was three guys at the door. She took so long in the bathroom, I was getting ready to leave. I hollered through the door, and told her I had to go because she was taking too long. When I tried to leave, they wouldn't let me.'

THE MEN? (pause: 17 seconds) AND THEN WHAT HAPPENED? (pause: 38 seconds) 'Excuse me. (pause: 13 seconds) All the guys, they raped me, and they kind of beat me up real bad. I didn't fight very much because one of them had a gun.' I REMEMBER YOU TELLING ME BEFORE WHEN YOU TOLD ME ABOUT THIS, THAT THEY HAD RAPED YOU VAGINALLY AND ANALLY. 'Yes.' WHERE WAS SHE WHEN ALL THIS WAS GOING ON? 'She was there in the room.' WATCHING? 'Yes. I can't remember how I got home. I didn't go to my mother's. I wasn't at my mother's when I came to myself. I was at the apartment where I used to live with my boyfriend. I don't know how I got there. I guess I must have went there because it was closer, being in the 1600 block of Forest Hill Avenue. Somehow I had gotten into my ex-boyfriend's apartment. I don't know how long I stayed there.' WAS ANYBODY THERE? 'No. I still had the key, but he very seldom was there.

'I think I waited a couple of days. I don't quite remember. I knew I went and had myself checked out by a doctor. I didn't go to a regular hospital. I went to the health clinic. I didn't tell them anything about what had happened. I just told that I thought there might be something wrong with me, and that I wanted to be treated, for venereal disease or anything like that.' AND YOU DIDN'T TELL THEM YOU HAD BEEN RAPED? 'No.' AFTER YOU WENT TO THE CLINIC, YOU WENT BACK TO YOUR BOYFRIEND'S APARTMENT? 'Yes, I still had a few things there. I got everything that was mine together, and then I went to my mother's. I remember at the clinic, the health clinic – I went there, and I can't remember if I was waiting for the doctor to see me or he had already seen me and I was in the waiting room or whatever, but I know I passed out for like 15 seconds. I was just out.' WAS THIS THE NEXT DAY AFTER THE RAPE? 'No, I think it was like a couple days. I'm not sure. I can't really remember. After I left the clinic, I know I got the rest of my things from the apartment, and then I went to my mother's. That's where I've been ever since. I was getting myself together to explain to her why I was going through all the changes I was going through.'

O.K., WHAT KIND OF CHANGES WERE YOU GOING THROUGH? 'I was afraid to even be in the house by myself without her being there. One time she was the only person that I would go out with, anywhere, for a very long time.' WAS THERE ANY TIME DURING THIS PERIOD WHEN YOU COULDN'T GO OUT, EVEN WITH HER? 'It was a while before I – when I first went home, I spent all of my time in my bedroom.' I REMEMBER YOU SAID YOU SPENT ALL YOUR TIME IN YOUR ROOM, BUT YOU HAD TO KNOW THAT SOMEBODY ELSE WAS IN THE HOUSE WITH YOU. 'Yes. She had to work and everything, but my sisters were there, my brothers. As long as they were in the house, the changes that I was going through when she would be out of the house, she would make them stay home with me, even though she didn't know what was wrong with me. She would make somebody always be in the house. I don't even think I went nowhere. The first time I came out with my mother, I think, she had made me come over here.' TO THE HOSPITAL? 'Yes, she made me come here. After that first visit over here, and I had talked to you all about what was wrong with me, it got a little bit easier but not much. I wouldn't go anyplace with anyone but my mother. Now if she's unable to go, I will go with my father. He'll go places with me, if I have to go somewhere.' BUT IT ALWAYS HAS TO BE SOMEBODY OLDER, I MEAN A GROWN-UP, AN ADULT? 'An adult, yes. Somebody that I feel like I can trust. If I feel like I can't trust them, I won't go anywhere with them. It's only just those people, my mother and my father, and I have a niece – well, she's more like a sister to me. She's thirty years old now. If I have someplace to go, she'll go with me, if can't nobody else go.'

In the aftermath of the gang rape, the patient's relapse into elective mutism for everything concerning it made it impossible for her to explain to her mother why she had returned to live at home, and why she was phobically housebound. After she had been at home for four months, she experienced the trauma of losing one of her brothers with whom she was quite close. Under circumstances that could not be explained, he was killed on the street by one of his companions. Seventeen months later there was another blow. Her favorite cousin, with whom she had shared the closest attachment since infancy, was killed as a consequence of an interracial love affair. His white girlfriend's brother fired a gun at his genitalia and killed him. The patient had foreseen this possibility, and had warned her cousin, but to no avail.

Following this tragedy, the mother observed that the patient became increasingly quiet and reclusive, and that she would act like a young child, playing with children and talking with them, but not with adults. For long periods she would sit and suck her finger. The mother's patience and tolerance gave out when she discovered that the patient had not been menstruating for two years during which time she had been secretive about not having replenished her supply of medication (prednisone) and about having been too phobic to keep clinic appointments. A month after the cousin's death, the mother herself took time off from work and accompanied her daughter to the clinic. The patient would undergo another seven years of undiagnosable mental and physical disablements, however, before her first words regarding the gang rape would be uttered. Only then would the multiplicity of symptoms begin to fall into place as all belonging to one syndrome, namely, of post-traumatic stress. Only then would she begin the tediously slow process of recovering which still has not been sufficient to allow her to lead a normal adult life socially and occupationally.

Lovemap biography

In childhood, while living as a boy, this child grew up as a participant of a black urban culture in which it is virtually certain that juveniles will learn about sexual intercourse, if not from catching sight of it, then from talking together. Large and well-developed because of his CVAH syndrome, at age 9, he was forcibly made aware of anal intercourse by an 11 year old boy at school.

At age 12:5, on the occasion of his psychohormonal intake interview, elective mutism was all-pervasive. His only responses were minimal nods and gestures, but he was attentive as the interview with his mother was designed for him as the listener. His mother had not talked to him about reproduction and was not aware of what he might know. Thus, with diagrams, the basic facts were presented.

Subsequently, I wrote: 'I spoke very straightforwardly, calling a spade a spade, as I explained his condition to him. I told him that he had female parts inside and that, if he would take cortisone medicine all his life, it would be possible for him, if he decided to live as a woman, to be able to go with a man and get a baby. By contrast, if he decided to live as a man, it would be possible for him to go with a woman and have a relationship (the word was supplied by the mother) which I spelled out as meaning the same as screwing or fucking or, in medical language, having intercourse. He would be able to have intercourse as a man, I explained, but he would not be able to let the woman get a baby. I explained that his doctors considered it very important not to admit a patient into the hospital and force him into being corrected as a male or a female without going in the direction of his desires. It was important for him to find out whether he would fall in love with a girl, or whether he would fall in love with a boy. He nodded his head to indicate that he had not yet fallen in love with a girl, and again nodded his head to indicate that he had not yet fallen in love with a boy.

'It was his mother's opinion that it would be easier for him to stay a boy. I took a sheet of paper and wrote two headings: man, woman. Under each was listed: yes, no, I don't know. After ten minutes of studying the page he almost furtively marked it and then, with embarrassment, allowed me to look. He had checked 'I don't know' in both columns. ... It is necessary for him to discover himself,' I wrote. 'He is at a very immature state, psychosexually. Today was the first step in self-discovery. He will require more interviews and psychological sessions.' They were scheduled weekly.

In sessions three and four, an attempt to break the barrier of elective mutism with the help of hypnosis proved ineffectual. In session five, at the outset, he spoke briefly about having attended school for the morning session only and having made an Easter basket. Then the trance-like state of silence returned. There was no mention of the written note (Fig. 9.4) that next morning I would find on the carpet, almost hidden under the sofa where he had been sitting. Though not quite literate, its message was unequivocal: 'Dear DR. I do not wemt to Be a Boy. I wemt to Be a girl Just my sisters. FAEM STAMLEY B. McGEE.'

Apart from this note, the only nonspoken evidence of femininity was adventitiously observed as, en route home from the hospital one evening, the boy was walking as though practicing a swaying, feminine gait. Otherwise, his appearance was that of awkwardness in ill-fitting garments appropriated from male relatives. Although he could not talk about being a girl, he had no diffidence in exposing his genitals in the physical examination. Thus it was possible to demonstrate that feminizing reconstructive surgery would entail complete loss of the phallus. (From age 7 or 8, he had masturbated up to three or four times a day, as would be ascertained 20 years later.)

As aforesaid, the story of how he had reached the verdict that he would change to be a girl was not ascertained until the patient was 24 years old and had lived for 12 years as a woman. This was when she said: 'When I couldn't talk about it, I used to be ashamed of it. But I'm not any more.' The change happened when she was 18 and had met Joan, bisexual and married like herself, with whom she had her first lesbian affair. 'She was telling me about all the girls she had been with. We were just laying around there on the bed. The lights and everything was out, you know, and we got to talking. And she was asking me a whole lot of questions about myself. ... I just had the urge to talk then. So I just went on and talked about it, and told her everything. And I felt free to talk about it ever since. I'm not ashamed of it any more. ... It took me a long time to convince her that it was true.'

Referring back to the age of 12, she said, 'I used to see the homosexuals dressed like women. I knew there was something unusual about me that there wasn't about other boys. And I had made up my mind that, no matter what I was supposed to be, I would never be a homosexual, staying as a man and wanting to be a woman.'

'I remember one time, I think I was about eleven years old, and I messed with this girl. ... We were in her house one day, and there was nobody there but me and her.

And I up and asked her. She told me no at first. And then, well, she made me come right out, and she said yes. For me it was kind of awkward, because I didn't know what to do. I didn't know how to go about it. I was excited, and I guess I was over-anxious. I just got some sort of a feeling from it, but it didn't last too long, and I didn't know exactly what it was, then. After that, we were supposed to, you know, be going together, boyfriend and girlfriend. When I had made up my mind to do what I was going to do, and when I came into the hospital, I told her about it. She said she didn't care what I was when I came back, so long as I came back. We never tried messing with each other again after I came out of the hospital. But we did remain friends for a long while, until she moved away.'

In view of this boyfriend/girlfriend experience, he did think about being a boy. 'I had a lot of thought about wanting to be a boy. And, well, mostly I knew that the doctor had told me if I had remained a boy and ever gotten married I could never make children. And this is what I wanted. I wanted to have children, whenever I was old enough to get them. When he told me that I did have a good chance of getting pregnant, if I became a woman, that is the reason why I decided. ... I did have another reason, too. I had already met my [future] husband.' This man, seven years older than himself, was a friend of the husband of his oldest sister. 'I didn't know him, but I had seen him quite often. And I used to find myself thinking of him the way a girl would think of a boy. I had thought to myself, if I stay this way, I'll never even get the chance to speak to him. So I decided then that I had made up my mind already, but had it made up more, after I had met him. When I came in the hospital, and had gotten out, he had gone in the service, and was gone for a long while. I didn't meet him again until I was about sixteen.'

Subsequently, at the age of 24, the patient recalled the feelings she had experienced at age 11 toward her future husband as compared with those for her girlfriend: 'The feeling I had for him was more than the feeling I had for her. I would have it for her when we were together, and when she was gone, it was gone. The feeling that I had for him was always there, whether I had seen him or not. I was always thinking about him.' She married him when she was 17.

Continuing to talk fluently, she responsed to a query about how difficult it had been to make the change from being a boy to a girl. 'I don't think it was too difficult,' she recalled. 'I did have a lot of problems with my stepfather and one brother because of the decision I made. If I had decided to remain a boy, I would have been the oldest of the boys. My stepfather always said he wanted to take all of my mother's kids and adopt them, and he wanted to name me after him. And I used to listen to him talk about how ugly I would be if I ever decided to be a girl. So one day, when everybody was out and I was home by myself, I decided to dress up like a girl and see how I looked. So I was messing around the house and I had on some of my mother's clothes, and one of her wigs, and I was standing there in front of the mirror, and everybody came in the house on me. They just stood there and looked and didn't say anything. And my mother told me how I looked, and if I looked all right to them, then I'd look all right to somebody else. So, I just stopped listening to my stepfather and did what I wanted to do.'

This recall, at age 24:3, accurately represented the changes recorded weekly prior to feminizing surgery at age 12:11. At first she had dressed as a girl only in private. Then, after the scheduled endocrine admission in preparation for surgery had to be canceled owing to a nursing shortage, she decided to appear in public dressed as a girl with a girl's name. Family counseling, specifically for the siblings, had helped them cope with their public relations roles with neighbors and friends. The patient herself partially overcame the problem of elective mutism by, so as to avoid eye contact, talking with her counselor on the office intercom telephone system. In this way it had become possible to deal explicitly with masturbation, erection, and the subjective cost of losing the phallus.

The recommendation to synchronize social and hormonal sex reassignment in advance of surgical sex reassignment was not adhered to, as the admission for surgical feminization took priority. It was scheduled for the 16th week after the initial psychohormonal evaluation.

During the hospital admission, and subsequently, the handicap of elective mutism had continued. It was aggravated when, on July 4th, at age 14:10, the girl was coerced into an unconsummated attempt at sexual intercourse by a disturbed 18 year old, the brother of her oldest sister's husband. After spending two years in prison on a rape charge, the boy apologized for having tried to prove to his own satisfaction that his brother's sister-in-law was really a girl.

At home, although the girl was able to converse provided she was not being interrogated, she did not initiate talk on personal and confidential topics with anyone except her sister Vera, two years her junior. She was $16\frac{1}{2}$ when she confided to Vera that she wanted to get married and have a baby, as another sister, aged 15, had already done. Vera passed on the information to the mother. The boyfriend was Ricky, for whom she had had, at age 12, the feelings that had helped her decide to become a girl. In the clinic, there were no verbal responses about Ricky, but with head nodding she indicated that they had been having intercourse, without birth control, and that he had had no trouble in getting his penis into her vagina (dick and pussy were the terms used). Additional nods added up to signify her assessment of intercourse as enjoyable and building up to a satisfaction.

Seven months later, when the clinic secretary commented on the patient's new diamond ring, she said that her boyfriend had given it to her, and that they would be getting married the next Saturday.

The couple stayed together, off and on, for two years, sometimes living with relatives, and less often in their own apartment. Though they both wanted a baby, she was unable to heed the advice, reiterated in the clinic, namely, that pregnancy would be contingent on compliancy in taking prednisone and bringing in urine samples for hormone measurements.

Five and a half years later, at the age of 24:3, she speculated as to why the marriage had failed: 'My problem, I think, was that, well, of course, I was a lot younger than he was and every place he went I wanted to go, and some of the places he went, I couldn't go. I don't like being in bars, anyway, and I just didn't like staying at home by myself, nobody else there but me. Even though we had had hard times and couldn't get along, we could enjoy ourselves as far as sex was concerned. I think our biggest problem was that, the whole time that we were married, I couldn't get pregnant, and this is what he was worrying about, because he loved children. This was all he talked about, and it seemed like this would always be our subject. Like he would always bring it up, if we were just sitting there talking, and he would ask why I haven't gotten pregnant, yet. ... So we came over here to the clinic, and the doctor checked me out and everything, and said there was no reason why I couldn't get pregnant, but maybe it just took longer. ... He accepted that for a while, and we tried to get closer together. But it just came up all over again, the same thing, and then, all of a sudden, like, he just stopped. He didn't stop having sex, but maybe two or three days a week he would stay away from home all night; and after two days it would get to be three and four, and sometimes he wasn't there at all for two or three weeks. This is when I found out he had gotten another girlfriend. I think she had something like three kids and, whenever I seen him, I would never see him with her. He would always have her kids, like she was never around, just the kids. He was always crazy about children. It was the same way with my sister's children. So I kind of think if I had had one, we would still be together.'

Childlessness alone did not explain the failure of the marriage. There was a problem of losing her feelings for her husband when they had sex, because he was, she said, too rough: 'When I'm in the mood and want to be touched, you know, it doesn't make any difference which part. If I want it, it all feels good to me. It's different being touched by a man than a woman, because men – well the only man I can say now is my husband, he was sort of lovable at one time and then at other times he was, you know, rough. And being with the other sex, it's the same way all the time. If her touch is soft one time, it's that way all the time. It's not rough. It doesn't change.'

The first sexual experience with a woman partner had been with Joan: 'She seemed to me that she had the same problem I had, because she was also married and wasn't pregnant. I don't know that she had ever did it before with a woman, but she had been talking to me like she had. So after me and her done it together for the first time, we just kept on seeing each other. ... Then she went in the air force. So I just went home to my mother. That didn't seem to do me any good, because it seemed like it was something I needed all the time. I was just in the mood all the time. So I just found another one.'

In the course of the next 20-odd years, she continued her life bisexually. Her allegiances were serial rather than concurrent, and were equally divided between men and women, with an identified total of 20, and possibly a few more. Some former lovers dropped out of her life completely, whereas others maintained a comradely relationship. Some relationships began with more intense passion than others, with the idealism of soul-mates forever, and none were trivial, one-night stands. The way of meeting new partners was through a network of friends and their friends, and to a lesser extent through pen-pal correspondence. The longest live-in relationships continued for between two and three years. In some instances the breakup was contingent on availability of income, employment, and housing, as well as the intrusion of personal incompatibilities. The patient's own greatest liability was the tyranny of her becoming electively mute in the face of anything idiosyncratically threatening or antagonizing, to which was added, after the incident of gang rape, the phobic tyranny of being unable to leave her place of residence, even to go to the corner store, unprotected by an accompanying adult.

The first occasion of being able to give a fluent interview, namely when the patient was aged 24 (Fig. 9.5), was also the first occasion when she gave details of her sexuoerotic functioning. She compared her feelings when making love to a man or a woman: 'It has to be different with a man and a woman because being with a man, there's a whole lot more there that you can get. With a woman, it's just the same as you. I can't say it feels quite the same. In loveplay, I would say that's the same, but when it comes right down to having sex, it's different. I don't know if I enjoy one more than the other, but I know, if I had to pick, it would always be a man. That should answer that, I guess.'

Answering an inquiry as to what sort of things she had done with her women

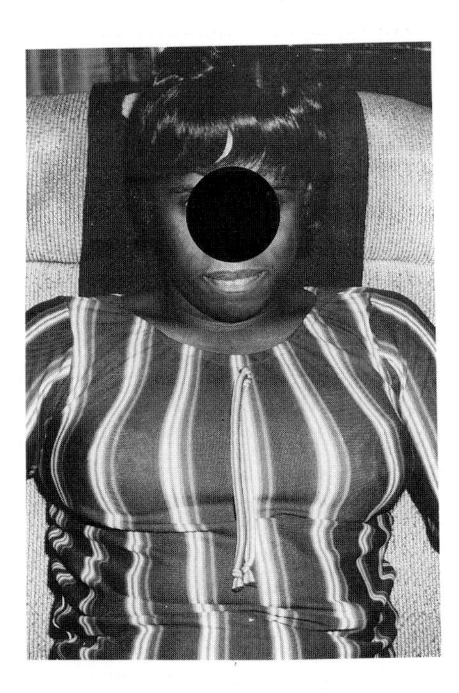

Fig. 9.5. Case #1. The appearance is feminine at age 24.

friends, she said, 'I didn't actually say or do anything. I was just there, you know. And everything that was done, they were the ones that were doing it.' Before telling what they did do, she requested that the audiotape be turned off, afterwards summarized: 'I like to have my breasts played with ... and sometimes they would make love to me the other way, I don't know the word, oral sex. There's not too much two women can do.' They had not used a dildo. Though she had enjoyed being the recipient of oral sex, she had been averse to having her genitals fingered, or to having them seen naked, regardless of the sex of the partner. She had not tried performing either fellatio or cunnilingus.

'You mean did I ever have one?' she replied to a query about climax (her own term was satisfaction). Then she responded to her own question, 'yes.' Describing climax, she said, 'Let me see. One thing, when I had sex with my husband, the feeling was good, but it seemed like it took him the longest time, and it took me the longest time to get satisfied, in other words to reach, what's the word? Climax. With a woman, it didn't take that long. With my husband, we would make love for something like half an hour. With the women it seemed less than ten minutes. ... With him, after we had a disagreement, it took longer, and sometimes it wouldn't happen at all. With a woman, if I wasn't in the mood, and just did it because she wanted to, there wasn't anything there. With the feeling, all I can say is that I enjoyed myself.'

She could not compare the feeling she now had with what it had been prior to feminizing surgery: 'I can't say, because, about me being so young, I wasn't even thinking about it, so I wouldn't know.' She did get feeling from the region where the greatly enlarged clitoridean organ had been surgically removed. The feeling from penovaginal insertion, however, 'would be a lot more enjoyable if, the whole time he was having sex with me, he would mess with my breasts, because that's where I have a lot of feeling.'

With respect to the sensory channels of sexuoerotic turn-on, additional inquiry indicated that smells, tastes and sounds were of no significance. Tactual stimulation, especially of the breasts, was the most potent source of sexuoerotic arousal, with explicit video erotica, and to a lesser extent narrative erotica, in second place. There had been only a few occasions when she had watched erotically explicit movies either in an adult movie theater or at a private party. She might project herself as a replacement for the woman in the movie. It was possible, though less likely that she would appropriate one of the figures in the movie for herself. 'One time,' she said, 'where I had seen the movie, I dreamed about it, and in my dream I was there with the fellow in the movie.' She had never had an explicitly sexual dream that was completed, with orgasm.

Thinking of questions that hadn't been asked and should have been, she said: 'I never did like to drink, but once in a while I would drink beer. If I'd drink enough of it to get me high, it seems like it would get me in the mood of wanting sex, and as long as I stayed high off the beer I could have sex all night and never be tired. When me and my husband was together, every night if I would drink beer, we'd make love in the night, and I was getting sort of out of hand for him, because he had to work next day. This is one reason why I stopped drinking beer. ... He would last sometimes like two hours, and then he would fall off to sleep. So I would just lay there, and try to go to sleep, if I could. And I'd just drink some more beer, so it would put me to sleep.' She had had no history of smoking pot, or using any other drug: 'I don't need it. I'm just not curious enough to try it.'

In retrospect, the patient had recalled that at age 12, she had not wanted to be a homosexual boy attracted to a male partner. Now, at age 24, how did she feel about being in bed with another girl? Her reply was: 'When I think about it, to me I didn't mind it too much, because it was something new and different; and to me it was sort of exciting, and I really enjoyed myself. ... Then, it seemed like to me it was really getting out of hand. Everytime we were together, we'd just want to, you know, go to bed. And, sometimes, I would see women that I didn't even know, and I would have this feeling. So I got sort of disgusted with myself, you know, and I stopped. I stopped going out. And I stopped seeing women, and I started seeing men.' In fact, what actually happened at this juncture was that she had quit seeing Catherine, her bisexual lover, and had begun having an affair with her male lover, Raymond, instead.

At this stage of her life, the lesbian part of her lifestyle was not openly declared at home. She lived in an apartment which she had constructed for herself in the basement of her mother's house, so that eventually she did reveal her bisexual story to her mother and one sister. This sister had two children and was the head of a single-parent household. When later the patient had no place in which to live with her new-found girlfriend, Rena, this sister provided accommodation for the two of them, until they found a place of their own, taking care of three motherless children of a working father.

Rena's introduction to the members of the psychohormonal clinic was on the occasion when, at age 26:4, the patient participated in a program for the freshman class of medical students. With Rena, she was enabled to defeat her phobia and negotiate the taxi ride; and, with Rena participating with her, she felt safer in appearing before an audience. Both women were very effective in their responses to the students' questions. One student asked the patient: 'Why did you definitely feel you did the right thing by having the operation to become a girl instead of a boy?' The response was: 'I feel that, by having the operation I did have, it gave me a more complete life, because I could feel like, if I had a bisexual life, I could have either male or female. By having the other operation, to become a male, I couldn't, I don't know, I don't think, even if I got the feeling, I don't think I could be with another man, if I was one myself. I mean, that's the way I felt then. It probably would have been different, if I had waited.' She did not regret the decision she had made.

After their presentation to the students, each woman did an audiotaped interview with a member of the psychohormonal staff. In her interview, Rena said: 'Having sex with her is almost the same way as having sex with a man, except that she doesn't have a penis. I get more satisfaction from her than a man because she makes me feel better than a man does. Where a man would make love to me, it feels alright to a certain extent, until I see the penis, and it scares me to death, you know. It inserts, and then it makes me feel miserable, and I can't stand it. In the past I had sex with men. It's like they don't care if they hurt you or not. Then, when you tell them they hurt you, it's like they want to hurt you more. They don't stop, you know, or anything.' With her girlfriend, it was 'much better.' Either one could take the initiative, and they could have orgasms. Nonetheless, her partner's phobic problems demanded too much social isolation, and she was already thinking that the relationship would not last. The disinformation between the two of them was such that the patient herself was reporting the possibility of following the example of three couples among their gay friends who had been married in a gay marriage ceremony by the pastor of a gay religious congregation.

Rena's problem of sex with men was, in the literal sense, deadly. She had spent time in jail on a charge of manslaughter, after a fight in which she slashed her lover for accusing her of being a lesbian and he then bled to death, so it was said, while awaiting emergency medical attention. Rena had, at the age of 12, been 'raped by my brother-in-law's godfather, and I was a virgin, and I just can't stand it. I've tried to make it with men, but it seems like I don't get no feeling in a sexual relationship. It hurts me. I don't do it anymore. And to be with a man that short a time [when she was raped] that's how come I can't have no children.' She became frenetic about becoming pregnant. Unaccepted for artificial insemination as a lesbian mother, she adopted a male as a sexual partner. A year and a half later, she still had not become pregnant. The erstwhile relationship between the two women continued as an agonizingly nonsexual friendship.

For two years, there was silence. Then, at age 28:9, the patient returned in June bubbling over with the good news that she would get married in December. Her husband-to-be had been released from an eight-year imprisonment in May. She had not known of his existence until the preceding January, when she had answered a newspaper pen-pal letter he had written from prison. She had visited him twice in prison and decided that she was in love, and that he could be released to live with her. The charges for which he had been locked up were a matter of indifference to her. So also was his financial responsibility for his three children. They were in foster care, and became known to her while he was still in prison. When he was released in May, he came to live with her. Three months later, he disappeared. She did not hear from him again ever, nor apparently did his mother, with whom she had established contact.

Following this debacle, the patient had an occasional relationship with Debby, through whom she met Tommy, who would play a very important role in her life for the next 10 years and more. On the credit side, Tommy was, as a male lover, an almost unparalleled Mr. Perfect. Sexually, he could keep pace with the patient who, estimating her own sex drive, said, 'I think I have a very high and strong one because, when he's not around, I can masturbate twice a week, or twice a day, and when he comes to see me, we always spend that time in bed. ... Tommy says I can never get enough, because I always want to do it whenever he's around. If he stays overnight, if he doesn't get tired and go to sleep, I'll want to do it about three or four times.'

With Tommy, she could enjoy sex with him on top, whereas with almost all other lovers of either sex it had to have been the other way round for her to be satisfied. With a woman partner, it had been 'better for me to be aggressive than for her to be that way with me. ... I have been with women who wear men's clothes, but I like women more like myself – women who look like women.' Tommy was one of her partners with whom orgasm became 'so great to the point where I get shaking in my body, and it takes me a long time to stop. It's always good.'

On the debit side, Tommy enlisted in the armed services and shipped out to Europe. They would not see one another again for six years. He had been gone three years when the patient next experienced a major love affair of great limerent intensity. The new lover was Molly who, at 27, was seven years younger than the patient. She had a 6 year old son and a 2 year old daughter, both fathered by the same man. He had been her only male partner, and she had gone with him so as to get pregnant. By contrast, she had since the age of 8 had five female partners.

Molly's introduction to the psychohormonal staff was a replica of Rena's, except that on this occasion, the patient would present her case not to the medical students class, but to the psychiatry department's weekly grand-rounds conference – another appearance before an audience in which, with Molly's support, she talked very openly about living in adulthood as a bisexual, after having changed from boy to girl at age 12. After the conference, each woman audiotaped an interview with a member of the psychohormonal staff.

The patient's own interview not only covered familiar ground but also added some new information. For example, no longer beset by a noncompliancy in taking prednisone, she was menstruating regularly, but was less concerned than formerly with becoming pregnant, partly because of her age, and partly because she had Molly's two children to care for. 'I don't like holding babies when they be first born,' she said, 'but after they get three or four months old, I can deal with them.'

She was quite sure that she had not, at age 12, made the wrong decision: 'I like myself just the way I am now,' she said. If able to start over again, she said, 'I would choose what I am now.'

There was new information on dreams: 'I have all kind of dreams, some nightmares, some happy, and some sad. For the past three nights I've dreamed that I've been rich. I've had all this money, you know. I ain't really had any bad dreams that I can remember – maybe falling off a cliff or something. In one dream that I had, I had found all this money. It was all change, like quarters, nickels, and dimes. I had picked it all up. I was stuffing it into my pockets. And, then, it was some more, like, laying on the street, and I had bent down to pick it up, and all these dogs leaped on me. I was running, and I was, it seemed like, I was on top of a building. Instead of on the ground, I was on top of a building, and I ran off the roof. I was in mid-air. But I woke up.'

The next query was about romantic dreams. She could not think of any, whereas of sexy dreams, her response was: 'I guess you would call it an orgy – more than three people. Everybody that was there, I knew them, people that I have met, some since I've been with Molly. ... I was there, but I wasn't doing anything. I was just looking.' The dream imagery was not derived from actual orgy experience, for there had been none.

There had been no wedding dreams, and no pregnancy dreams – 'None where I was pregnant, but I did have a baby. I wasn't pregnant. I'd already had it. ... It was a good-sized baby, so I know I probably would have taken care of it. In my dream, the baby was mine. ... That was about three weeks ago.' There was nothing to report about daydreams – 'I've been watching too much television during the day, when we're not dealing with the kids.' There was no evidence of any type of paraphilic imagery or ideation.

Regarding sexual diversity with Molly, she said, 'There are a lot of different things I should have talked to her about, but haven't yet.' Oral sex was mutually enjoyable. Of finger sex, she said, 'I don't like it. I don't know if she does. I mean, I'm not, you know, I haven't tried it.' They had not tried using any sex toys, like a vibrator: 'We've been talking about it, but that's it.' With Molly, she was able to have an orgasm, 'forceful, hard and strong' from prolonged breast stimulation alone. She and Molly were readily affectionate with touches, hugs, and kisses; and she was 'very huggy and kissy' with the children also.

Queried about anything in her behavior that other people might consider strange, she allowed that she sometimes had a bad temper, 'but I don't fight or anything. Molly tells me that I'm mean, but I don't see it. ... When she tells me I'm being mean, I don't talk to her. I don't give her no attention when she wants it, when she tells me that I'm mean. I won't talk to her. ... She tries to get me out of it. Sometimes it works, and sometimes it don't. It all depends on how mad she's made me. We don't fight. I don't like fighting. I used to see so much of it when I was coming up, I don't do no fighting. ... If I get mad to that point, I'll just leave. I did leave one time, and went to my mother's. I stayed home about three weeks, but might as well not, because she was there every day.' The patient had absolutely zero insight into the aggravative passive-aggressive power of silence and absence as weapons, nor of their ultimately destructive effect on, one after another, the relationships that she most treasured in her life. Incredibly enough, from the ashes of each relationship lost, there would arise the phoenix of a new relationship, extolled as more perfect than any of its predecessors.

In the course of her own audiotaped interview after the grand-rounds conference, Molly gave information that either confirmed what the patient herself had said or in some instances expanded upon it. She allowed that there was no evidence of the patient's birth defect of the sex organs from looking at her. She knew from having sex with her that something was different about the inner lips. She and another woman who had had sex with the patient agreed that when she was getting ready to nut (that is, to have a climax) there was a hard part of her lips (where the clitoris should be) that got harder: 'And she asked me was that the right thing. I told her, yes. She said she didn't know that's what she was working with when she was having oral sex with her. So I said, yeah, that was the right thing, and left it at that. ... We have oral sex quite a bit, but mostly we both like getting on top of each other. Because, being that her vagina is different from mine, when she is getting ready to nut, just that hard part of her lips, you know, freaks me out. Right? And that's the best part to me. But I love making love with her. I like her breasts a lot. Everything turns me on. ... Her orgasms are different than mine, because she can go at least three times behind each other, where I might go only twice. It's a whole lot different. After a little break, she's ready for seconds. ... Altogether, she's a lot different from anybody I've ever been with, and very exciting.'

The intensity of Molly's own lesbian possessiveness notwithstanding, she considered her friend to be bisexual, 'even though she tries to tell me that she's not. You know how you can just feel it? And then it's some of the men that comes around her, they're always giving her money and stuff, you know. And they're always asking her. Right? I think she don't, you know, say anything to them, because of me. But I believe before I came along, it was a lot of different things, you know, going on.'

Before the month had ended, the patient was no longer on a live-in basis with Molly, but had moved back to her own quarters in her mother's house. She and Molly continued to be lovers for another two years. They separated on a friendly basis, and the two children maintained a visiting relationship with her, the woman whom they had for three years adopted as their aunt.

It was at this time that her former friend, Tommy, had re-entered her life. 'I wouldn't mind settling down with him,' she said, 'but not to the point of getting married [he was not divorced from this third wife]. I do have the feelings in my mind that I want to settle down and leave this other life, this gay life alone. But, knowing Tommy, he's not going to let me leave it alone, because he's, he's – he likes to be sexually involved with more than one woman at a time. He likes to be – to party heavy. Sometimes. And I think this is the very reason why he liked coming to see me, so much. But when it had gotten to the point where I stopped getting involved with him and another female at the same time, because I didn't want it any more, you know –. He likes these things. And I know he wouldn't let me stop completely. ... That's why he stays in the street [instead of with his wife].'

Tommy's age was 38, and the patient's 36. The synchronization of their social life together was limited by the continuing handicap of her agoraphobic inability to leave the house, unless accompanied by a family member or familiar companion. Their relationship continued, however. Four years later they decided to move into an apartment together on an indefinite basis.

CASE #2: SEX REASSIGNED AS BOY

Diagnostic and clinical history

Following extensive clinical attention and treatment from birth through early childhood, this patient became lost to active follow-up until 11:0 years of age. At that time the county director of maternal and child health, having been alerted by the school nurse, referred the child to the Johns Hopkins pediatric endocrine clinic for evaluation and treatment of sexual ambiguity. At school, at home, and in the community, the child was known by a girl's name, Andrea. Except for short plaits in her hair, one on each side, her dress and appearance made no concessions to femininity. She was afflicted with elective mutism that precluded her talking about anything concerning her status as a boy or girl, and much else, as well.

It was only from the records of the Bureau of Vital Statistics that it was ascertained, 12 years after the fact, that the child had first been registered as a male, Joseph, and then two weeks later reregistered as a girl, Andrea Marie. In retrospect, the mother failed to recall the male name. She also had no recall of uncertainty as to the baby's sex at the time of birth. The delivery had been at home.

The earliest retrieved medical record indicated that at age 13 days the baby had been admitted to the county hospital with a two-day history, typical for the CVAH syndrome, of vomiting and dehydration. At 25 days of age, she had been transferred to the nearest university medical center. There the diagnosis was correctly established as the salt-losing variant of the syndrome of congenital virilizing adrenal hyperplasia (CVAH). The corrective hormonal treatment for this condition, newly discovered only five years previously, was begun, namely with the adrenocortical glucocorticoid hormone, then generically known as cortisone, plus extra dietary salt and a salt-retaining hormone. Following discharge, the treatment was supervised by the local physician.

At the age of 21 months, the child was admitted to the nearby national medical center for evaluation and study of her CVAH condition. At age 2:6, she was readmitted for control of persistent symptoms of vomiting and diarrhea. During this admission an exploratory laparotomy was done. It revealed normal ovaries and fallopian tubes, bilaterally, and a uterus. It was decided to defer plastic surgical reconstruction of the genitalia, but the reason was not recorded. Following discharge, the child was carried as an outpatient for an unspecified period of time, and then became lost to follow-up.

The local physician supervised the child's hormonal therapy until he came to the conclusion that, insofar as it was having no effect on her masculinized external genitalia, the case was one of a congenital deformity for which nothing could be done. This, he explained, was his reason for discontinuing treatment.

At the time of the child's admission to Johns Hopkins at age 11:11, the discrepancy between chronological age and physique age was apparent upon physical examina-

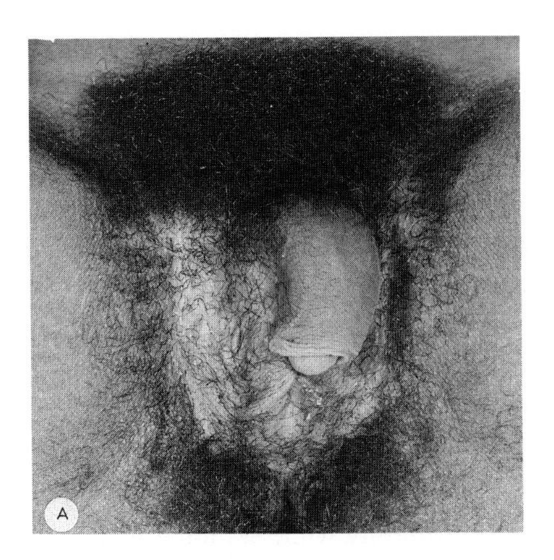

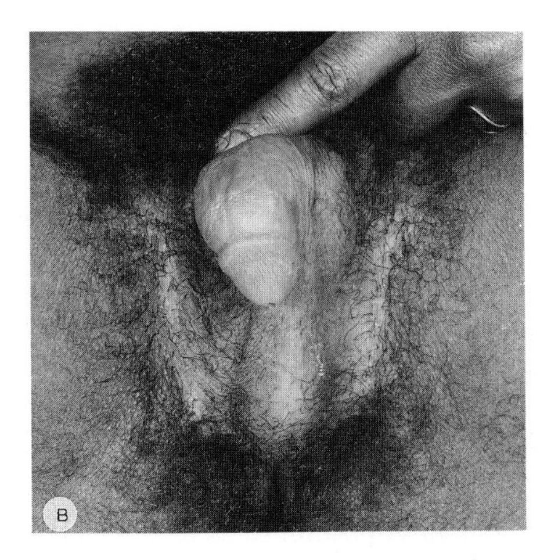

Fig. 9.6. Case #2. Two views (A and B) of the genitalia prior to masculinizing surgical reconstruction.

tion, for she had the morphological maturity of late male adolescence. There were no signs of adolescent feminization of the genitalia (Fig. 9.6). The body build was recorded as 'normally proportioned, extremely muscular masculine habitus with broad shoulders and narrow hips.'

Staturally, there was a discrepancy between the advanced degree of virilization and the height age, which was that of a 10 year old (Fig. 9.7). The actual measurement was 141.5 cm (4 ft. $7^1/_2$ in.). The weight, 40.75 kg (89.5 lb), was congruent with the height. There was no likelihood of additional increase in height, as excessively rapid early bone growth had been followed by early and irreversible closure of the epi-

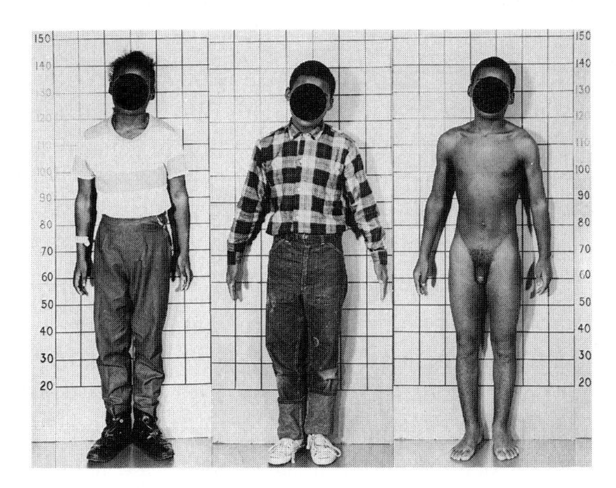

Fig. 9.7. Case #2. (Left) Appearance at age 11, when admitted as a girl. (Center and Right) Appearance seven months later, living as a boy.

physes. The bone age was read from X-rays of the bones as being 17–18 years.

The signs of adolescent development were enumerated, thus: 'Slight amount of dark hair on upper lip and chin. Abundant axillary and pubic hair extending around anus and onto abdomen around umbilicus. Body hair was not increased. Slight facial acne and much seborrhea. Breasts completely infantile.'

In the formation of the external genitalia the midline of the perineum had completely fused so as to form a rugated, empty labioscrotum instead of bilateral labia majora. In the place of a clitoris, there was a large clitoridean phallus held bound down in chordee by what would otherwise have been either labia minora or the external skin covering of a penis. The length of its outer curvature measured 9.5 cm, and its diameter 2.5 cm. At the tip of the glans there was a dimple not connected with the urethra. There was a tiny urinary opening hidden below the glans on its underside, and so positioned as to be compatible with urinating while standing which, although difficult, the child would attempt to do.

Laboratory tests confirmed the previously established clinical diagnosis of congenital virilizing adrenal hyperplasia (CVAH) with hermaphroditism (synonym: pseudohermaphroditism) in a genetic female. From the genetics lab, the report of the buccal smear test as chromatin positive, as for a female, was confirmed with a chromosome count which produced a female 46,XX karyotype. The hormonal assays showed, as expected, elevated levels indicative of excessive hormonal masculinization: urinary 17-ketosteroids, 37–44 mg/24 h; pregnanetriol, 50.8 mg/24 h; and dehydroepiandrosterone, 4.5 mg/24 h. During the course of a prolonged inpatient evaluation, it was demonstrated that glucocorticoid (cortisone) treatment would effectively normalize these measurements. To correct the salt-losing propensity, unrestricted additional salt in the diet proved to be therapeutically sufficient.

Even though the chromosomal and hormonal data gave confirmation of the patient's diagnostic status as a female CVAH hermaphrodite, they were insufficient for a complete sexological prognosis and plan of treatment. To obtain the patient's own input on whether the prognosis would be better as boy or girl, it would be necessary to surmount the obstacle of elective mutism. This obstacle had been identified in the report of the physical examination under the heading of general appearance: Sullen, withdrawn appearance, often sucking on middle two fingers, usually nonresponsive to questions, occasionally nodded her head and responded if at all only with single words. Words tended to be blurted out harshly, with the pitch of the voice masculine.

The setting of a date for corrective surgery was put on hold until such time as the patient had become able to talk and provide information relative to whether the correction should be feminizing or masculinizing. Six months later, after the verdict, namely masculinization, had been declared, the first-stage operation was done. It comprised a hysterectomy with bilateral salpingo-oophorectomy, an appendectomy, and a release of the clitoridean penis from the chordee that held it bound down and restricted the mobility of the urethra.

The second-stage operation, to construct the urethra, was performed one month

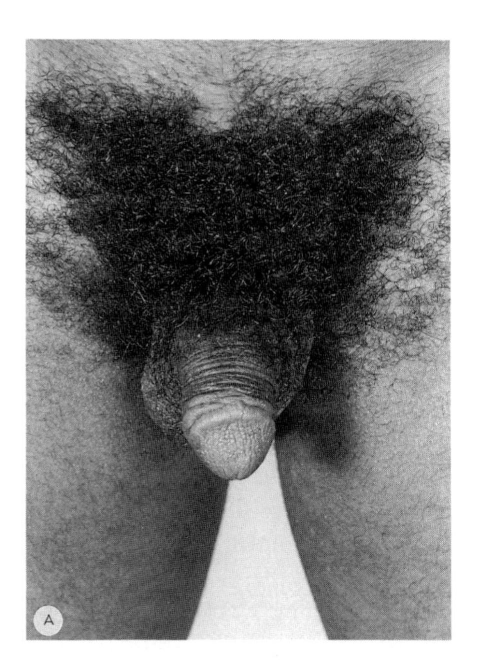

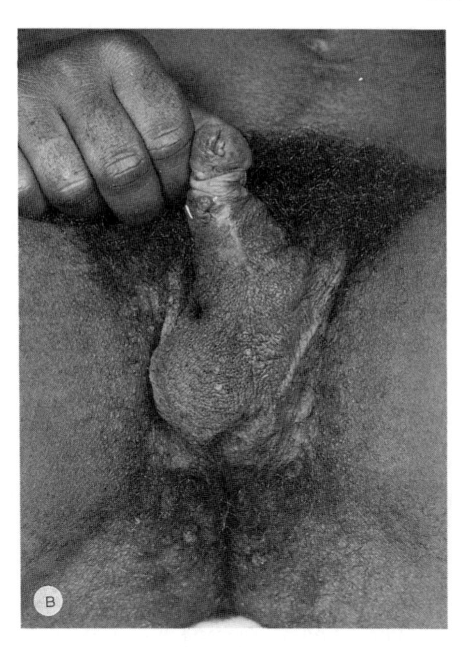

Fig. 9.8. Case #2. Two views (A and B) of the genitalia after masculinizing surgical reconstruction. Left prosthetic testis was awaiting replacement.

Fig. 9.9. Case #2. Appearance at age 15:6 living as a boy.

later, with the expectation that it would be final. The grafted tissues healed imperfectly, however. After four months, there was a second attempt to reconstruct the penile urethra. A temporary urinary opening was left in the perineum, to be closed at a later date. When the patient was discharged one month postsurgically, the total length of the hospital stay had been 11 months.

Readmission for the fourth operation was in three months. Five days postoperatively, it was apparent that the reconstruction had broken down. Once again the patient was discharged, still able to urinate only in the sitting position.

Fourteen months passed by before the next readmission – for operation number five, a final attempt at urethroplasty so as to enable urine to pass through the end of the penis. During the same admission, a separate operation was performed to implant prosthetic testes in the empty scrotum, to improve the cosmetic appearance of the genitalia. Within six days, a revision of this operation was necessary to anchor the left prosthesis in place. It did not stay anchored, however. Its replacement, the eighth and final operation for this patient, took place 16 months later when he was 14:10 years of age.

Evaluated on the criterion of urination, the surgical outcome was less than satisfactory. The urinary opening was not at the tip of the glans penis, but approximately 2 cm lower down on the shaft of the penis (Fig. 9.8). From this opening urine sprayed out and could not be aimed properly, and afterwards it left a dribble. The boy's own words were: 'Where the water come out, on the end, it sort of spray out all over the place. The doctor told me when I, before next time, try to hold it a little, sort of up from, sort of, so it just come right out.' It was more convenient to sit and urinate. There continued to be a problem of 'mild enuresis, almost nightly.' It had antedated surgery. No explanation could be found, other than that it might possibly be syndrome related, since it had been noted in other CVAH patients also.

In preparation for each surgical procedure, the patient had been placed on high-dosage glucocorticoid (prednisone) replacement therapy. This was a protective measure, standard in all CVAH cases, insofar as it mimics nature's normal adrenocortical hormonal defense against stress. The rationale for glucocorticoid treatment at other times of no stress is to prevent masculinization in CVAH females. In this patient's case, prednisone would have suppressed the masculinizing hormone secreted by the adrenocortices, thus requiring testosterone replacement therapy to maintain masculinization. The logistical problems of ensuring compliancy in taking medication when not under supervision in the hospital were, in this particular case, insurmountable. Thus the endocrinological decision was to allow the patient to be self-maintained on his own hormones, with the recommendation of supplementary prednisone in times of illness, injury or surgery.

By age 15:7, there was a beginning growth of hair on the upper lip, chin and chest, and a more dense growth on the arms and legs. The voice was masculine. So also were the gait, posture, gestures, mannerisms, and clothing style. He had his hair cut in the fashionable Afro style, which increased his visible height by three inches or more (Fig. 9.9). 'Height don't mean nothing to me,' he said, when asked about it. 'Nobody say nothing. They don't pay much attention to it.'

At age 19:10, the patient was referred by his social worker for an endocrine evaluation, as he seemed to have no energy or interest at work. At this age, the height (141.5 cm) and height age (10 years) were unchanged since age 11:0. The weight was increased from 40.75 kg to 45 kg, without his being obese. The stretched length of the penis was 11 cm. Overall, no impairment in health could be detected.

Gender-coded social biography

This patient was a descendant in a colonial Maryland lineage whose forebears had been slaves, possibly on the same land as that now farmed by the patient's kin. Also descended from a colonial family, one of the major landowners of the region was the local physician. When the patient was 11 and the feasibility of sex reassignment was being evaluated, he foresaw no problem in the community if she returned home reassigned as a boy. 'Everyone here,' he reported, 'already knows her as the boy with long hair.'

The family lived in a seven-room farm house on an eight-acre farm said to be a subdivision of the land worked by the paternal grandfather. He had had 10 daughters and three sons, of whom the eldest was the patient's father. This son and his wife had had eight children, the oldest born when he was 26 and his wife 16. The patient was the sixth child. The fifth was a boy with normal male genitalia. He may also have had the CVAH syndrome, insofar as he had the tell-tale symptom of vomiting since age three days, and died at age one month, reputedly of pneumonia. The surviving siblings were five brothers whose age ranged from 19 to 8, and a sister, aged 15. This sister seemed to have a closer relationship with the patient than did any of the brothers. The oldest brother was employed away from home.

The household included a 9 year old girl, an orphan, who was the patient's aunt. There was also a paternal great-uncle, aged 60, who had always occupied a place in his younger brother's household. Newcomers within the household for three years were two teenaged girls and their mother, Mavis, a woman of 50. This woman assisted in the domestic affairs of the household and to some extent substituted for the maternal grandmother who had died two years earlier.

The grandmother had been the effective head of the household, as her daughter had supplemented the family income by working outside the home as a domestic, in the most recent four years in the home of the local physician. Before that, for three years, she had been the live-in housekeeper for an elderly woman, formerly a neighbor, who had moved to another county, three hours distant by bus. From this job she had been able to return to her family for one weekend every two weeks.

While the grandmother was alive, she had been the one who took responsibility for the running of the household, and supervised the patient's medical affairs. The child had a close relationship with her grandmother who was, according to the father, 'the only one she would really talk to.' After the grandmother's death, the child expressed her grief only by becoming very withdrawn. Her father recognized that she was more deeply affected than other members of the family.

The woman, Mavis, knew of the anomaly of the patient's genitals because her mother had required that she show them to Mavis 'so that she wouldn't get upset when she saw her later.' Neither Mavis, nor the mother herself, took over any of the grandmother's involvement in the child's health care. The attitude of the adults in the household towards the child's anomaly was one of laissez faire which at times became scarcely distinguishable from indifference and neglect.

During the child's 11-month hospitalization, there were weeks at a time when there was no written word from home, no phone call, and no visit. The family lived on the edge of poverty, and were dependent on a relative or neighbor to drive them to Baltimore. But that was not the whole story, for the availability of a travel reimbursement or a collect phone call did not eliminate the problem of unfulfilled promises to visit, and failure to keep appointments for both outpatient follow-up and inpatient readmission. This extreme degree of negligence on the part of the parents could not be attributed to mental retardation or psychosis. Nor was illiteracy an explanation, for the mother, at least, was able to write a letter literately. Drinking may have been a factor. For example, when the mother's employer was contacted on the occasion of a readmission appointment not kept, he said she had not been at work for several days, and conjectured that she and her husband 'both might be sitting somewhere drinking.' There were several notations in the hospital record of a smell of alcohol on the mother's breath when she had an interview. The father also had a reputation for drinking, according to the notations of a visiting teacher which included also: 'pretty rugged home life, crowded, poverty-stricken; tenant farmers.'

At home, the child had a consistent history of being taciturn. After her grandmother died, her elderly great-uncle was the only person who could engage her in sociable conversation. 'If other people overhear her talking to him, she doesn't mind,' her mother said. 'Anybody else, she just can't be bothered talking to them, I reckon.' She did not speak with anyone, including her great-uncle, about her sexual anomaly.

In the same way that 'she never do talk much' the mother said, 'she don't want to have too much to do with nobody. But she will play games, outside, things like that, outside. Anything else, she just don't bother. ... She likes to play with mostly things boys play with. Like the rest of them play ball, all the rough things. Climb trees. She loves to do that. ... When she was a baby, she liked dolls and all the other things for girls. In the last three or four years, it must have been around the time she started school, she started changing.'

The father's observations were concordant with those of his wife. 'Well, in the way she acts, it reminds me more of a boy,' he said. 'Like boys will play ball and go fishing, and things. She likes to be out with the boys more than be with girls in the house – won't play with girl things, you know, like dolls, nothing like that. She'd rather be out in the fields with me, and the boys, rather than in the house with other girls.

She help out right good.'

'When you got a whole lot of children,' the mother said, 'it's a lot easier to wash pants than to be washing and ironing a whole lot of dresses. So at home everybody wears pants.' Like the other two girls, the patient did not object to wearing dresses to school. 'She don't like nothing that fit her real tight. And certain colors she likes, like blue, red, and yellow, but she won't wear pattern colors.' The father's statement about style of dress was brief and to the point: 'Well, I think just like a boy.' The clothing she brought with her to wear in the hospital confirmed this statement.

'As far as dressing or anything, she'll always be the last. Dressing, taking a bath, or using the toilet, she would never go in with any other girls. She just wait until they get through, and then she'll go in. She does most of her dressing in my room.' She would not risk being seen unclothed except by her mother and no one else.

The girl's tendency to recoil from association with children of her own age did not carry over to young infants. 'Oh, she loves them,' her mother said. 'She likes to play with them. She's very gentle with them. Of course it's seldom that she see any. But when we go over to their [paternal] grandmother's, there's a house full of babies over there.' It was not exactly as if she played being mother to them, but it was different than what a boy would be doing with babies, according to the mother's opinion.

'She don't like school,' the mother said. 'I guess at school she's the same way she is at home, not wanting to have to bother with nobody.' She attended the local public elementary school. There she was placed among third graders despite the discrepancy between their chronological age as 8 year olds and hers as an 11 year old. There was an even greater discrepancy on the criterion of physique age, with the development of the 8 year olds being juvenile, and that of the patient being adolescent and virilized, like a boy.

Academically, her highest ratings on standard achievement tests had been at the second grade level. These ratings were in agreement with those obtained after the child had been admitted to the hospital and was evaluated in the reading clinic. The summary from this evaluation is as follows: 'Although the patient's oral reading rate and accuracy were poor at the Grade 1 level, her comprehension, on both oral reading and listening, were good. Her sight vocabulary was very limited and she evidenced absolutely no ability in word analysis skills. The same difficulties apparently prevented her from reading silently. However, she was able to learn an auditory analysis skill in a very short time. She did not spell most of the Grade 1 words. She answered simple addition problems but did not subtract, multiply or divide. In view of the foregoing findings and of reports from the hospital school, it is my opinion that this child's academic disabilities are consequent on personal and cultural factors other than mental retardation. The child was bright and eager. She answered my inquiries about her schooling without hesitation and read to me willingly. This was in great contrast to my previous contacts with her. During the first month she was in the hospital she neither looked at me nor spoke to me, and as recently as two weeks ago she only nodded or said yes or no to me.'

As an inpatient, the patient was enrolled in the Baltimore City public school situated in the Children's Medical and Surgical Center. The report from this hospital school, 10 weeks postadmission, covered the following: 'She does not get emotionally upset about things and controls her temper well. In a group, she gets along well with others, is considerate of their rights and wishes, and seems accepted by all. She can be depended upon to begin work promptly. She requires much supervision to compensate for her deficiencies of understanding. Once she understands, she does more than expected. She now expresses herself more adequately, and is more at ease in speaking. She likes to paint, draw, and make things beautiful, and is very good at this. She works well with her hands. She loves to put puzzles together and will work for hours doing them. She does superbly in physical activities. She loves to go to the gym, and enjoys a game of basketball or other sport. She excels. She always appears neat. She has a problem with body odor.'

Nine weeks after this report had been written, the patient at age 11:4 was tested with the Wechsler Intelligence Scale for Children (WISC). The IQs were: Verbal, 77; Performance, 80; Full, 77. These figures were, like those for blood pressure, declared as viable for no longer than two years, at which time the test should be repeated. The logistics of scheduling a retest, however, proved insurmountable. Whether or not there had been a degree of intellectual catch-up growth, and a corresponding IQ increase following sex reassignment remained unascertained.

When the patient returned home as a boy following the almost year long hospitalization, he was refused readmission to the school he had formerly attended, on the assumption that his sex reassignment would be disruptive for other children. Official inertia and bureaucratic neglect conspired to deprive the boy of regular schooling, except during hospitalizations, for 18 months until he was 13:6 years old. The impasse created by noncoordination among service agencies responsible for the child's welfare was finally broken by the expedient of having him taken legally into protective custody for a year. As a ward of the county social service department, he continued to live at home, but he was no longer reliant on a noncompliant family to ensure either his schooling, his transportation, or his medical follow-up visits.

At the age of 12:5, and a year after the medical and surgical declaration of a change of sex, the change had been made legal by a court order. It was obtained with the assistance of a special assistant to the Attorney General in the State Department of Health, and through the *pro bono* efforts of a county attorney. It authorized a change of name from Andrea Marie to Andre, and required that the birth registration 'on file with the State Board of Health and Mental Hygiene and Vital Statistics ... concerning his sex be, and it is hereby amended to read "male".'

On the occasion of the patient's outpatient visit and progress report at age 14:0, he was accompanied by his oldest brother, aged 22, who gave a brief report. His reaction toward the sex reassignment was: 'Well, if that what he want to be, I got no objections. I think it was the best thing, the way that he, that it turned out. From the way he turned out he did better. ... He seems to be getting along pretty good.

A new person. He get along with people. ... He talks. Before he wouldn't say nothing. Just suck his fingers. Now he don't even suck his fingers no more. ... He go out working with my father a lot.' Regarding the reaction of other family members, he said: 'They don't, well if they had anything to say, they ain't said nothing. They get along with him just as good as they would anybody else. Better than some kids, I reckon.' People in the neighborhood had not gossiped: 'I don't know why, but they didn't. I ain't heard nobody say nothing.'

This older brother's report was in conformity with what the mother had to say about the child's social improvement. Everybody was happy with the outcome, she said. 'The whole house is happy. ... He's reacting very well. As a matter of fact, I think he was more happier than I was, to tell the truth. But it was his decision, so. I really do think he was very successful. He do have more friends now than he had before. And he associate with people a lot more than he used to. He used to stay off by hisself all the time. ... Even at home he didn't do very much talking. He would just answer if you call him, and that was about as much as you could get out of him. He never did talk much until here lately. And now it's just sixty miles an hour!'

At the time of his 17th birthday, the boy was in an ungraded 'special center' in the public school system. His attendance record was excellent. The public health nurse who supervised his case obtained a report from this center: 'He has worked in a work/training program as an aide to the physical education teacher for the past three years. He is functioning at a fourth grade level and is highly motivated. He has good comprehension and progresses at a faster rate than the other children here. He benefits from the social relationships with them. His size limits his acceptability for employment. I think he would be best suited for a teacher's aide, an orderly in a hospital, a messenger, a xerox operator, or in a sporting goods shop. This student is very intelligent; he has stated that he enjoys this school and hopes that he will never have to leave.'

The public health nurse added her own observations. 'He engaged actively with the boys and girls in the physical education class. He participated in a wrestling match with the boys and was quite aggressive. He seemed to fit well in the interaction with the other boys. He spoke freely with the girls in the class and they seemed to accept his direction. His teacher said that the other students respect him and often look to him as the leader. He said also that he might be sheltering the boy too much, as he has hesitated to release him for employment out in the community.'

Lovemap biography

On the day when the child was admitted for evaluation at age 11:0, the parents had time only for the routine admission interviews. It was a month later that they were first interviewed by members of the psychohormonal staff. The policy decided in advance was that their child would not be pressured or coerced into undergoing either feminizing or masculinizing surgery, since the outcome, either way, would be irrevocable. The aim was to reach a decision that would be compatible with her entire future life, and that she would never have to say that she had been the victim of a mistake. The most prevalent opinion among specialists in her type of case was that she should have been surgically corrected and hormonally treated as a female from the first week of life onward so that she would never have had reason to get ideas of herself as being a boy. It was still not too late for her to undergo surgery and take hormones to be a woman, with a possibility (subsequently refuted in salt-losing CVAH) even of carrying her own pregnancy (Mulaikal et al., 1987). But, for that to be successful, she would have to be sexually attracted to a male lover. If her attraction would be only to a female lover, then it would be better to undergo corrective surgery and hormonal treatment to be a man who would be able to have sex as a man, but with no pregnancies.

In accordance with standard procedure, each parent first gave an interview alone, and at the end of the day joined in a group conference. The mother, in her interview, said that she had passed on to the child, at an age unspecified, the information that she had had an exploratory operation, at age 2:6, which showed that she had female organs inside. The child's reaction was: 'She's puzzled; she really don't know.' The mother had no explanation for the girl of why she had an organ that looked like a penis. She had no explanation for herself of why no operation had been done in infancy. 'I was willing to go all the way to have her normal,' she said, 'like everybody else.' The medical record itself was silent on why the child had not been surgically corrected as a female before age 3, during her long-term hospitalizations in two of the nation's prestigious medical centers.

The mother was the only person at home who had seen what the child's genitalia looked like at age 11. 'Well, it's the same as before,' she said, 'only it seems to have gotten bigger. And you know, more hairy.' She concurred that somebody could mistake it for the organ of a boy. She had been unable to ascertain how the child herself might have been thinking about it: 'Like I say, she don't express her thoughts,' she said. Her guess was that the child must have thought she was really supposed to be a boy, 'after she started taking up with the boys, playing around with rough games. It's hard to tell.'

Asked to use only her own common sense, and to think whether the child would be better off as a boy or a girl, the mother's response was: 'A girl, I think. I don't know why, but I still think a girl. She likes to do girlish work in the house, like washing dishes. She don't mind that at all. Make the beds. Anything like that. But as soon as she finishes, she's outside. She stays to herself most all the time. Once in a while she plays with the rest of the children. Then she begins to give up, and sit down, and put her fingers in her mouth. That's when you can tell she's thinking. She puts those fingers in her mouth. But what she's thinking, I don't know.'

One thing that the mother did not know was: 'Well, what happens, when inside is a girl, and outside is a boy, when the monthly period times rolls around?' After the explanation of CVAH treatment had been given at length, with detailed reference to the child's own unpreparedness to make a decision, the mother's response was, 'Well I wish she'd hurry up and make up her mind,' and with a laugh she added: 'It costs too much to be running back and forth. It ain't like we don't want to get up here. It costs too much.'

Contemplating the possibility that the child would come home to live as a boy, the mother's response was: 'It wouldn't bother me at all. As a matter of fact, I'm used to boys. That's all I had except for her and her sister.' People talking wouldn't be a problem: 'They do that anyway, no matter what happened. Down there, a little town, no matter what happens, they always add to it, anyway, so no reason to bother about that. Them people down there do talk. It wouldn't bother me.' Of those who already knew about the child's anomaly, the mother said: 'The family knows. And more friends of mine. They don't talk about it at all. A whole lot of them down there don't know it.'

The mother confirmed that she thought it really would be better for the child to come home as a girl. 'I don't know,' she said. 'I don't have but that one other girl. The rest of them all boys. And practically everybody around there ain't got nothing but boys.' At this point, the mother listened once again to an enumeration of all the points in favor of the child's being a girl, with the final proviso that none of them would help her if, a year later, the child returned saying, 'You cut it off, and I really want it back.' The mother agreed that she would tell the child in plain words that to return home as either a son or a daughter would be acceptable.

In his interview, the father's response to the alternatives of having a son or a daughter was: 'To tell the truth, I don't know much about it. I ain't never seen her with her clothes off.' Nor had he seen his other children unclothed, even in infancy, except for a couple of the older sons. He had never heard the patient say anything about wanting to be a boy, and was noncommittal about trying to talk to her about it. He didn't know how he would feel if she would come home as a boy instead of a girl, but if he had to make the change over, he said: 'I could try. I reckon so.' If he told her his own feeling, it would be that he'd rather for her to be a girl because 'it's been that's what I thought it was all along. I reckon.'

During the first weeks of the child's hospitalization, it was not possible to predict the occasions when she would be able to utter even a few words, nor the topics that would evoke them. As early as the sixth day, in a short conversation on the ward, she answered, 'boy', when asked which she thought it better to be, a boy or girl. She gave the same answer when the question applied to herself, but had no words to explain why.

The next day, since there were again no words, the Draw-A-Person test was resorted to. The first request specified nothing more than to draw a person. The drawing is a classic of naive expressionism for it symbolizes not only a person, but also a screaming glans penis on a long neck with no torso, but two hefty arms and two skinny, footless legs (Fig. 9.10). The drawing of the self was less symbolic as it had the front view of a face as well as both arms and legs, but the legs had been drawn downwards from the armpits and then covered with a dress, the hemline of which created a gaping square hole where the genitalia would be. The self-and-friend drawing (Fig. 9.11) depicted herself dressed in a skirt and her sister in pants, a depiction of the unrelatedness of sex to style of clothing.

After having done these drawings, the child listened to the explanation that she was in the hospital so that she could be 'fixed up, down here' so as to become completely a girl, or to change over to a boy, and eventually to be either a father or a mother. She did succeed in saying that her sister had said 'nothing' to her about her genitals, and that she would 'fight' anyone who teased her about them, as she had fought her youngest brother.

On the 13th day of hospitalization, she had a session of an hour or more with two members of the psychohormonal staff, a man and a woman. Her nods, shrugs, and rare monosyllabic responses were sufficient to indicate that she comprehended the explanations and illustrations of making a baby, including the function of 'the pussy and the dick' in sexual intercourse. In her case, she was told she would have her 'un-

Fig. 9.10. Case #2. First drawing of self.

Fig. 9.11

Fig. 9.12

Fig. 9.11. Case #2. Drawing of self (left) and friend (sister).

Fig. 9.12. Case #2. The 'Bomb Shoot' made of modeling clay.

finished' sexual parts 'finished' surgically. Then she would be able to become, respectively, a woman or a man in sexual intercourse. As a woman she might carry a baby, but fatherhood would be by the sperm bank or adoption.

She next heard a story, told as a parable applicable to herself, of a child (Case #1 of this chapter) with the same condition as her own, and the same handicap of not being able to talk about being a boy or a girl. The precept of this parable was that the child had solved her problem of not talking by writing the message that she wanted to change and be a girl instead of a boy.

Two days later, using modeling clay, the patient fashioned a cannon which she called a 'bomb shoot,' that had the surrealistic and symbolic appearance of a hypospadiac penis, like her own (Fig. 9.12). The cannon was manned by a little creature, whom she identified as herself. It was all head, mounted on a flat plate. Behind the cannon was 'a kitten on a piece of mountain.' It was added after she had told a friendly secretary that the animals she had on her family's farm were a dog and a cat. Something which looked like a ball atop a post, which the child called a statue, was placed as a target at which the cannon was aimed. In response to a series of questions, she identified the scene as a man shooting at a male statue, with a kitten watching the whole thing.

From modeling, she turned her attention to a box of well-worn toys and created a scene of three men, one on a horse. All three had guns, shooting at each other. She added a flexible rubber-doll fireman, and tried to add a mother doll and a child doll, but could not make them stand up.

She stacked the toys back in their box. Sitting in the chair, she nodded to signify, yes, she had again been thinking about being a boy or a girl. She shook her head clearly to indicate that she did not want to stay a girl, and nodded that she would rather change to be a boy. Another series of nods and shakes of the head indicated that she really had made up her mind; that she did not want to be like her sister; that she did want to be like her brothers; that she had made up her mind not to have babies as the momma; and that she wanted to become a daddy.

Despite its major significance for the child herself, the foregoing exchange was not substantial enough to convince all of those who carried the ethical and medicolegal responsibility for a sex reassignment. Thus the child was reminded, from time to time, that it was essential for her to help by talking more. After such a reminder on the 25th day of hospitalization, there followed a period of free-choice modeling and drawing. The model was a 'duck egg' in a bowl. The drawing was, after three false starts, a self-portrait with three words enclosed in a balloon as in a cartoon: 'You Andrea Boy' (Fig. 9.13). Twice she read the words aloud: 'Me Andrea boy.' (She had made the same dyslexic error, transposing you for me in the reading clinic.) She made other drawings of her sister, of herself renamed Andre with her brother, and of Easter eggs under a rainbow. With the drawings finished, she was visibly stimulated and aroused, as well as smiling and happy. Where she had been sitting, there was a small patch of wetness where urine had leaked out.

Fig. 9.13. Case #2. Self-declared as a boy for the first time.

Fig. 9.14. Case #2. Self-declarations as a boy, one for each parent.

Fig. 9.15. Case #2. Final reaffirmations of being a boy, separately for each parent.

Following this breakthrough, there was a relapse into elective mutism. From the patient's vantage point, her answer had been final. However, in a week, after she had been in the hospital for a month, her parents would return for their first interview (see above) about whether their child would be sex reassigned, or not. They carried the responsibility for signing the operative permit. They would encounter professional discord regarding sex reassignment of a potentially fertile female hermaphrodite as an infertile male. Thus it was absolutely necessary for them to hear their child express herself.

The patient's only sister, aged 15, accompanied her parents on this occassion. There was no role into which she could be coopted as she said she did not know why her sister had been hospitalized, and had no idea what was wrong with her. She had never seen her undressed.

The parents returned home without reporting on the outcome of their talk with their child about whether she should be a boy or a girl. The child's own report of the outcome was nonverbal. It consisted of three drawings (Fig. 9.14). Each was a drawing of herself, and each was labeled, 'Andrea Boy.' The first signified her own decision. The second had the word 'Mother' added and the third, 'Father,' to signify her parents' decisions.

While waiting to begin the session in which these drawings were produced, the child had occupied herself with drawing a house and beside it two shapes, which, some time later, she identified, when asked, as a 'flower' and a 'water hole.' These were the only three words actually spoken during the entire session. There had been only gestural responses to other inquiries indicating, inter alia, that her sex organ did sometimes become hard and stiff, and that she didn't like to touch it with her hands; that she expected to be able to talk more if living as a boy; that she did not have either a boyfriend or a girlfriend; that she thought that, living as a boy, she might have a girlfriend and get married; and that she was not afraid of operations, nor of requiring more of them to be a boy than a girl. To the information, illustrated with an X-ray film, that her bones were already fused, so that she would be a very short man, her reaction was impassive.

Another two weeks passed by before the parents returned. They were together before having separate interviews. The mother reported having asked the child, 'What do you want?' Her reply was, 'The operation.' Then, together, they said 'girl' was what she wanted to be. In her solo interview the mother added details. The girl had been silent for about 10 minutes after her mother asked her what she wanted to be, and then said girl. 'Is that what you really want?' the mother asked, and the reply was 'yes.' She would get used to being a girl once she was back home, the mother said, and would not want to be a boy. The father maintained his position that it would be better for her to remain a girl.

The patient was now in the invidious position of being beset by conflicting messages. One message was that her life was her own, and that her own self-conception as male or female, now and for the future, would be an essential consideration in her prognosis and surgical treatment. The other message was that her parents thought it would be better to remain a girl than change to a boy, and that the current consensus of medical opinion was the same, not in her case, specifically, but in all children with her diagnosis.

Perhaps the child interpreted another visit to talk with her parents as aggravated harassment, since she had already circumvented the constraints of elective mutism and declared herself a boy. Whatever the explanation, she at first objected to leaving the ward to participate in a joint interview with both parents and the two psychohormonal staff members, a man and a woman, who were her personal counselors. After her nurse's aide offered not only to accompany her to the psychohormonal unit, but also to wait for her, she agreed to come on over.

Without her parents in the room, she listened to the explanation that there had been two conflicting messages, one from her parents that she wanted the operation to be a girl, and one from her drawings for us that she was, 'Me, Andrea, boy.' She sat, head bent over, playing with a pencil. Provided with paper, she began to doodle: first, three small open squares and loops joining them; second, letters to form a word in three trials, each underneath the other – Bo, Boo, Boy; third, spaced further down the page in much larger script – Boy, with AN, the beginning of her name, under it; and fourth, shading scribbled all over the page. While scribbling, she was once again reminded that, despite the arguments in favor of being a girl, she would not be coerced. It would be a good idea to take advantage of having her parents here, she was told, to put her message in writing, one addressed specifically to her father, and one specifically to her mother.

Her father was called into the room. She shook her head with a little smile to indicate that she'd not be able to answer questions from him, then nodded that she would write. Slowly and with corrections she wrote: 'I got to B a Boy,' then wrote it again (Fig. 9.15). As requested, she wrote 'Father' at the top of the page, and the date. She was not able to read the message out loud, but nodded confirmation when it was read for her. The father was then asked to add his signature as proof of having received the message in person. He said that he would like her to change her mind, as he would rather have a girl, but he agreed that he would go along with her decision if she did not.

The father left the room and the mother came in. Only a clean sheet of paper was in sight. Again the child nodded that there would be no talking, only writing. This time, she began with the date, and with the name, Mother, followed by the message in duplicate, with a spelling change: 'I gut to B a Boy' (Fig. 9.15). The mother appended her signature and commented that she would like her to change her mind. If not, she would cope with having the child as a boy.

Week followed week after this session in which the parents did not visit and, within the hospital, the management of the case stagnated. With the knowledge of hindsight, it became evident that it was not sufficient to put copies of clinical reports and case notes in the child's pediatric medical record, for these materials took too much for granted regarding the fundamental etiological and diagnostic principles on which prognosis and treatment are based. Explication of these principles was precisely what was necessary for many of those who, at all levels of the health-care hierarchy, were involved in the patient's care. One or more tutorial seminars would have helped at least some of them deal with the medicoreligious and medicomoral issues of sex reassignment for which their conventional wisdom about male and female was deficient.

As the child became increasingly disgusted with procrastination and inaction on her case, the freedom of conversation she had been achieving suffered from more frequent relapses into elective mutism. She told her favorite teacher in the hospital school that she was sick of being asked what she wanted to be, as everybody already knew her answer. So she did not reply to the redundant question. Various people construed her silence to mean that she was having second thoughts. One alleged piece of evidence was that she was wearing pink pants. Her own explanation was that her blue ones were torn. The pink ones had been given her by one of the ladies who was a clerk at the desk. Another example was that she was using perfume. Her explanation was that a student nurse gave it to her, and she didn't have it any more. Another example, of concern to a member of the pediatric house staff, was that she had been playing with a doll from the playroom, and kept it. In her gruff, explosive way of speaking, she said that it wasn't her doll, and she hadn't asked for it. She had taken the doll to keep, because she wanted to give it to her cousin who didn't have one. She had taken five dollars also because she had no spending money of her own, but had given it back.

She had been an inpatient for three and a half months when a conference was convened to resolve the impasse regarding surgery. It was attended by representatives of social agencies from the child's home county. The idea had been mooted that she be placed in a foster home for a year as a boy and then re-evaluated for surgery. This idea was dropped as logistically not feasible. The child herself was averse to it. What she wanted was to stay in the hospital 'so I can get my operation.' The only condition on which she would not go back home, to the farm, was if her parents demanded she live as a girl. Instead of that, she would go to a foster home. Her clothing was already masculine. She had not yet had her hair cut, and she had not yet settled on a boy's name. Eventually she would accept the suggestion of masculinizing Andrea to Andre.

A couple of weeks later, and following one more intramural conference on the case, the way was cleared for the child to be known officially as a boy, and for plans to be laid for masculinizing surgery, and also for change of legal documentation, including the birth certificate which had been changed from male to female at age two weeks. It was necessary also to negotiate various logistical business with the county health and rehabilitation services, one item of which was to obtain a signed operative permit from the parents, who did not keep hospital appointments.

Finally, all the hurdles were cleared. Six days before the six-month anniversary of her admission to the hospital, she had an interview in preparation for surgery, to explain what would be done, and to confirm yet once again, the correctness of having it done. The next day she had a haircut, the official symbol and declaration of becoming a boy, one day before undergoing the first stage of masculinizing surgery. Six weeks later, the second-stage operation followed. He went through both operations alone. For months he had been totally neglected by his family, as if abandoned.

Six weeks after the second operation, in a follow-up operation he was minimally conversational. He said he felt all right. He would not want to go through the operations again, but he would not stay living as a girl. He didn't know what advice he'd give to another child just like himself. Other people did not make comments or tease him when he changed. In the hospital, the person who helped him the most was 'my surgery doctor, the doctor that operated on me.' He never had a feeling of being sorry that he had changed.

It was arranged that he would celebrate Christmas day in the hospital, so as to be present at the charitable gift-giving. He went home the next day for a five day visit. There were no presents for him there, and no Christmas dinner. He enjoyed his visit. People called him Andy, the same as they always had done. They couldn't remember Andre. Sometimes they said she and her. It didn't upset him. He did not report that only his father consistently called him Andrea, she and her (and would persist in doing so for a couple of years). No one made any comment about his haircut, clothes or being a boy. Mostly he stayed indoors and looked at television. He showed that laparotomy scar to his mother and some of his siblings, but he didn't show the surgery on his penis. He wasn't too shy; he just didn't want to.

There was an entry in his chart on the ward that his mother thought the most remarkable change in him was his ability to converse. Asked if he could explain this change, he replied, 'Because I had something else in my mind. I said to myself, when somebody talks to me, I'll talk back to them.' He did not know if changing to live as a boy had made any difference to his talking. The following day, though still laconic, he was able to answer each question with a short sentence when being interviewed before an audience. It was his second experience of Grand Rounds.

Following the disappointing failure of the third operation, it was decided to allow at least a couple of months before the fourth attempt. He was attentive to the explanation of future plans, but his responses were minimal. At the next meeting he talked more. He said he'd be attending a new school and was not afraid of being teased: 'Who's scared of them children? I know somebody who had an operation worse than I did. He stuck something in his mouth and his whole mouth blowed up. ... I'll just hit them, that's all. ... I ain't worried about what they call me. ... Sure I know what I wanted to be when I came here. I knew what they were gonna do when I came in here.' Was he glad about it all? 'If I wasn't, I wouldn't have told them I wanted to be a boy.'

He was ambivalent about having more surgery: 'I wish I could get that while I'm here, so that when I go this time, I go for good. ... If it's not done here, now, I'm gonna tell them I don't want it anyway. Only got two things to put back on there.' He was referring possibly to bilateral prosthetic testicular implants. He did not mention that, without further surgery, he could urinate only in the sitting position.

He had less to say about getting a girlfriend later in teenage: 'I might have one,' was all he said; and, of children, he might want her to get, 'maybe a few.'

The Saturday following this interview was posted as the day of discharge, the conclusion of the 11-month hospitalization. The parents failed to appear to take him home, and could not be traced by phone. Wiping away a tear of betrayal, the boy questioned his parents' fidelity: 'Don't you think I have a right to know when they come?' he asked.

He was three weeks short of his 12th birthday when on the following Saturday he did get discharged. This was the occasion when, while the housestaff officer was on the phone settling details for their discharge interview, the parents abruptly departed with the boy, under the pretext of taking his luggage down to the car.

It would require almost three more years before, at age 14:10, the genital reconstructive surgery would be completed. During much of this period, it required an inordinate amount of effort to ensure that the patient returned for follow-up. Such information as was obtained included nothing new with respect to sexuoerotic development.

Subsequently, in follow-up visits between the ages of 14 and 19, inquiries about having a girlfriend or being at the girlfriend stage produced responses signifying that at school he socialized with girls, and perhaps with one in particular, as well as with boys. When he was 15, his mother said, 'I know he likes girls. He's got his little telephone book with all the girls' names in it. But everybody lives so far away, they get together, I guess, when they have a party that everybody go to.' There was no report of dating, and no evidence of any romantic involvement with anyone, male or female. Nothing was disclosed about masturbation, nor about masturbatory fantasies or erotic dreams. The only information he gave about dreaming was that 'I very seldom ever dream.'

Reluctantly, at age 24:2, he returned for a general checkup. He had long ago had his fill of doctors and their hospital, and wanted to be left alone. Even though he had never doubted the correctness of his decision to be a man, he could not think of any special advantages to being a man – 'None that I can see' – or a woman – 'I wouldn't know about that, I don't have their feelings.' Self-rating his appearance, he said: 'I think I look all right myself. I don't know about the rest of the people. I look better than some of the people I've seen.' He did not want to change anything about his appearance. He did not take hormonal or any other medication.

His sports interests were playing horseshoes, baseball and basketball with his brother and neighbors, and sometimes going to a baseball game with his brother. Indoors, he put together puzzles and model cars, Chevrolets and Pontiacs only. He rated himself not as a leader, but one who generally goes along with everyone else. Most of the time he would be by himself, but not lonely. He wouldn't change anything in his life: 'If you change it, it wouldn't do no good.'

His reply when asked about having friends who are male was: 'All I fool with are kin [brothers, cousins, uncles]. I don't fool with no one else.' Likewise, female friends: 'No more than my kin. All the girl friends I got are my sister, cousins, and my aunt.' He had been friendly with a few girls when he was in school, but no longer saw them. He had never had a girlfriend, and never bothered to look for one: 'I ain't got nothing about no woman or girlfriend on my mind.' He said he'd never had any experience of hugging, kissing, or sexual intercourse. Although he had previously, as an inpatient, been given explicit information on reproduction and a sex life for himself, he disclaimed having knowledge of where babies come from, or of having ever heard talk about sex from his brothers or anyone else. He rejected marriage: 'After seeing what these married people down in Maryland do, I don't want to get married.'

He allowed that although he got good feelings from his penis 'every now and then, maybe about once a week,' he didn't do anything about them. He could not say what would start the feelings, but it wasn't thinking about girls because he never thought about them.

He did not want to have children, nor to have 'to put up with them.' Though he did do babysitting for his sister's five children, ages 4 to 11, he didn't like it, because 'they have gotten out of hand. Their father and mother let them get away with too much.' He would rather sit with the older children, as 'I don't have to tote them everywhere I goes.'

The fluency of his responsiveness was no greater than in times past, even though he had known the interviewer, a black woman member of the psychohormonal staff, for years. She switched, therefore, to giving the Sacks Sentence Completion Test, orally. The noteworthy items were, by number:

11. My idea of a perfect woman... 'I don't know nothing about no woman and don't want to. I don't want to hear anything about anything.'
12. When I see a man and a woman together... [no reply].
25. I think most girls... 'Skip it altogether.'
26. My feeling about married life is... 'It's stupid.'
30. My greatest mistake was... 'That I came up here. Wasted my time. Got to wait till next week, now, to pick up my check.'

Thereupon, he left the remaining 30 items unanswered. He had had enough. Subsequently he did not return.

EXPOSITION

The likelihood of finding a pair of cases doubly matched, initially on the concordance of prenatal history, and subsequently on the concordance of undergoing a sex reassignment, is not absolutely unique, but it certainly is extremely rare. It is all the more rare, insofar as the two cases were discordant for sex of rearing, one with a boy's name and the other with a girl's name, and hence for sex of reassignment, one from boy to girl, and the other from girl to boy.

This multiplexity of concordance and discordance can be attributed in part to disjunction in the delivery of modern medical care when there is no one agency or clinic responsible for the conjunction of all clinical findings and long-term case management. Such lack of conjunction is not specific to the children of any one socioeconomic class. Nonetheless, it is exacerbated for the children of poverty, as in the present cases, when the parents lack not only the free time, transportation, and finances, but also the sophistication to obtain coordinated case management.

In both cases, the postponement and neglect of corrective genital surgery early in life cannot be attributed to lack of parental sophistication. The program of treatment, surgery included, was a responsibility of specialists within the health-care system. There is no retrievable information by which to explain, during the period of infancy, the neglect of corrective surgery concordant with the sex in which each child, after the inconsistency of having a neonatal redeclaration of sex, was being reared. The juvenile years, the patients' own elective mutism, misconstrued as either indifference or mental incompetence, provoked still further neglect not only of corrective surgery, but also of corrective hormonal therapy.

The consequence for both children was that the sex in which they were named and were being reared was not confirmed by the appearance of their ambiguously malformed genitalia. Both were able to construe themselves, and to be construed by others, as belonging partly to both sexes. Hormonally untreated, both of them outgrew their infant age-mates in size, and underwent premature pubertal changes so as to have the masculinized appearance of an adolescent boy. In Case #2, these changes further increased the evidence of her genitalia that she was more boy than girl.

For both children neglect of treatment was equivalent to avoidance of treatment, and it was consistent with neglect and avoidance of explanations or, indeed, even of any explicit reference to their boy-girl, genital ambiguity by those responsible for their well-being. They responded in kind, not only by avoiding to talk about their sexual dilemma, but by being overcome with elective mutism regarding it. It became the unspeakable monster in their lives about which, literally, they could not speak. They could not escape the entrapment of being damned-if-you-do and damned-if-you-don't talk about the monster.

Since the outcome of the Catch-22 is to be damned by talking or not talking, either alternative might, hypothetically, be equally available, or one may take precedence over the other. There is no principle of human behavior by which to predict which alternative will prevail. Nor is there a principle to explain why, when elective mutism takes over, its strength and persistence are intense and unyielding to the logic of common sense regarding the self-sabotage of not speaking up.

The child afflicted with elective mutism is not, like a brain-damaged patient with receptive aphasia, incapable of following the logic of common sense. The pathology of the electively mute is encapsulated and applies specifically to their being incapable of responding in words. That is why their suffering is made worse by those who, their patience exhausted, admonish, chastise, humiliate, punish or abandon them. By contrast, it is why their suffering is alleviated by those who resort to alternative modes of communicating and transmitting messages. In the present two cases the alternatives were messages transmitted by the child in gestures, three-dimensional models, drawings, and in writing. These modes of communication did not require person-to-person eye contact, which could also be circumvented by talking on the telephone, even on the intercom between offices. In primates averting the eyes is associated with avoidance and submission, and staring with assertion and dominance-threat. In both patients eye contact became unthreatening and elective mutism lifted after the genital deformity had been surgically resolved and the sex reassignment accomplished.

Didactically, the special significance of the sex reassignment in these two cases devolves from the fact that, whereas they were concordant for prenatal developmental history and diagnosis, one elected to be reassigned from boy to girl and the other from girl to boy. If there had been some factor or factors already present in prenatal life that had preordained how their G-I/R would differentiate postnatally, then, irrespective of the sex of rearing, both should have differentiated either a female G-I/R or both a male G-I/R. Had they both been prenatally preordained to differentiate a female G-I/R, then a sex reassignment would have been needed only in Case #1, from male to female, and in Case #2, the child would have remained living as a girl. Vice versa, had they both been prenatally preordained to differentiate a male G-I/R, then a sex reassignment would have been needed only in Case #2, from female to male, and in Case #1, the child would have remained living as a boy.

Insofar as neither of these two alternatives actually materialized, the hypothesis that the differentiation of the G-I/R was totally preordained in the course of prenatal development does not hold up. The alternative hypothesis is that the differentiation of G-I/R is ordained in two stages, one prenatal, and the other postnatal, and that the postnatal determinants are not only an extension of the prenatal determinants but may, under specified conditions, override them. The present two cases lend support to this hypothesis. They do so insofar as the two children grew up to differentiate a G-I/R that embodied the unresolved ambiguity of their genital status which failed to confirm the correctness of the sex to which they had been officially assigned.

In the society in which we live, and in which these two children grew up, sex is a binary system of male and female with no in-between position for a hermaphrodite. Thus, for anyone, hermaphrodite or otherwise, who experiences himself or herself as a misfit in one sex, the only logical alternative is the other sex.

The evidence in support of being a boy who had been misassigned as a girl was undeniable in Case #2. She could look at her genitalia and see that they looked like those of a boy, and no one ever talked to her to tell her otherwise. Her body matured prematurely to look big and strong, with the build and the muscular strength of a boy, while she was still very young. She grew body hair and had a deep voice. She liked the sports and outdoor activities stereotyped as being for boys. She wore the clothes that, even though in fashion for girls, were those that a boy might wear. The two short plaits in her hair alone gave other people the message that she was classified socially as a girl. She did not discuss her situation with anyone, but silently developed the conviction that she had to be a boy. After the sex reassignment which allowed her to be officially proclaimed a male, there were no signs of androgyny. Socially, in adulthood, the transformation was complete, except for holding back from romance and a sex life with a partner. There was no evidence of homosexual attraction as a male. His disclaimer regarding marriage was likely to be contingent on being sterile.

In Case #1, the evidence of being a girl misassigned as a boy was, in the early years of childhood, exclusively genital: the organ that was supposed to be a penis was far too small, it was not equipped to use for urination while standing, and there were no testicles. No one talked to him about his predicament of being a boy without the penis of a normal boy, and no one ascertained whether, as a small child, he might have had a conception that his genitals would develop to become more masculine like those of his brothers, or more feminine like those of his sisters. What is known is that he became a juvenile recluse, hiding himself and his genitals from others, as his body matured too early, but in the manner of a male.

Whatever reassurance the bodily evidence of masculinization may have given him, it was undermined if not cancelled, at around age 10. That was when his mother, following doctors' orders, told him that he was a female. He was at the same time given corrective hormonal treatment with cortisol which released the ovaries to secrete female hormones, so that his breasts began to enlarge. The feminizing changes of the body were not spontaneously accompanied by feminizing changes of the mind. It would, in fact, require another two years of searching for a solution to the issue of sex reassignment before he would reach the verdict that he should change and live the rest of his life as a woman.

The substantive evidence of beginning breast enlargement was a reassurance that medical predictions of further feminization, including menstruation, could probably be relied upon. There was no prediction, however, that feminization of the mind would be included, for there is no psychohormonal evidence of a direct, point-for-point correlation, between the male or female hormones of puberty, on the one hand, and the male or female orientation of the G-I/R on the other.

The failure of G-I/R feminization by estrogen was well illustrated in another similar case of the CVAH syndrome in an individual with two ovaries and a 46,XX chromosomal karyotype (Money, 1988; Money and Daléry, 1976; Money and Ehrhardt, 1972, Ch. 8). The child in this case had been born with external genitalia of normal male morphology except that the scrotum was empty. The sex was declared as male. From age 3:6, he was correctly treated endocrinologically to correct the CVAH hormonal anomaly, but on the assumption that he was a chromosomal (46,XY) male with undescended testes. The error of this assumption was recognized at age 11 when he began to develop breasts. For another two years, until menstruation was about to begin, the hormonally feminizing treatment continued, presumably on the assumption, also in error, that the mind would feminize in synchrony with the body. The boy's reaction was to be mortified, as is typical for normal males afflicted with the gross breast enlargement of adolescent gynecomastia. His masculine G-I/R did not change. He continued to go hunting and fishing with his father, to go dirt-biking with his friends, and to have a profitable farm-club hobby of raising steers. 'He has a sister,' his mother reported, 'and they are completely different. He does not think like a girl, and he does not have the same interests. Right now, the one thing that made me very sure is that he has a girlfriend. And that to me was a relief, because that was just the clincher that he wasn't a girl.'

When the boy first heard of Johns Hopkins and the possibility of a surgical mastectomy to flatten his chest, 'I really couldn't wait until I could get here,' he said. 'I was very overjoyed.' For him, feminization of his body was incompatible with his G-I/R as a male. With the incompatibility surgically and hormonally resolved, he resumed his life satisfied and successful, as a man. In his case, female hormones secreted from the ovaries, and circulating in the bloodstream, did not feminize the G-I/R. Rather, the masculinized G-I/R dictated that female hormones be rejected, and replaced with male hormone.

So also, in the case of the boy who underwent sex reassignment, female hormones secreted from the ovaries, and circulating in the bloodstream, did not feminize the G-I/R. Rather, female hormones became progressively more compatible with the G-I/R as its femininity took precedence over its masculinity. Socially, the transformation became progressively more convincing. Except for the skeletal structure, the bodily evidence of masculinity disappeared. So also, to a large extent, did the behavioral evidence, the notable exception being sexual attraction, which was toward not only males, but also females. Sexuoerotically, the masculine component of the G-I/R from the first 10 years did not disappear, but coexisted with the superimposed feminine component, and was manifested as serial bisexuality, with many love affairs, one at a time, in sequence, and a continuously active sex life.

BIBLIOGRAPHY

Abramovich, D.R., Davidson, I.A., Longstaff, A. and Pearson, C.R. (1987) Sexual differentiation of the human midtrimester brain. *Eur. J. Obstet. Gynecol. Reprod. Biol.*, 25:7–14.

Migeon, C. (1989) Diagnosis and treatment of adrenogenital disorders. In: *Endocrinology*, 2nd edn., Vol. 2 (L.J. DeGroot, Ed.). Philadelphia, Saunders, pp. 1672–1704.

Money, J. (1988) *Gay, Straight, and In-Between: The Sexology of Erotic Orientation.* New York, Oxford University Press.

Money, J. and Daléry, J. (1976) Iatrogenic homosexuality: Gender identity in seven 46,XX chromosomal females with hyperadrenocortical hermaphroditism born with a penis, three reared as boys, four reared as girls. *J. Homosex.*, 1:357–371.

Money, J. and Ehrhardt, A.A. (1972) *Man and Woman, Boy and Girl: The Differentiation and Dimorphism of Gender Identity from Conception to Maturity.* Baltimore, Johns Hopkins University Press.

Mulaikal, R.M., Migeon, C.J. and Rock, J.A. (1987) Fertility rates in female patients with congenital adrenal hyperplasia due to 21-hydroxylase deficiency. *New Engl. J. Med.*, 316:178–182.

White, P.C., New, M. and Dupont, B. (1987a) Congenital adrenal hyperplasia (First of two parts). *New Engl. J. Med.*, 316:1519–1524.

White, P.C., New, M. and Dupont, B. (1987b) Congenital adrenal hyperplasia (Second of two parts). *New Engl. J. Med.*, 316:1580–1586.

CHAPTER 10

Bisexually concordant, heterosexually and homosexually discordant: a matched-pair comparison of male and female congenital virilizing adrenal hyperplasia (CVAH)

SYNOPSIS

The matching of the two cases in this chapter for concordance is like that of Chapter 2, for in prenatal life their concordance was partial, but not complete. On the criterion of diagnosis, they were concordant, for each had inherited from each parent the recessive gene for the syndrome of congenital virilizing adrenal hyperplasia (CVAH), also known as the adrenogenital syndrome. As a consequence of their CVAH inheritance, in prenatal life both fetuses were hormonally flooded with masculinizing androgen erroneously secreted by their adrenocortical glands as a substitute for the correct glucocorticoid hormone, cortisol, that should have been secreted.

At the time in prenatal life when they were being masculinized with adrenocortical androgen, the discordancy between the two cases was in chromosomal and gonadal sex. In Case #1, the chromosomal sex was 46,XX, and the gonadal sex was ovarian. In Case #2, the chromosomal sex was 46,XY, and the gonadal sex was testicular. Thus, the masculinizing effect of the excess of adrenocortical androgen was to masculinize a female fetus in Case #1, and to hypermasculinize a male fetus in Case #2.

In Case #1, the prenatal masculinizing effect of CVAH was by no means as extreme as it might have been, which makes her case all the more instructive as one of the pair matched here with respect to degree of bisexuality in adulthood. Had she received the full flood of prenatal CVAH masculinization, she would have been born with a penis and empty scrotum, instead of a homologous clitoris and female vulva. In her case, however, the flood held off until after both the internal and the external sexual organs had already differentiated as female, except for insignificant enlargement of the clitoris, and insufficient distance separating the orifice of the urethra from that of the vagina.

Prenatally, the timing of the flood of masculinizing adrenocortical hormone, though too late to masculinize the sexually dimorphic fetal differentiation of the external sexual organs in Case #1, would not have been too late to masculinize the sexually dimorphic differentiation of the brain, and so increase its dualistic potential for a further degree of masculinization in postnatal life.

In Case #2, by contrast, the prenatal flood of adrenocortical androgen would not alter the dimorphic differentiation of either the internal or the external genitalia that were already destined to be those of a male. Likewise, it would not alter the masculinization of the brain, already destined to be that of a male. However, it might hypermasculinize the brain, and so preordain further masculinization and the exclusion of demasculinization in postnatal life.

At birth the two infants were discordant for assigned sex, in Case #1 as a girl, and in Case #2 as a boy, on the basis of the visible evidence of the external genitalia. The sex of rearing was discordant, as girl and boy, respectively. In Case #1, the rearing was not complicated by uncertainty as to the sex of the external genitalia until age 4:8, when the CVAH diagnosis was first made. In Case #2, there was never uncertainty about the sex of the external genitalia.

In both cases, the hormonal status in infancy and early childhood differed from that of other children, and was concordant for continuing high levels of adrenocortical androgen in the bloodstream. In both cases, the androgen level was sufficient to produce the bodily signs of precocious onset of puberty. On this criterion, the two cases were again concordant, for in each one the signs of puberty were those of virilization.

In Case #1, the signs of virilization were correctly recognized at age 4:8. They were accelerated growth in height, oily hair and skin, acne, enlargement of the clitoris, and the beginning growth of pubic hair. In Case #2, the signs of virilization were precocious in onset, but otherwise were the same, except that enlargement was of the penis, not the clitoris. By age 3:6, the appearance was that of a boy well into adolescence.

Concordance for the CVAH diagnosis was established at different ages, 4:8 in Case #1, and 3:6 in Case #2. Following diagnosis and the onset of replacement treatment with the glucocorticoid hormone, cortisol, the masculinizing error of the adrenocortices should have remained continously suppressed in both cases, indefinitely. In Case #1, it did do so. In Case #2, it did not, owing to maternal noncompliancy based, as she later confessed, on her conviction that her son's condition was untreatable, as she attributed it to divine retribution for the sin of his conception in sibling incest.

In adulthood, the two cases were discordant for hormonal maturation and bodily appearance. In Case #1, because of compliancy with treatment, virilization of the girl's body ceased, and her ovaries functioned to induce normal feminine maturation and the onset of menses. The consensus of those who saw her was that she was quite stunningly beautiful, with the qualities of a fashion model. In Case #2, the chief penalty of noncompliance with treatment was accelerated bone maturation and short stature in adulthood. Masculine maturation of the body and its appearance was effected during the period of noncompliancy by adrenocortical androgen, and during subsequent compliancy in adulthood, by testosterone secreted in normal fashion by the testes.

In the everyday social life of adulthood, the two cases appear to exemplify the polarity of masculine and feminine, and to be wholly discordant for gender-identity/role. Sexuoerotically, however, there is an unexpected concordancy, for in each case there is a history of having tried to engage in a relationship with a male (more than one male in Case #1), and of having failed. Both the woman and the man have been romantically and limerently attracted only if the partner is a female. Each has been able to fall in love only with a female. Each has established long-term relationships only with a female. On the criterion of falling in love, the two are discordant, for society defines her love as homosexual, and his as heterosexual. Bisexuality for her meant trying to be heterosexual, and for him, homosexual, in both cases without success.

CASE #1: CVAH FEMALE

Diagnostic and clinical biography

This patient was adopted at six days of age, and is an only child. There is no available information concerning the birth or the birth parents. Although the adoptive parents had noted clitoral enlargement not long after the child came to live with them, the family doctor assured them that the size of the clitoris was within normal limits. It was not until the girl was 4 years old that he decided otherwise, and made a referral to the Johns Hopkins pediatric endocrine clinic.

The initial endocrine examination, at age 4:8, revealed a clitoris measuring 2.8×1.2 cm with a well-formed glans and prepuce (Fig. 10.1). There was a urogenital sinus at the base of the clitoris and fusion of the labioscrotal folds. There were 10 coarse black pubic hairs. The child's height was 123 cm (48.5 in.) and the height age 7:9 (Fig. 10.2). Hair and skin were oily. Acne had been present for six months. The diagnosis was established as hyperadrenocorticism (CVAH), nonsalt-losing type, and treatment with hydrocoritisone was begun. A week later, surgery was performed which included clitoridectomy, separation of labioscrotal folds and plastic repair of the introitus. On a dosage of 12.5 mg of hydrocortisone three times a day, the patient's hormonal condition was always well controlled from infancy onward, into adulthood.

The earliest sign of the onset of pubertal feminization, namely breast enlargement, was first recorded during the eighth year. Onset of menarche was at 10:4. At age 13:0, the height was 162 cm (64 in.). The final adult height of 164.5 cm (65 in.) was attained by age 17:0. Weight at age 17:0 was 59.5 kg (131 lb). A gynecologic examination at age 15:2 revealed a normal cervix and a vagina which admitted one finger. There was no constriction of the vaginal outlet. The uterus was of adult proportions, but was

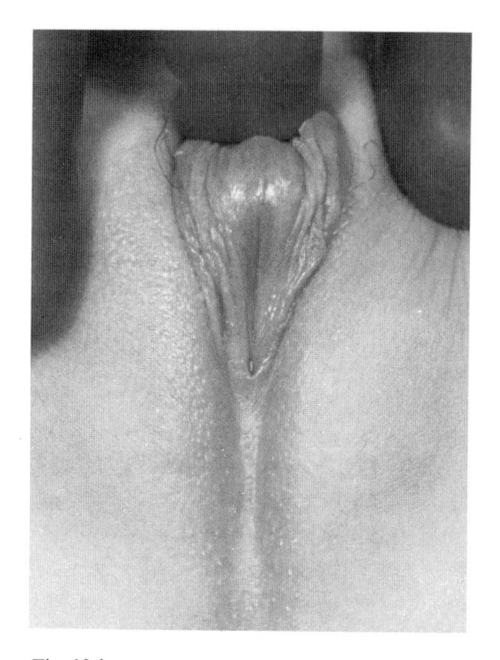

Fig. 10.1

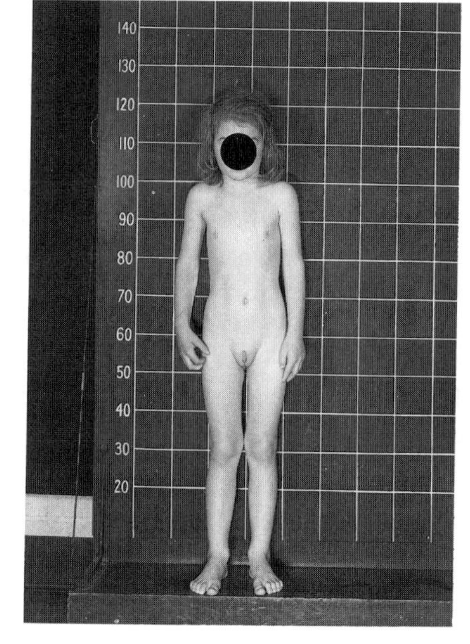

Fig. 10.2

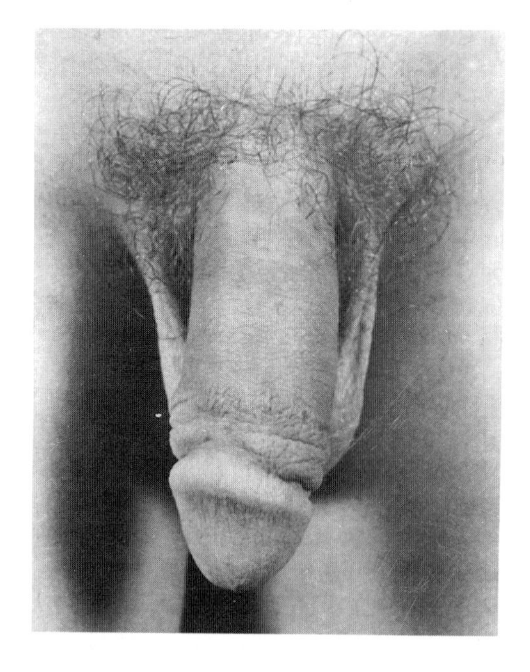

Fig. 10.3

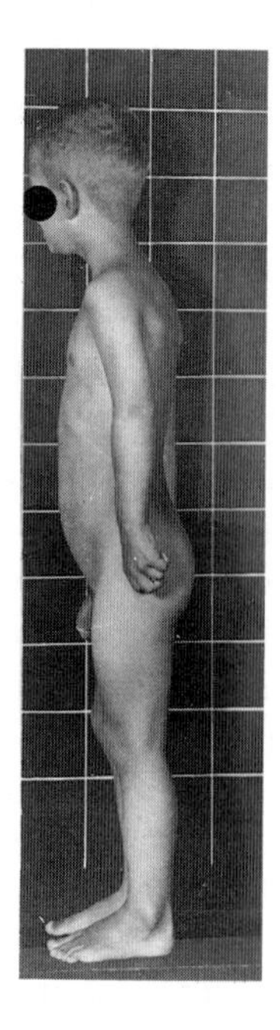

Fig. 10.4

somewhat smaller than normal for age. At the age of 18:0, the young woman's endocrine care was transferred from Johns Hopkins to an endocrinologist in her home town. Psychohormonal follow-up was maintained in the Johns Hopkins psychohormonal unit until age 31.

During childhood and adolescence, the girl was given age-appropriate self-knowledge regarding her diagnosis and clinical history. As an adult she had an adequate working knowledge of the etiology and prognosis of the CVAH syndrome.

Gender-coded social biography

At age 4:8, this girl's IQ on the Revised Stanford-Binet Scale (Form L) was 98. At age 7:0, on the Wechsler Intelligence Scale for Children, the findings were: Verbal IQ, 105; Performance IQ, 117; Full IQ, 112.

School achievement was always high. In junior high school, the girl made the honor roll. She was reported by her mother to always want to be the best in everything she did. While in junior high school, she played a stringed instrument in the local symphony orchestra and also tutored 6th-grade students once a week. In interviews, she always maintained that she wanted a professional career. The choices of career varied over the years from teacher to doctor. She went on to college and obtained an undergraduate degree in business. She has worked in the insurance and investment field from graduation to the present.

In childhood, she was reported by her parents to have tremendous energy and to be somewhat tomboyish. They did not consider the tomboyish behavior unusual in light of the amount of energy. In retrospect, at age 21, the girl corroborated that she had indeed been energetic when younger. She emphasized, however, that although she had been a tomboy and had hung around with boys, she had also hung around with girls.

Doll play was late in onset. According to the mother's memory, which differed on two occasions, the age at which the girl played with dolls was somewhere between 5 and 11 years. She enjoyed playing with boys' toys and was reported to remark that they were more interesting.

Despite her tomboyish reputation, she also cried easily and did not take up for herself when there were peer conflicts in childhood. Tomboyism did not extend to

Fig. 10.1. Case #1. Appearance of the genitalia at age 4:8, before treatment.

Fig. 10.2. Case #1. Appearance of the body habitus at age 4:8, before treatment.

Fig. 10.3. Case #2. Appearance of the genitalia at age 3:6, before treatment.

Fig. 10.4. Case #2. Appearance of the body habitus at age 3:6, before treatment.

being active in sports in a formal way. She was not well coordinated and did not excel in physical education or sports.

She has always had a good relationship with her parents, summed up in adulthood as follows: 'They are two of the neatest people, and they have a good relationship. They love each other so much, and it's so visible. They've been married for so many years, and they still hold hands. They have always gotten along well. It's great to be brought up in that kind of atmosphere. I can't imagine being brought up where a father or mother drank or fought.'

She has lived on her own since she was 22. 'After I graduated from college, I was at their house for maybe a few months. I just couldn't deal with that. ... I think I would have felt bizarre bringing lezzies and faggots there.' At age 31, she conjectured that her parents knew that she lived in a homosexual relationship, but avoided mentioning it: 'I think they do know, but it's never really been spoken. I think they've known I'm gay maybe five years, maybe longer. My mother asked me, a long time ago, if I was gay: "Jenny, you would do something like that," she said, "just to be antisocial or whatever." That's what she used to think. At the time, I think I said no. I didn't want her to know.'

She has lived with her current lover for two years, and they recently purchased a home whithin an hour's drive from her parent's home. She sees her parents at least once a week. When her job requires her to be in their town, she invariably visits them for lunch. She does fieldwork as an insurance adjuster, greatly enjoys her work, and is satisfied with her salary.

So far as she can recall, there has been no period in her life when the fact of her adoption was guarded as a secret. Her parents at the time of the adoption expressly requested to have no knowledge of the baby's birth parents. However, they supported her, when at age 31, prior to a change in state law, she 'sent away for my adopted birth certificate, and I got it. And I got their names.' Though the birth mother lives only a few blocks from her adoptive parents, she has not yet attempted to make contact. The birth mother is not someone that she would recognize, she said, even though she lives in the same neighborhood.

Lovemap biography

There is a note in the girl's history, when she was aged 4:8, that she was pretty and appropriately feminine in appearance and bearing. When she was seen again at age 7:0, her mother was positive about the girl's behavioral development without mentioning anything specifically psychosexual. The parents were emancipated with regard to sexual learning. They took an interest in providing age-appropriate sex-educational materials. They were given guidance by the psychohormonal staff regarding the adaption of sexual information for a girl who had had feminizing surgery for CVAH.

When the girl was 12, she confided to her mother that she had been having a good feeling in the genital area and thought that that made her bad. The mother reassured her daughter that the feelings she was having were normal. If there was a history of childhood sexual rehearsal play, the mother was not aware of it.

As a 12 year old in junior high school, the girl spoke of an interest in boys. Retrospectively, however, at the age of 21, she recalled that she had been aware of a sexual attraction toward girls since the latter part of elementary school. 'I used to have crushes on girls. I used to think they were really neat, you know. Older girls I'd look up to. And this summer I had a sexual experience with a girl, and that was oral sex. I really had a good time.'

Dating was one of the topics of an interview with the parents when their daughter was 14. They agreed that she was not dating — which they favored as they were against going steady at age 14. Though the mother considered that the 'boy stage' was already in evidence in her daughter's development, the girl's own opinion was that, though she did some boy-watching, she did not find it to be sexually arousing. By the time she was 18, she told her mother that she hadn't met the right boy yet, and would rather be with girls than boys. Looking back, at the age of 21, she said that she had never been boy crazy, because: 'I like older people. I could always relate more to older people. And the guys in my high school, there were a few that I had crushes on, but not many. I thought most of them were assholes. ... I thought other girls who were boy crazy were very silly.'

At age 14, she had discovered a focus of feeling in her sex organs in response to reading sexually explicit pages of works of popular fiction which she designated, not exactly correctly, as 'dirty books.' But she had no subsequent urge to masturbate.

At age 18, she kidded a boy into taking her to a homecoming dance, after which they began dating, though not seriously. They got into 'petting and stuff like that, but not going the whole way.' She discovered that she was erotically turned on primarily by way of the sense of touch, the breasts being especially sensitive. The tactile primacy of the breasts gave way to the genitals by the time she was in her twenties, at which age she had also discovered an erotic zone around the neck and ears, on the abdomen, and around the perineum and anus.

Self-touch, as in masturbation, was not erotically satisfying. She had masturbated occasionally, but without orgasm: 'I need someone else, I guess. I can't do it myself.'

Visually, she did not get turned on by pictures: 'Magazines don't turn me on. No. No way.' The same applied to movies. Seeing a man who was attractive to her was not an erotic turn-on, unless he was wearing tight clothes, especially tight pants or swim trunks. The sight of a woman's buttocks was especially erotic, and she could be turned on by the sight of her woman lover.

She preferred oral sex to vaginal intromission, and enjoyed it with either a male or female partner. With a male, she mainly enjoyed cunnilingus. She rated her sexual experiences with each of two woman lovers as different from and more satisfying than those with a man. She would not go to bed with a man whom she did not like. A man she liked would be 'intelligent, really sensitive, and good in bed. He would also have to be open, tolerant, honest, and somebody I can change with, and that can change with me.'

Her sexual experience with men had been as a freshman and sophomore in college. 'I just can't remember them,' she said. 'They were like one-night stands, or a short-lived relationship, maybe a month. But I just never enjoyed it. It never turned me on.'

With a woman partner, she said, 'That's entirely different. Exciting someone else excites me. I get excited when my partner gets excited, and it always results in orgasm — well, most of the time.' The history and personal knowledge of neonatal clitoridectomy did not adversely affect orgasm. Usually it was her woman partner who would initiate sexual activity. Her current partner did not like to receive digital-vaginal sex, but she did like to perform it as well as oral sex. 'And that really does turn me on,' the patient added. She described her orgasm as being: 'Just my whole body. It's like I'm in a different state. It starts in my genitals, and it just feels so good. It's so sensitive. ... It radiates, and my whole body just gets limp, you know, afterwards, and I feel all relaxed.'

Rating her sex drive, she said: 'I think it's pretty high, because I love sex. I have always. I think I'm a very physical sexual person. At least that's what I'm told. People tell me that I'm very seductive, and I'm told that I'm sexy, but I don't think I am. People tell me that I flirt a lot, but I don't think I do.'

She said that while having sex, 'I don't like to do it in the dark, because I like to see the person. I like to see what I'm doing, you know.' She considered herself a bit experimental, using, for example, whipped cream as an adjunct to oral sex.

Queried about maternal interest, her response was: 'I've said I don't like kids, but I do. I get along with them. I've always been able to relate to children. ... Not little babies, you know. I'm afraid I might drop them. I think I like them maybe between the ages of five and ten. ... Yes, I do think I'd like to have a child someday. ... Yes, my own, definitely. I'd like to have one of my own. I would like to have it by a woman, but I know that will never happen.'

She considered how she would go about picking out the father. 'He'd have to be a very — he'd have to be a man that I loved, you know; and he'd have to be gay also. I can think of one or two right now — three right now that I truly love, and they truly love me, but we just — I mean, sex is not even, it's not there, and it won't be there. But I think we could have a child. I don't know if they would want to take care of it, you know, with me. ... Maybe a couple, two women that liked each other, and two men. The man and woman would be married, you know — two marriages where the child can be raised, but the women would be lovers, and the men would be lovers.'

On a more realistic note, and with the wisdom of additional years, she said: 'I do think about having children from time to time. There is the obvious act of doing that which I don't really care to participate in. What I think I might do is to become a Big Sister. I think that would be really fulfilling.'

CASE #2: CVAH MALE

Diagnostic and clinical biography

At age 3:6, this patient was referred by way of the state health department to the Johns Hopkins pediatric endocrine clinic because of premature onset of puberty. The initial clinical examination revealed a penis (8 cm × 3 cm, flaccid) and scrotum of a 'boy well along in adolescence,' although the testes were infantile in size (Fig. 10.3). The height was 120 cm (47 in.). The height age was 7 years, and the bone age 11 years. He was easily mistaken for a 6 or 7 year old (Fig. 10.4). The skin had an unusual bronze color, and the blood pressure was elevated. The diagnosis was hyperadrenocorticism (CVAH) of the hypertensive type. Both the skin color and elevated blood pressure responded to glucocorticoid treatment which was stabilized at 37.5 mg, intramuscularly, every third day. Subsequent treatment was erratic or nil, owing to maternal noncompliance.

By age 3:10, under the influence of glucocorticoid treatment that suppressed adrenocortical androgen, the patient's testes had begun to enlarge. By 5:1, they were at the 'borderline of adolescence,' and the prostate was palpable. Only with the knowledge of hindsight were grossly elevated ketosteroids at age 5:4 attributable to parental noncompliancy. When the boy was seen again at age 6:6, the penis measured 12.5 cm × 3 cm, and the testes 4 cm × 2 cm. The scrotum was of adult configuration with pigmentation and wrinkling.

The patient was then lost to follow-up for 21 months, after which he was located through the State Board of Health. He was admitted to the pediatric service to reinstate treatment. His chronological age was 8:3. He was well-muscled and looked like a pubertal boy. The skin again had increased pigmentation. The mother could not specify how long it had been since the last injection of cortisol. Thenceforth, the boy was lost to follow-up in the psychohormonal unit until he was aged 13. In pediatric endocrinology, even though his mother said he did not need an endocrine follow-up, he was seen again at age 10:6 after an otolaryngologic evaluation. She reported that during the interim, she had obtained medication 'locally' and had been giving it to the boy 'on and off.' On clinical examination, however, it appeared that his ketosteroids had been inadequately suppressed, consistently. His height was 152 cm (60 in.). The epiphyses were closed, and he did not grow any taller.

The mother had a fatalistic attitude toward the boy's condition. His increased coloring was, for her, the evidence that the doctors had little to contribute. Nonetheless, she went along with the reinstitution of treatment. When the boy was seen three months later, at age 10:9, it was evident that he had been receiving and was responding to medication.

He was lost to endocrine follow-up until, at age 13:1, he was admitted in an adrenocortical crisis. To escape what later proved to have been child abuse, he had run away from home and for several days and nights, with the help of a school friend, went into hiding in a farmer's barn. There his feet had become frostbitten. Necrosis necessitated amputation at the transmetatarsal level on the left foot and the metatarsal-phalangeal level on the right foot.

At this time, the mother revealed that her son was a product of sibling incest. Her reaction to this secret in her life had been an intense love/hate relationship with her son. She interpreted the boy's adrenogenital condition as divine retribution for incest. This was her rationale for having not sought treatment for the child early in infancy, and for subsequently having withheld treatment regularly, except when he was at risk for coming down with a life-threatening illness. In such an emergency, she would sometimes surreptitiously obtain a supply of glucocorticoid medication from the physician's supply in the medical institution where she worked.

There had been periods from babyhood onwards when she abusively abandoned, neglected, and punished the child. At age 13, he recalled episodes when she had whipped him with an extension cord, after which he was required to brush her hair as she lay on the bed clad only in scanty lingerie, until she fell asleep.

After a 10-week admission, the boy was discharged to live for a year in a children's recovery center, then in a church-run residential children's center where he stayed until age 18. After leaving the center, he lived a haphazard lifestyle for about a year, and during this time was neglectful of medication. From that time to the present, he has become always reliably responsible about his medication, with only occasional lapses. Hormonally, he has stabilized on oral prednisone 5 mg twice a day.

Gender-coded social biography

At age 3:6, this boy's Stanford-Binet IQ was 140. At age 8:3, on the Wechsler Intelligence Scale for Children, the findings were: Verbal IQ, 124; Performance IQ, 111; Full IQ, 120. On the same test, administered at age 13:1 during the admission for frostbite, the IQs were: Verbal, 119; Performance, 100; Full, 111.

Because of his size and advanced maturation, the boy entered first grade approximately half a year prior to his 6th birthday. He attended a public school and made passing grades. At age 8, he said he wanted to go to college and become a lawyer, the profession of his mother's then current paramour.

After residential placement in the children's center at age 13, he continued to attend public school. His grades were mostly C. A month before he graduated, he left the center to live with married friends, alumni of the center. With this move, he forfeited the complete support that he would have received for four years as a college student. Instead of further education, he tried his hand at stereotypical masculine trades and service jobs that put too much strain on his amputated feet and small body build. Eventually he settled into an apprenticeship and became a registered electrician. The heavy work of climbing and hauling, required for the wiring of new high-rise buildings, proved too strenuous. At age 27, he enrolled in a vocational rehabilitation program, went to undergraduate college, and after four years graduated with a bachelor's degree. He continued with graduate training, periodically interrupted because of financial need.

Coincident with the career change, he also changed his living arrangements. His marriage at age 20 had lasted for only two and half years, after which he set up housekeeping with a woman friend. It was when this friend became explicitly concerned about childlessness, having used no contraception for four years, that the relationship became unsteady. He had always had other partners, but now became involved in a love affair with another woman who soon became both his college partner and his room-mate as well. Before he discontinued completely the prior live-in relationship, that woman's pregnancy did materialize. The baby was born after the couple had permanently separated. The patient had favored an abortion, as he could not reconcile fatherhood with being a college student. Subsequently, he claimed to doubt his ability as a father, because of his own history of child abuse. He also claimed to be in doubt about paternity. He saw the baby, a girl, in early infancy, and since then has evinced no interest in her. Shortly before this baby was born, his new lover became pregnant, and at two months had an abortion. There have been no subsequent pregnancies.

The way in which the patient ended his marriage and his first live-in relationship was without resort to physical assault, but with verbal argument and by simply leaving. His partners agreed that he was predominantly even-tempered, and passively rather than actively aggressive — a type of behavior, common in CVAH males, that disaffirms the popular folklore that androgen is the aggression hormone, since CVAH boys, as in the present case, have a history of elevated serum androgen.

Despite his superior size and strength in the early years of his pubertal precocity, the boy did not get into fights with his playmates. He was even-tempered at school and at home. He met the crisis of domestic abuse at age 13 not with destructive violence, but by running away. His reaction to being rescued by being residentially relocated was typical in cases of child abuse, namely unrequited attempts to appease the abuser, a fantasy of paradise being regained, and blame or ambivalence directed toward the rescuers. It took years before he finally reconciled himself to the fact that absolutely nothing he could do would satisfy his mother, so that she would treat him humanely. He then established a more trusting relationship with the professionals who had rescued him.

It was not until adulthood that he became cognizant of the circumstances of his birth, namely that his maternal uncle, his mother's brother, was actually his father of conception and that his mother had been 15 years old at the time. She was exceptionally good-looking and so had no difficulty in covering her tracks by having intercourse with other local youths to whom she might attribute paternity. One of them, a good-natured farm boy, was so infatuated with her that he jumped at the opportunity of marriage, though he knew quite well that the pregnancy could not possibly be attributed to him. He did not know that for his wife the marriage was an expedient, and that she expected to end it in divorce after the baby was born. She stayed married, however, for nine years, having one son by him and a sequence of many casual affairs, before her demand for divorce hit him like a thunderbolt. This is the only man whom the patient has ever named as his father. After the breakup, the mother began an extremely tortured relationship with a local politician who sacrificed his family and career on her behalf. The storminess of this relationship contributed to the patient's institutionalization after losing half of each foot to frostbite. It contributed also to the placement of his younger brother in another residential institution.

As an adult, the patient has a good masculine appearance and looks his age, despite the fact that, on the basis of height alone, he might be judged younger. He walks as if being cautious, guarding his balance, and protecting his feet.

Lovemap biography

Ordinarily, in a case of precocious onset of hormonal puberty, one expects greater visibility, if not greater frequency of erections, and increased incidence of masturbation. In this case, the reports of the parents were inconsistent. On the initial visit, when the boy was aged 3:6, the father (his actual status as stepfather was not known until 10 years later) said that he had not noticed many erections except upon waking, and no masturbation. Six months later, after the beginning of glucocorticoid therapy to suppress adrenocortical androgens, both parents agreed that they no longer observed masturbation, whereas formerly they had.

On the first visit, the mother expressed some perplexity about her son's affectionateness. Once he sat beside her looking at television, hugged and kissed her, and told her that he loved her. She noticed that he had an erection. Otherwise, she thought that erections were no more frequent than in her infant second son.

On this same visit, in the examining room, the boy related to staff members, all men, in a friendly, puppy-like way, by hugging their legs and kissing them. He kissed and cuddled against the examining physician, especially during the genital examination. He was too young at this time to be consistent in providing answers to inquiries, but he did seem to understand that he had a big peewee with hair around it that other boys did not have, whereas his father did. Sixteen months later, aged 4:10, he was able to say that he sometimes had erections and sometimes would play with his penis.

Four months later, at age 5:1, his vocabulary allowed him to say that sometimes 'my mommy has to whip me ... because I'm bad ... for playing with my penis.' To a query about love dreams, he did not differentiate dreaming from waking activity. 'One time I kissed Linda Bowen. That's about it,' he said. 'I was home, watching T.V. Daddy told me not to kiss her, it wasn't nice. When we grow up, then we can kiss each other? You can get married, can't you? But you don't have to, do you?' In response to a question about play, he said his daddy would not allow him to tell Linda to undress when they were playing doctors.

Around this age, according to retrospective recall in adulthood, there were many times when he would be at home while his parents were working. Then a neighborhood girl, herself in early puberty, 'used to talk to me about hooking school. She used to come over and play with me and try to get me to have sex.'

There is a note in the record when the boy was next seen at age 8:3, and had been off treatment for some time, that he sometimes had erections — in school, he said, and in bed, but he did not know how often. They appeared to be spontaneous, and not triggered by erotic thoughts or perceptions, and not accompanied by masturbation or ejaculation. He seldom had a love dream, he said, but quoted one example: 'Well, I didn't win. The other boy did. He got the girl. ... We was in school, you know, and then we went out with this girl. I dreamed that she was going to pick one of us and marry us. And she picked the other guy, and they got married and everything.' There was no associated ejaculation.

At this same age he said that if his mother would scrub him down in the bathtub, 'I don't like it. I don't want her to see my penis.' He was shy, but less so, if his father or younger brother saw him naked.

He wasn't sure whether he had a girlfriend or not, because 'I've never asked her.' Asked about the marrying age, he said, 'I'd like to get married when I'm nineteen.'

At age 8, he already had some knowledge of where babies come from: 'I've seen in the encyclopedia where they come from. I don't know how to get a baby. The woman borns the baby. I've seen pictures. I didn't read about it'. To augment his knowledge he was given, with diagrams, the missing information regarding the baby egg; the swimming race of the sperms from the father; the penis as a sperm pump fitted into the vagina where the race begins, only one sperm in 300 million being the winner; and the vagina as a baby chute for delivery.

It is no easier for a 9 year old with precocious hormonal puberty to disclose the circumstances or the content of his erotic ideation and imagery than it is for a child whose age of pubertal onset is normal. By age 8, both the proscriptions and the prescriptions of sex in our society have been assimilated. The penalties of self-incrimination send sexual self-disclosure into the closet. It emerges only guardedly, after there has been sufficient time to establish trust.

In the present instance, trust became established when the boy was 13 and convalescing from plastic surgery of his frostbitten feet. Without the benefit of intimate disclosure, it was evident during this hospitalization that he was enamoured of the teenaged sister of a fellow patient. The two did carry on a correspondence, for a while. It was by no means easy, however, for the boy to give any information about the role of this girl, or anyone else, in the ideation and imagery of his erotic fantasies and wet dreams; nor was it easy to find out if he had masturbation fantasies and wet dreams with ejaculation. With the knowledge of hindsight gained in adult years, it eventually appeared that the boy's sleeping history had, indeed, been negative for ejaculations with or without a dream. By contrast, at age 13 he had a waking history of one or more ejaculations a day. The accompanying masturbation fantasies did not go all the way to intercourse: 'When it discharges, that's the end. I don't go any further. I get my bad feelings, and then I don't want to think about that anymore.'

While asleep, he did at age 13 have some erotic dreams, minus ejaculation. They were not mentioned until the topic was made safe to talk about by means of the Parable Technique, as exemplified in the following excerpt from his psychohormonal history.

'He told me that once when his father found him masturbating, while under a blanket looking at television, he told him simply that he should go to his room to do what he was doing. I added that I had heard all kinds of parental reaction, some enlightened like his father's, some not. I quoted one parable of a precocious three year old whose parents dealt with bathtub masturbation by teaching the boy to do it in private. Then I made up a composite parable of a boy of five whose father had recently died. When he masturbated, his mother got very disturbed. Could he [the patient] guess why? He guessed that it was because it got her aroused. He was right, I told him. This, then, is the background against which, a short time later, he revealed his own Oedipal dreams on the tape. I had asked him to guess what his mother might have said had she caught him masturbating, as his father had done.'

'I don't know. I've dreamed about it though,' he replied, 'her catching me. I dreamed that she jumped right on in. And I dreamed she started right on in with sexual intercourse. That was odd. ... This was a sleeping dream. It had no climax. Why should it? Because I didn't have no wet dreams.'

Subsequent inquiry revealed that there would be no actual discharge of semen to accompany the dream. However, in the dream imagery, fluid would begin to flow and at that point the female would jump on, as the flow continued, to catch it. The boy had always believed that ejaculation in intercourse was different from masturbatory ejaculation, the latter being short-lived, and the former being continuous.

There had been an estimated six to ten Oedipal dreams. Only one more was reported. It had happened during a visit home. Then, following residential placement away from home, there were no more. The possibility of getting rid of these dreams had been a primary consideration in the boy's acceptance of the offer of residential placement away from his mother.

Upon contemplation of the hypothesis of a connection between having no father in the home, having himself declared to be the man of the house, having these dreams of intercourse with his mother, and having run away, hiding, and getting frostbitten, he decided: 'There's no connection there. No. You brought up a pretty good point there, but I don't believe there's any connection.'

The mother did not express her own viewpoint, but her boyfriend said she had told him that it disgusted her if she would see the boy get up in the morning with an erection in his pants. He quoted her as saying, 'I could kill him,' as if she seemed to take it as a personal insult and a sign that he had unchaste thoughts in his mind. Elsewhere it is recorded in the history that the boy was a look-alike for her older brother, his father of conception, who had had sex with her since she was 8 until she married at 15 — and for which she blamed herself, since she would accept his chaperonage, trading sex as the price of transportation to teenaged social events in the rural area where they lived.

In the residential center where the boy lived from age 13 to 18, there were the usual limitations on a girlfriend relationship. His first experience with sexual intercourse was at age 15. At 18, he had a brief live-in relationship, followed by various casual affairs. With his marriage, at age 20, he began the first of a sequence of three long-term relationships of living together.

Though it was always clear that his primary orientation was heterosexual, from boyhood onward he had had some peripheral homosexual contacts. When he was only $6^1/_2$ years old, his penis was of adult size. Because of its size, the mother of a neighborhood boy became so alarmed that, on the basis of suspected homosexual play, she took her own son for a medical examination. Though he did not disclose the information until talking about bisexuality nearly 20 years later, as a 9 year old he had played around with sucking on his younger brother's penis while masturbating his own.

In preparation for sex-segregated living in the residential center with boys who were postpubertal like himself, but two or three years older, he had a psychohormonal counseling session on strategies for living with teenaged sexuality. This session paid off when, at age 15, while he was waiting for a bus at the corner of the hospital, he was offered a ride home by a smart-looking young man driving a Mercedes. The driver's technique for sounding out his passenger as a potential sexual partner was to converse about sex and, from the glove compartment, to hand him a sexy narrative to read. Obtaining an inadequate response, he produced another story from his briefcase on the back seat. For a moment, this move scared the boy, for he thought the driver might have a weapon in his briefcase. Recognizing that the actual agenda was homosexual, the boy maintained his cool and requested a stop, en route, at the auto dealership where his uncle worked. The driver circled the block and waited, but the boy did not reappear. This story was authenticated by the driver himself. His identity as a paramedic was established from the description of his appearance and automobile. His escapades did not always end so well. Within two years he was dead, the victim of teenaged homosexual murder.

Within the residential center there were, as was to be expected, some reports of homosexual tomfoolery, and of some residents who were predominantly homosexual in orientation. Recalling one of these boys, years later, the patient said: 'I had a roommate who tried to put the moves on me, and wanted to have anal sex with me one night, all drunk. He had greased himself all up with Vaseline and kept on saying, "Coates, give me some butt." I look back now and laugh about it, but I distinctly remember he came over on my bed, trying to get me in position. I told him no, and hopped out, off the bed. That was enough to stop it.'

He did not know then that the day would come when he would experience intense erotic feeling if his woman partner would digitally stimulate his anus and massage his prostate while stimulating his penis — or else if he would do it himself. He had originally discovered anal eroticism as a young child when he would use a bobby pin anally, while masturbating. In adulthood, he had had a fantasy of anal stimulation by a man, but had not ever realized it.

Once, however, he had tried not only receiving fellatio from a bisexual male friend, but also performing it on him. It lasted all of thirty seconds. 'I didn't continue long enough for him to come. I was amused at myself, sucking him — yeah, like "What am I doing in this man's crotch?" You know? I didn't get turned on. I felt bad for the guy that he was turned on, and I wasn't into it. ... And it didn't culminate in anything other than an erection for him, which he had before he started. I was thinking about his ex-wife [asleep in the other room]. She and I had been flirting, and we had slept together. ... He knew I had the hots for her, and had given me his verbal permission, if anything did happen between the two of us, that it was okay.' Thereafter, he still was curious to find out what it would be like to have an orgasm with a man, but never did.

The significance of the foregoing experience to the patient was that it took place when he had quit his macho job as an electrician and, as a member of a group belonging to the Human Potential Movement, was 'getting in touch with my physical and emotional being ... and my homosexuality. ... I felt a genuine love and caring from this man I had sex with.'

The experience did make a difference: 'One of the major things or points would be that I've learned it's okay for a man to feel attraction towards another man, and okay to have sex, and it does not mean that I'm condemned to a life of homosexuality. ... I think I'm pretty much a Kinsey zero or a Kinsey one. That one incident satisfied my curiosity. I find that I like gay guys as friends. I find that they're more sensitive guys. I can express my feelings with gay guys a lot better than I can with most other men.'

Having discovered that he could be both heterosexual and masculinely sensitive, he was ready to quit being a handicapped Hercules, vocationally, and ready to resume academic training for a professional career, using his mind. This was the time that he left his partner of five years, and took up with a new one. He was 27 years old.

One of the products of getting in touch with bisexuality was a fantasy of a threesome with two women or, rather than two men, a two-couple foursome. Now free to express her own bisexual inclination, his lover did get into a threesome with him and another bisexual woman. It didn't work out as planned. The two women became totally engrossed. 'If there is such a thing as a ghost,' he said, 'I mean I was a ghost whom nobody saw.'

Thinking about his sexual energy over the years until he was 35, he said: 'I think it has increased, if it's done anything. I've learned more about how to put it into other areas than sex. I still do masturbate a fair amount [in addition to intercourse]. I believe that an orgasm a day keeps the sexologist away. I've noticed that I'm not as promiscuous as I used to be. I'm not letting myself get involved. When I do have sex with women, I tend to get involved. ... But I miss the excitement. Sometimes I think I need to fall in love again to finish up my dissertation, because I like the excitement, the motivation, the drive. Yeah. Life looks fresh, new and alive, invigorating. If I could stay in a love state all my life, that's one of the things I'd really like.'

Recently, in adulthood, he responded to a checklist concerning the sensory channels of sexuoerotic arousal. Visual arousal was clearly pre-eminent: 'To get started? Sight. I say most definitely seeing. The combination of that and the voice. I mean if it would be what I would consider to be any serious opportunity to do any fucking, it would have to be started with sight and some sound. ... I used to get more turned on by thinking about it. And now, I think it's more hearing and seeing. Pictures, or people — women in the shopping mall or at school, or something of that sort.' When he spoke of being turned on by sound, he explained: 'It's mostly sight, but if I know somebody a bit, then it's sight and sound — more the content of the conversation.' Erotic stories and writing are not a turn-on for him. Touch becomes important after the initial turn-on: 'Some talk, and shortly thereafter, some touch. I like being touched.' Taste and smell also are secondary: 'Taste? I think that's pretty important too. The mouth, and mouth to female genital type thing. After I've been involved with somebody, I would say that their body odors are important to me.'

He and his lover both wanted and did not want to get married. 'I don't have any visions of being wealthy, but I would like to be comfortable, and would like to have children, maybe just one or two. ... I would really like to have some of the things that normal people have in life, instead of being a student all the time.'

EXPOSITION

In experimental psychoneuroendocrinology, there is an expanding body of laboratory animal evidence that sexual dimorphism of the brain is differentiated under the influence of the presence or absence of steroidal sex hormones (Kelly, 1985; McEwen, 1976). In the human species, psychoneuroendocrine laboratory experiments are, for ethical reasons, not possible. The alternative is to investigate nature's own experiments. One such natural experiment is the adrenogenital syndrome (CVAH). The prenatal excess of adrenocortical androgenization characteristic of this syndrome occurs irrespective of whether the fetus is karyotypically either 46,XX or 46,XY. Hyperandrogenization does not alter the anatomy of the genitals of the 46,XY baby, whereas it masculinizes those of the 46,XX baby. In addition, it may be inferred that prenatal hyperandrogenization of the 46,XY brain precludes the possibility in postnatal life of the differentiation of a homosexual or bisexual gender status (Money and Alexander, 1969; Money and Lewis, 1982); whereas, by contrast, prenatal hyperandrogenization of the 46,XX brain facilitates the possibility of a subsequent homosexual or bisexual gender status (Money and Daléry, 1976; Money et al., 1984).

Among 33 CVAH males followed in the Johns Hopkins psychohormonal service since 1951, there is only one instance of a consistent orientation as a homosexual male. This man had an attenuated form of the CVAH syndrome, sufficiently asymptomatic that his case would not have been ascertained except for a pedigree study. By contrast, three (10%) among a sample of 30 CVAH women in long-term follow-up proved to be homophilic lesbians with a history of a pairbonded love affair with only a woman, never with a man. The case of one of these three is used in this chapter.

If bisexuality is behaviorally defined strictly in terms of the sex of the partner in a sexual act, then both members of this matched pair have had a behavioral history that is bisexual. By contrast, if bisexuality is defined in terms of a pairbonded, limerent love-affair relationship with both a man and a woman, then neither of the pair is bisexual for, on this criterion, the woman is lesbian, and the man is heterosexual.

Either a homosexual or a heterosexual experience may be contrived, perfunctory, expedient, or opportunistic. The alternative is an ecstatic, abandoned, passionate peak experience, often involving limerence or love-smittenness, and love requited in a prolonged attachment. In each of the two cases that have been presented, homosexuality and heterosexuality were not evenly balanced. One was more perfunctory and contrived than the other: heterosexuality in the woman's case, and homosexuality in the man's.

There is no final, sufficient and total explanation of the sexuoerotic orientation or status of the woman and the man in this comparison pair. If each had been differentially affected with respect to juvenile sexuoerotic rehearsal play, then the evidence is insufficient to permit a firm conclusion. The greater amount of information on record for the boy than the girl shows that, despite some early male sexual contact, and despite the capriciously abusive and seductive relationship his mother had with him, he was clearly sexually attracted toward females from a very young age. This attraction toward females survived the early teenaged challenge of having to cope with incest dreams and, later, the challenge of sex-segregated living.

Irrespective of the role of postnatal and social factors in influencing the differentiation of heterosexuality and homosexuality, in the present two cases it is very likely that prenatal hormonal masculinization of the brain also had a determining influence on subsequent sexuoerotic status. In the case of the boy, it may be hypothesized that the flood of prenatal adrenocortical androgen ultramasculinized the sexual brain, with a possible reinforcing effect, postnatally, because of the discontinuities in endocrine therapy to suppress androgen. Ultramasculinization predestined the brain to resist the social forces of postnatal life that, in his particular home life, might conceivably have disrupted the development of a heterosexual status in a boy not so ultramasculinized.

In the case of the girl, adrenocortical androgen did not ultramasculinize the sexual brain, but masculinized what would otherwise have been a nonmasculinized and feminized brain. The influence of this masculinizing effect survived postnatally until, at age 4:8, enlargement of the clitoris finally made an endocrine consultation imperative. It was only then that the parents became aware that they were rearing not an ordinary little girl, but one with a mild degree of genital anomaly and with an excess of masculinizing hormone that needed to be corrected. Thus they had not been covertly educating her to be like a boy. Her early life does not support the feminist argument that dismisses prenatal and neonatal hormonal influences on G-I/R formation, and accepts only the hypothesis of the parents' covert expectations that their CVAH daughter will be like a boy.

It was at age 4:8 that hormonal treatment with hydrocortisone was commenced to suppress further masculinization. Treatment has since been maintained continuously. The early lesbianizing effect of brain masculinization has persisted despite the social forces within the family and community that favored heterosexual differentiation. In other words, the prenatal and neonatal hormonal effect survived and facilitated the differentiation of a homosexual status that eventually manifested itself in falling in love and pairbonding with a female partner. The sex of the partner in a love affair, not genitosexual practice per se, is the ultimate and definitive criterion of homosexual and heterosexual status.

In this woman's case, homoerotic love manifested itself without other masculine behavior, and without any masculinization of the body in adulthood. The masculinized counterpart of this case (Money and Daléry, 1976, and unpublished records) is one in which a baby with the same 46,XX CVAH diagnosis is declared and reared as a boy. The rearing and clinical management – endocrine, psychologic, and surgical, if genital surgery is needed – are then directed toward habilitation as a male. The end result is then an adult who looks indisputably masculine, and acts and thinks as a man. He falls in love as a man. He has sexual intercourse as a man. Nonetheless, he has the 46,XX chromosomal pattern typical of the normal female, and was born with a uterus and two ovaries. Had he been habilitated as a girl, he would have been fertile and could have become pregnant.

The paradox of a 46,XX man, as in the foregoing paragraph, challenges the very definition of lesbianism. Common sense dictates that he not be called a lesbian when he falls in love with a woman, because he looks like a man, acts like a man, and copulates like a man with his penis. His counterpart, as in Case #1 of the present study, whom common sense does call a lesbian, looks like a woman, acts like a woman, and does not copulate as a man does with a penis. Nonetheless both are chromosomally 46,XX, and both were born with ovaries and with the capability, given proper clinical treatment, of becoming pregnant.

Such a pair of cases shows rather dramatically that the prenatal hormonal history works not alone but in concert with the postnatal clinical and social history in producing not only the final sexuoerotic product, but also society's definition of it as heterosexual, bisexual, or homosexual. Although current technology has shown no difference in the adult hormonal status of homosexuals as opposed to heterosexuals, there is an abundance of evidence for the influence of prenatal hormones on the development of brain pathways, which is discussed in full in Money (1988). The two cases give dramatic support to the hypothesis that the prenatal hormonal history is implicated in the genesis and ultimate differentiation of a homosexual, bisexual, or heterosexual status. This hypothesis gains further support from the 46,XY CVAH male herein exemplified in Case #2. This man was prenatally ultramasculinized and heterosexual in status, though not homophobic, and not totally unfamiliar with homosexuality at first hand. Current technology does not yet make it possible to measure the relative weighting of the prenatal hormonal history nor to differentiate it from the postnatal history in the genesis of sexuoerotic status as homosexual, bisexual, or heterosexual.

BIBLIOGRAPHY

Kelly, D.D. (1985) Sexual differentiation of the nervous system. In: *Principles of Neural Science*, 2nd edn. (E.R. Kandel and J.H. Schwartz, Eds.). New York/Amsterdam, Elsevier.

McEwen, B.S. (1976) Steroid hormone receptors in developing and mature brain tissue. In: *Neurotransmitters, Hormones and Receptors: Novel Approaches.* Neuroscience Symposium, Vol. 1 (J.A. Ferrendelli, B.S. McEwen and S.H.Snyder, Eds., G. Gurvitch, Assistant Ed.). Bethesda, Society for Neuroscience. pp. 50–66.

Money, J. (1988) *Gay, Straight, and In-Between: The Sexology of Erotic Orientation.* New York, Oxford University Press.

Money, J. and Alexander, D. (1969) Psychosexual development and absence of homosexuality in males with precocious puberty: Review of 18 cases. *J. Nerv. Ment. Dis.*, 148: 111–123.

Money, J. and Daléry, J. (1976) Iatrogenic homosexuality: Gender identity in seven 46,XX chromosomal females with hyperadrenocortical hermaphroditism born with a penis, three reared as boys, four reared as girls. *J. Homosex.*, 1: 357–371.

Money, J. and Lewis, V.G. (1982) Homosexual/heterosexual status in boys at puberty: Idiopathic adolescent gynecomastia and congenital virilizing adrenal hyperplasia compared. *Psychoneuroendocrinology*, 7: 339–346.

Money, J., Schwartz, M. and Lewis, V.G. (1984) Adult erotosexual status and fetal hormonal masculinization and demasculinization: 46,XX congenital virilizing adrenal hyperplasia (CVAH) and 46,XY androgen-insensitivity syndrome (AIS) compared. *Psychoneuroendocrinology*, 9: 405–414.

CHAPTER 11

Amphoteric bisexual pathology evoked by solitary confinement and imprisonment in two syndromes: 47,XXY and 47,XYY

SYNOPSIS

Under conditions of solitary confinement in prison, two men became diagnostically concordant for manifestations of gender transposition related to transexualism. The noteworthiness of their concordance for gender transposition is the discordance of their chromosomal status which in Case #1 is 47,XXY (Klinefelter's syndrome) and in Case #2 is 47,XYY (supernumerary Y syndrome).

The term for the type of masculine/feminine inconstancy or pendulum-swinging manifested in the two cases is amphoteric (from Greek, *'αμφότερος*, both). Conceptually amphoteric is more inclusive than bisexual. Bisexual duality signifies sexuoerotic attraction that is homosexual as well as heterosexual in either fantasy, actuality, or both. Amphoteric duality includes not only bisexualism but also masculine and feminine attributes and stereotypes other than sexuoerotic — as, for example, in gender-crossed impersonation, cross dressing, and sex reassignment.

Amphoteric duality is not synonymous with androgyny, the difference being that androgyny signifies a constant synthesis of masculine and feminine, whereas in amphoteric duality masculine and feminine are inconstant and take turns. They may take turns cyclically, on a more or less regular basis, or irregularly on a more sporadic basis. Whether regular or irregular, the back-and-forth amphoteric alternation may persist over many years of the life span. Conversely, the alternating may come to a standstill. Its final position may be permanently either gender-uncrossed or gender-crossed. The gender-uncrossed position, being defined as heterosexual and normal, receives little or no scientific attention. The gender-crossed position is that of either gender mimesis or transexualism, and receives intensive scientific attention. In the vernacular, gender mimetics are drag queens and diesel dykes, and transexuals are sex changes, or those who have undergone surgical sex reassignment.

There are no data against which to make a statistical evaluation of how prevalent amphoteric duality may be in the XXY and XYY populations at large. However, it is known that the supernumerary X chromosome puts its possessor at risk for any one of a wide range of neuropsychopathological disorders, sexological disorders included (Nielsen, 1969; Nielsen et al., 1969). It is also known that the supernumerary Y chromosome puts its possessor at risk for psychopathological disorders, sexological disorders included, associated with impulsivity, social disobedience, and criminal offense (Money et al., 1970, 1974).

Each of the two cases in this chapter shows that sex reassignment surgery, because of its irreversibility, is contraindicated until the amphoteric pendulum of masculinity and femininity has quit swinging. The Two-Year, Real-Life Test is imperative (Money and Ambinder, 1978).

Each of the two cases in this chapter also shows that gender-transposition syndromes in general, including transexualism in particular, are fundamentally dissociative syndromes in which sex between the legs is dissociated from sex between the ears. The full implications of sexological dissociation have yet to be incorporated into the theory and practice of scientific sexology as applied not only to medicine, but also to the law.

CASE #1: PATIENT XXY

Diagnostic and clinical biography

The patient was referred to the Johns Hopkins Psychohormonal Research Unit for a consultation two months after his 15th birthday. He was at that time serving time in the Maryland Training School for Boys (MTSB), having been sent there following an evaluation, at age 14:5 years, at a state psychiatric facility. He had set fire to populated suburban woods, and had a history of dangerously ungovernable behavior. The consulting psychiatrist at MTSB was the first to recognize that the boy had gynecomastia, which is prevalent in the 47,XXY (Klinefelter's) syndrome. Therefore, he arranged for his transfer to the state university hospital for endocrine and surgical evaluation, prior to the psychohormonal referral.

The diagnosis of Klinefelter's syndrome was established on the basis of seminiferous tubule dysgenesis and aspermatogenesis, as determined by testicular biopsy, plus a positive sex chromatin determined from a buccal smear. The chromosome count of 47,XXY was confirmed at a later date.

The findings from the physical examination were typical for the 47,XXY syndrome. The penis was within normal limits for an adult, but small. The testes were abnormally small and soft; each measured 2 cm × 1 cm. The prostate was half normal size. Pubic and axillary hair had developed, but there was no facial hair. There was mild facial acne. The breasts were 8 cm in diameter, and they protruded 2 cm. The nipples and areolae were not grossly enlarged. The height was 175 cm (68.75 in.), the span 170 cm (66.75 in.), and the weight 50 kg (110 lb). Androgen level was, in that era, inferred from the urinary 17-ketosteroid level, which was low. Urinary FSH (follicle stimulating hormone) level was high.

Ten months later, the breast enlargement had not diminished. The patient was, therefore, admitted for surgery and a bilateral mastectomy was performed, with the cosmetic result marred by a sunken left nipple.

On postsurgical follow-up a year later, at age 16:8 years, the height was 178 cm (70.25 in.), the span 175 cm (69 in.), the sole to symphysis measurement 96 cm (37.75 in.), and the weight 53 kg (116 lb). Pubertal masculinization was considered to be adequate, and supplemental treatment with testosterone, sometimes recommended in Klinefelter's syndrome, was not considered.

At this time, the symptoms of behavioral pathology had not remitted. Because he had attempted stabbing a guard, the patient had been transferred from the reformatory to serve time in a regular prison, while still only 16 years of age.

When he was 18 and in the prison psychiatric institution, an EEG was read as showing 'a slightly slower than average dominant frequency, but no specific abnormalities.'

In the first few years following the onset of behavioral pathology, all of the professionals in the psychiatric and correctional bureaucracy construed the patient's symptoms in terms of a chronic and continuous state of sociopathic personality disorder. Since no one took professional responsibility for long-term care and follow-up, no one recognized the evidence that irrationality and pathological antisocialism occurred episodically. The parents were the first, apparently, to make this observation, which was first put on record when the patient was 22 years old.

At this time, the mother, contacted by phone, spoke of cycles her son went through, every eight weeks or so, of depression followed by uncontrollable tendencies, followed by normalcy. His sister, the next youngest sibling, six years his junior, also had behavioral cycles, but on a monthly basis. They coincided with the menstrual cycle, but were more pathological on alternate months. They began at age 14 and had necessitated residential psychiatric treatment episodically for most of the preceding two years.

At this same time, the father gave an interview in person. He compared the behavioral cycles of the two children. The daugther would, he said, alternate between apathetic depression and being high, with no good period in between. When high, she got into sexual affairs and eloping. The brother, by contrast, might have a long period of good behavior before an outbreak characterized by what he listed as running away, reckless spending sprees, lying, disobedience, destructiveness, arson, assault, or violence.

There was no corresponding cyclicity of behavior in the other five children, though the youngest brother, aged 13, was showing some signs of undisciplined reactions.

In the family pedigree, there was a brother of the father's father who had been institutionalized in middle life because of a 'nervous breakdown.' The diagnosis was complicated by an earlier history of heavy drinking.

With the knowledge of hindsight, the diagnosis for the patient's sister should have been periodic psychosis of puberty. Today this condition is still extensively unknown or overlooked, diagnostically. It does respond, at least in some cases, to treatment with a progestinic hormone, for example, medroxyprogesterone acetate (Berlin et al., 1982).

In girls, cyclicity in the periodic psychosis of puberty synchronizes with the hormonal cycle of the menses. Logically, therefore, one would not expect to find an exact counterpart in a male, unless perhaps a male might cycle in synchrony with a female companion. In the case of XXY (Klinefelter) males, behavioral cyclicity is certainly not the rule, and the possible existence of rare cases of hormonal cyclicity is terra incognita.

As a boy of 15, the patient himself had formulated no concept of cyclicity or periodicity of behavioral symptoms, nor of why he did the things that got him incarcerated. 'I don't know the reason,' he said, for setting fire to the suburban woods across the street from his house, 'I can't think of any. ... I did in a way do it for the excitement. ... I told my sister, and she told my family.'

The reaction of other people was, as might be expected, judgmental, moralistic, and punitive, on the basis of the premise that his behavior was a product of free will and voluntary choice. Turning that same logic inside-out, to give a rational explanation to the irrationality of his own acts, he came up with the hypothesis that: 'I'd always wanted to go to military school, but didn't have the money, so I got myself sent up. ... I wanted to learn military life so I could join the army. I like the discipline ... military precision. ... I love drill.'

Five weeks later, he had rationalized a new hypothesis: 'I was hoping I would get sent up. I just wanted to get a record, so I could join the gang, the Wolf Gang in the city. I thought it would help me to join. Most of them have a record for going around and stealing cars, or something. Having a real great time.'

He could not predict whether he would get into performing more illicit acts. He thought that he had gotten such behavior out of his system, but then added the qualification: 'Maybe — if I stay up here at the Training School a little longer. ... I think I've gotten most of it out, but I could have a little more to get out, like stealing some more.' His own clinical prognosis was accurate, as accurate as if he had comprehended the periodic inevitability of his psychopathological disorder. But it was not comprehended by those responsible for his treatment at that time. He was destined to spend the next third of a century, until the present, in and out of psychiatric institutions and prisons — chiefly the latter.

When he was 20 years old, the patient resumed contact on his own initiative, writing from the institution for defective delinquents: 'I was sentenced to three years for assault with a deadly weapon; and I was so deeply shocked as to what I had done that I was unable to testify. ... I could remember that I had used a weapon, but I did not have the facilities to remember why I had used it. If I had not been so emotionally disturbed, I could have entered a plea of not guilty by reason of temporary insanity. Now that I have come to my full senses, I am sure that I went temporarily insane. I have had these fits of temporary insanity before. In fact I went legally insane in June two years ago.' No one believed him.

Gender-coded social biography

The social history on record in this case was obtained retrospectively from the mother when the patient was 17, at which time he had been reinstitutionalized as a repeat offender. A few points could be cross-checked with the patient's own recall.

The patient was born, together with a twin sister, in the early 1940s. His mother was 26 and his father 30 years of age. There was a brother aged 6, and a sister aged 4. After six years, a second brother was born and after three more years, twins, again a brother and sister.

Both parents had been college students, the mother for two years until she married, and her husband until, as a graduate in engineering, he obtained permanent employment as a civil servant in the administration of water resources. The mother's occupation was full time at home. The family income and residential location met the criterion of middle class.

On the basis of such evidence as is retrospectively available, it appears that the developmental history in infancy and early childhood was not recognized as unusual. The bond between the boy and his father was not a good one, partly because the father was, like his father before him, rather a loner and social-distancer; and partly because this son did not measure up to his older brother's brilliance, academically. The mother-son bond was positive. The mother leaned toward self-blame for having been inadequate to help her son.

Though the age of onset of first symptoms of behavioral disorder is not on record, it appears to have coincided with the onset of puberty, between the ages of 12 and 13. It was at the age of 13:10 that the boy had his first serious run-in with the law. The offense was setting fire to 12 acres of suburban residential woods. Thus, he entered the revolving door of mental and correctional institutions where, subsequently, he was destined to spend the major part of his life.

The first IQ on record, on the Wechsler Intelligence Scale for Children, was obtained at age 14:6 years: Full IQ 98; Verbal IQ 108; Performance IQ, 87. Eight months later, on the Wechsler-Bellevue Intelligence Scale I, the corresponding figures were 102, 112, 92. Eight years later, when the patient was 23 years old, the Wechsler Adult Intelligence Scale was administered and the IQs were: Full, 117; Verbal, 116; Performance, 116.

Early academic achievement was consistent with IQ. In teenage, however, achievement was sporadic, and dependent on periods when the symptoms of behavioral pathology were in remission. For the same reason, there has been no long-term history of employment. In the course of his various institutionalizations, however, the patient has developed a high degree of literary proficiency, as exemplified in an autobiographical fabrication, a tale of pseudologia fantastica, which begins thus:

'I was born on the Txzuchami (pronounced Chookahameye) Reservation in Brazil, South America on 27 February 1942, but my birth records were falsified to show that I was born in the Sutter Memorial Hospital in Sacramento, California on 18 May 1942, because my 'father' had to buy me and his other six children from the Mafia baby stealing racket, due to the fact that he was born sterile. A close look at his seven children will show that we are all of different ethnic varieties to prove it! But I celebrate my birthday on May 18th because I am so used to doing so. All of the Txzuchami women go nude all of the time because of the excessive heat in the Amazon River Basin, where the reservation is located. So, seeing a naked woman is nothing unusual to me.

'When I was three months old I was stolen off the reservation by the Mafia, taken to California, and sold to my Caucasian parents. I had not been in the Amazon River Basin long enough to develop the light brown skin color of the Txzuchami Indians. But I have a light brown birthmark on my left leg, in the exact configuration of the Txzuchami Reservation. It is the royal birthmark ... my real father is Chief of the Txzuchami Indian Nation.

'I was raised in Sacramento, California, the first seven years of my life. I spent first grade at Trinity Lutheran Parochial School, since my 'mother' was of the Missouri Synod Lutheran Church. I very diligently learned the catechism, the sacraments of the altar, etc. Second grade I spent at Fair Oaks Primary School, and we lived in Fair Oaks, a suburb of Sacramento.

'I'll never forget that house in Fair Oaks. It was a big Spanish style adobe house with an open courtyard in the middle of it. There were six olive trees in the front yard, two almond trees in the side yard on the driveway side, two orange trees and two fig trees in the backyard, a big weeping willow tree further back, and in the other side yard there was a grape arbor, and all the way over next to the boundary fence between our yard and the next door neighbor's yard was a pomegranate bush. I really love to eat those pomegranates. You cut them in half and eat the red seeds out of them. They are really sweet. Another thing I like is fresh figs right off the tree. And eating fresh oranges, and all the almonds I wanted was a real treat! And then of course there was nothing quite like picking those ripe black olives off the trees in the front yard. And there was a small stand of bamboo at the entrance of the driveway that we kids used to play hide and seek in!'

Lovemap biography

When he was 15, the patient in a sex-history interview gave responses consistent with the relative inertia and apathy that commonly typifies erotosexual life in adolescents and adults with the 47,XXY (Klinefelter's) syndrome. In content, the responses otherwise were male stereotypic and in conformity with sex-restrictive orthodoxy: masturbation mostly outgrown, aversion to homosexuality, heterosexual imagery in dreams and fantasies, and sexual intercourse not yet experienced. There had been no romantic experience with either a girlfriend or a boyfriend; and no sexual activity of any type with a partner, according to self-report.

In the reformatory, other inmates had tried to feel his breasts, and they had called him a morphodite. 'I just try to ignore it,' he said.

After the workup at age 15, the patient was not again seen at Johns Hopkins in person until he was aged 23. The next contact was six years later, in a letter written from a state psychiatric hospital. 'My behavior pattern runs in cycles,' he wrote. 'At the acme of my cycle I am extremely homosexual, whereas at the base of my cycle I am quite anti-homosexual. When my anger has been aroused deeply, I become extremely powerful, so much so that I am almost omnipotent. This omnipotency lasts for a period of two months, and tranquilizers have no effect on me. Then without any warning, I become normal again. ... I want to know if there is some way possible to correct the cycle.'

A month later, he wrote: 'I know I must have an endocrine imbalance because most of the time I am an active homosexual, whereas when I have become emotionally disturbed I am very anti-homosexual. ... I've lost many of my best friends because of my homosexuality, mainly because there isn't anything I can do to control it. To make my problems worse, I broadcast my homosexuality! I've lost many jobs because of it. I almost kept myself from being discharged from here because I told the staff that "to me homosexuality is not a game, it's an obsession which can only be satisfied by frequent sexual intercourse." That went over like a lead balloon.'

At the age of 35, after another relapse, he wrote from a state psychiatric hospital in another state: 'Psychoactive drugs have had no effect on controlling my personality functions, and I would say that the probable reason is the excessive amount of adrenalin which is secreted when I become intensely angered during the masculine phase of my behavior cycle. There is also the possibility of a pituitary secretion which causes schizophrenic behavioral changes when the endocrine changes occur. These, of course, are just educated guesses from observing my behavior over the years since you first told me I am a Klinefelter syndrome. Now what course of action would you suggest to halt my behavior cycle so I can permanently maintain my masculinity? Can it be done?'

It was typical in the history of his correspondence that the patient seldom was able to maintain continuity of follow-through. Eighteen months went by before he wrote again. 'I really want to know,' he wrote, 'especially since I'm facing eighty years Federal incarceration, whether there is any truth to the theory that being a Klinefelter causes a natural tendency toward criminal insanity. ... I've noticed I have a compulsion to commit crimes; and that I fantasize a lot about crimes I've never yet committed, but want to commit. My current fantasy is that I own a cocaine factory that supplies five Mafia families with cocaine. ... I am beginning to believe that being a Klinefelter doesn't mean I have a natural tendency toward homosexuality. But I do believe I'm a true bisexual. Although I prefer women, I do enjoy performing fellatio, especially if the man is circumcised. And I still enjoy anal intercourse.'

Prior to his next letter, four months later, he had read more about his own syndrome and also the XYY syndrome (Money et al., 1974), and came up with a new theory about himself based on the comparative psychopathology of XXY and XYY:

'I have discovered that every year from late March to early May, I change genetically from 47,XXY to 47,XYY. I have known for many years that I have a genuine Jekyll-Hyde personality syndrome, and that this syndrome manifests itself most during the same time period that I am in Power, i.e., usually from late March to early May, although I can induce the condition at any time of year from just going seventy-two hours without sleep. I deliberately induced the condition in the first week of January (six weeks ago) this year because I wanted to be in Power sooner than usual. But the usual genetic change will still happen next month anyway, just as it always does. Beginning tonight, I'm deliberately gonna start losing that 72 hrs. sleep just so I can get full of pure, cold, deadly anger, and cold deadly hatred to begin killing some people again. Some people are getting on my nerves. ... My Jekyll-Hyde personality syndrome has also been able to manifest itself at times when I'm not expecting it ... like when I'm in jail awaiting trial ... like now! So I've been writing letters to the mafia trying to get them angry enough to kill me. ... It's not that I'm suicidal ... just testing my theory that they can't kill me. ... I've always thrived on danger. ... It also makes a monotonous life more exciting. ... If you don't hear from me again, you'll probably read my obituary in the newspaper.'

Three days after writing this letter from the county jail, he was put 'in the hole' for 100 days. It was some months subsequent to his experience in solitary confinement that he attributed to it a reversal of the phases of his cycle, which had become nine months of deadly rage as XYY and three months of being 'a simpering faggot' as XXY. 'It takes a full two weeks for the complete change to occur,' he wrote. 'I'm not overjoyed to be XYY for nine months, and I've learned that Prolixin changes me back into XXY.'

In response to a specific query, he replied: 'My name during my bisexual personality is Tiger Lily. My name during my heterosexual personality is Grasshopper. All of the inmates call me Grasshopper because of my high proficiency in karate and kung fu. The change is not connected with sexual activities per se. The sexual dreams, thoughts, and fantasies concern past heterosexual experiences.'

The next week, in another letter, he further explained: 'When I'm XYY, I go through sixteen different personality changes ... with a different identity for each different personality.' Some of these personalities featured in the autobiographical novella that he wrote the following month, though Tiger Lily did not.

Some of his correspondence at this time was answered by a woman postdoctoral fellow. To her he wrote: 'I don't want you to get your hopes up about me. Even though I prefer women, I'm a bisexual. I'm queer as a three dollar bill when I'm locked up with nothing but men, as I am now. That is due mostly to the nonavailability of women. To put it bluntly, "I'll suck a dick in a heartbeat!" That's not to say women don't turn me on. I just can't get to any women here in the penitentiary (and in the isolation cell). Right now, I've got several men vying for my attentions. Quite a few have already claimed me as their bitch.'

In his next letter, and in response to a specific inquiry, the patient gave additional information on his developmental sex history as follows:

'I received my sex education from friends both my own age and somewhat older. I also received it from bisexuals, homosexuals, and lesbians I've lived with at various times. I got some of it from books, movies, and magazines. I had my first sexual encounter at the age of six with a girl from across the street. My first homosexual experience was at the age of eight when I sucked the penis of a boy who was fifteen. My next homosexual experience was at the age of thirteen when another boy and I fucked each other. I've lived with homosexuals, bisexuals (both male and female), and lesbians at various times. The only time I am not gay is when I am in 'Power.' That is when I have a complete Chromosome Reversal: from XXY to XYY, and I become ultra-heterosexual. I am now in 'Power' and have been in 'Power' since 27 April [seven weeks]. Although I am not yet so deep into 'Power' that I don't still have subliminal urgings to be gay once in a while. That will completely discontinue in another couple of weeks when I am deep into 'Power.' When I am in 'Power' I gradually increase both Mental Power and Physical Power, until after about seven months I pass my Peak and go Power Mad from too much Power. This usually lasts about two weeks then I lose Power with accelerating speed till my chromosomes change back to XXY, and I become bisexual once again!

'Now of course, due to the fact that I'm in prison locked up away from females (which I've always preferred), I settle for whatever else is available for sex. I give you three guesses what that means. The first sixty-nine don't count!'

In response to a query, written by a male graduate student, about sensory channels of erotic arousal, he wrote:

'I usually get turned on by seeing a foxy lady and/or a very handsome dude. By the way, how do you look Bill honey? I hope you are hung like a horse. Maybe we can get down sometime! I also get turned on by just discussing sex. And I always get turned on by a well hung dude, no matter what his looks! But when it comes to women, I appreciate the whole body! My tongue gets hard just from seeing a well built foxy female. I wouldn't say I eat pussy ... but then I guess the Pope isn't Catholic either. When I eat a lady, I like to start at the top and work down very slowly. I'm looking to give her the utmost in sexual pleasure! I want her to have just as many multiple orgasms as she possibly can! I want to make her eyes glaze with ecstacy if I can. I like to take a shower with her, then lick her dry. That way I find all of her erogenous zones, and I make sure I concentrate on them! Also, that way she is gonna want me back for more sexual fun. I try to make sex as pleasurable to a woman as I possibly can. That way I guarantee I never do without a woman when I can get to them! Of course that's not to say I can't also satisfy a male sex partner. I've never had any complaints when I've really tried to satisfy a man. When I'm really in the mood, I give such good head, that I've been able to make a man's eyes glaze with ecstacy! I remember the last time I accomplished that. I gave him such good head that he was just sitting there like he was in heaven. He had a big smile on his face and his eyes were literally glazed with ecstacy. I give head better than Linda Lovelace when I'm really trying! And many a dude has found that I've got some damn good booty too! Some of them have said I'm much tighter than their wife! I know how to keep it tight. I hope I've enlightened you with my sexual mores (and not many lesses either). I've been as truthful and straightforward as I can be. All of what I've written here is the absolute truth, I will swear to God. I'd get this notarized, but the Notary Public might die laughing, and I wouldn't want to be responsible for a death I didn't mean to cause!'

In view of the periodicity of psychopathology that had been ascertained, an attempt was made to arrange for an evaluation of its possible relationship to manic-depressive cycling, and to the so-called periodic psychosis of puberty. Bureaucratic red tape intervened, so that the patient did not get the benefit of a trial treatment with either lithium or medroxyprogesterone acetate, each of which would have been appropriate. He continued life as a prisoner.

CASE #2: PATIENT XYY

Diagnostic and clinical biography

At age 14, a court-ordered psychiatric hospital report noted that the patient 'must fight a constant battle with impulsivity.' This report was the first notation in his record to indicate that impulsiveness was one of his problems. A survey of XYY boys and men subsequently showed that impulsiveness might be considered pathognomonic of the XYY syndrome (Money et al., 1970). In the present case, the patient was in a correctional institution because of impulsive and antisocial behavior when, at age 16, genetic screening led to the discovery of his supernumerary Y chromosome. Subsequently, he was followed, along with other XYY patients in the PRU. He enrolled in a study on the effects of the antiandrogen, medroxyprogesterone acetate (Upjohn, Depo-Provera), in controlling impulsive, aggressive, and paraphilic behavior in XYY men (Money et al., 1975).

At age 18 (Fig. 11.1), a routine physical examination was given. The height measured 180 cm (71 in.), the weight 80.6 kg (179 lb), and the span 188 cm (74 in.). The flaccid penis measured 12×4 cm, and the testes 5.3 cm in length (Fig. 11.2). There was a normal amount of pubic hair. Facial hair was sparse. There was an unexplained hand tremor. Hormone levels were within normal limits: testosterone, 806 ng/100 ml; LH, 44 ng/ml; FSH, 190 ng/ml. The sperm count was 19 million per ml, volume 3.5 ml; and a repeat count two days later was 34.75 million per ml, volume 1.4 ml (norms: 60–120 million and 1.5–5 ml, respectively); with on each occasion normal viscosity and 80% normal morphology. An EEG done earlier, at age 16, had revealed no evidence of seizure discharge or focal abnormalities, though when he was 15 there had been an EEG report of a 14/6 abnormality. An updated chromosome count found no additional chromosomal abnormalities.

A geneticist informed the patient of his XYY status when he was 18. The report stated that he 'seemed to understand, but to be unimpressed.' A year later, during one of his weekly follow-up interviews in the PRU, he expressed interest in learning 'all about the XYY chromosome and how it affects the person, and if this XYY chromosome makes a person dangerous to other people and himself.' The interviewer talked with him and gave him journal articles on psychologic research pertaining to XYY men. Several times later in his life, while in prison, he requested and received more to read on the subject, including psychology and genetics textbooks.

Depo-Provera treatment was started when the patient was 19, at a dosage of 400 mg per week. He was more enthusiastically cooperative with the treatment program than might have been expected from his prior history. He had himself admitted to a nearby state psychiatric hospital where the effects of Depo-Provera were confounded by additional medication, notably Haldol. Depo-Provera treatment was continued for 15 months, until the patient returned to prison, on this occasion

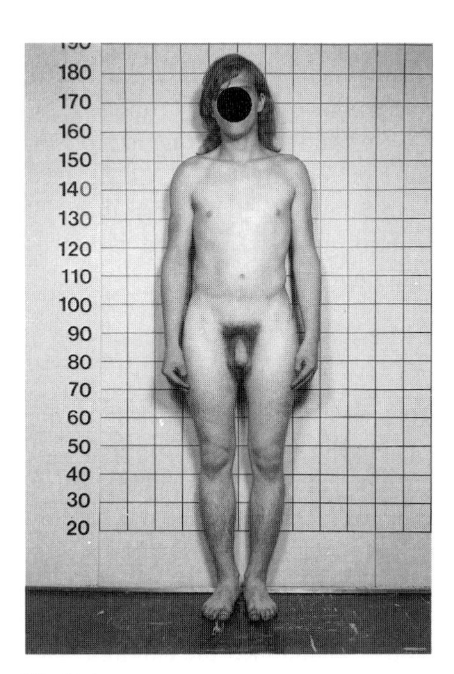

Fig. 11.1

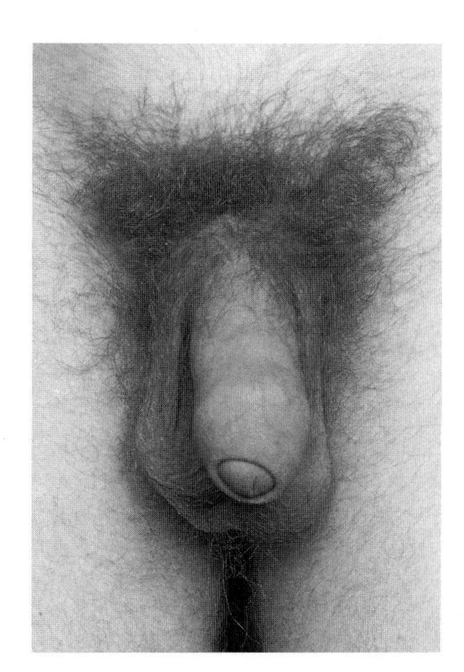

Fig. 11.2

Fig. 11.1. Case #2. Appearance of the body habitus at age 19:3.

Fig. 11.2. Case #2. Appearance of the genitalia at age 19:3.

charged with being a passenger in a stolen car. For many years after he was taken off Depo-Provera, he occasionally stated that he thought his behavior was controlled better with the hormone treatment and requested that it be resumed. There was one attempt to begin Depo-Provera therapy again when, in his late twenties, the patient was in prison serving an extended sentence. For bureaucratic reasons, it lasted less than a month.

The patient's behavior, with or without Depo-Provera, was extremely difficult for him to manage in or out of correctional and mental institutions. He recurrently threatened suicide. On some occasions he followed the threat with an actual suicide attempt, e.g., by overdosing on prescribed valium which he had saved up for this purpose, or by cutting his wrists and arms with razor blades that he kept hidden in his mouth. Both arms were scarred with multiple slash marks from the inner elbow to the wrist. In a prison incident report, at the age of 25, self-mutilation and suicide were interpreted as follows: 'While in the office of this psychologist ... the inmate removed a single edge razor blade from his mouth and proceeded to inflict several superficial wounds on his left forearm. This incident resulted after he was told he would not be placed in the prison infirmary because he does not want to be on lock-up. ... This man is not suicidal, but is extremely manipulative.'

When the patient was 28, he was convicted of aggravated assault on two prison employees and sentenced to six and a half years of solitary confinement. Prior to this conviction, he had been scheduled to be released from prison in less than a year. After serving the first two years of this six and a half year sentence of solitary confinement, he requested a transfer to the prison hospital so as to avoid participating in a prison riot in the section of the prison near his cell. 'Or else,' he threatened, 'I'll be dead in 30 minutes.' Despite the standing order that he be under constant supervision, 30 minutes later he was found, hanging from a ventilator grill, strangulated with bedsheets. He was brain dead at the age of 30, and could not be revived.

Social biography

The social history of this patient was obtained from social service agencies and from retrospective court investigations. The patient was born in the mid-1950s, the oldest of four children. His mother was 17 at his birth. She had married the patient's father when she was 13 and he was 24. The father worked as a carpenter. He was said by his wife to spend his earnings on drinking rather than supporting his children. For this reason, the patient's mother requested, for the first time when he was 3 years old, that the local department of welfare place all four of her children in adoption. After an unexplained delay of 18 months, the mother abandoned the children. The patient was $4^1/_2$ years old. He and his three younger siblings were placed in foster homes, while awaiting adoption. The two youngest children were quickly adopted, but the patient and one brother each rotated through a different series of blue-collar foster homes for the remainder of their childhood. The patient had no further contact with his mother after his first foster home placement and was 15 before he saw his father again.

Information about the patient's childhood was sparse. He was said by his father to have had severe temper tantrums when he was very small. Other than that, the father remembered little about his son's behavior. According to his first foster mother, the boy was 'an unusually sweet but moody child who had a bad temper.' He seemed to behave normally and get along with the other children. Since no adoption was found for the boy, he was removed from his preadoption first foster home after 13 months. Next he spent 18 months in a group home, considered marginal by the social service agency. The foster father there said that, owing to the large number of children, he could remember nothing about the boy except that he seemed fairly normal.

Between the ages of 7 and 12, he lived in three foster homes and was removed each time at the request of the foster parents because of behavior problems listed as 'bulliness, stealing, running away, compulsive eating, bed-wetting, public masturbation and exhibitionism.' The secretary at his elementary school remembered him, six years after he had attended the school, as having been an academic and a discipline problem to every teacher in every grade.

The patient was 12 when he was arrested and put in juvenile detention for the first time, on the charge of breaking into a grocery store after he had run away from a church-sponsored institutional home for boys. Between the ages of 12 and 18, he was sent to juvenile correctional institutions seven times and psychiatric hospital centers three times. He spent only three and a half months of these six years not institutionalized.

After age 18, he had five psychiatric admissions for a total of eight and a half months. One hospital disqualified him from returning, except on court order, because of his history of signing out against medical advice. In addition to being hospitalized on court order, there were seven different times that he was imprisoned between the age of 18 and his death at age 30:6. He spent 10 of these 12½ years in prison, and only 21 months free.

During each imprisonment, he was disciplined for infringing prison rules, disobeying guards, and explosive fighting with other inmates. The result of these disciplinary actions was that he spent a great deal of time, from 30 days to two years at a time, in solitary confinement. He had expressed his reaction to solitary confinement at age 19 during a period when he was not in prison. It was in the form of an unsolicited written instruction which he left, undelivered, on a seat in the waiting area of the Johns Hopkins Psychohormonal Research Unit (PRU). It read: 'What to do when I get upset: Don't ever leave me in a room by myself. I would rather be put in a strait jacket and shackles for my feet and then tie me to the bed and then give me a needle, but don't leave me alone until the medicine takes effect because I get very scared when I am left alone.'

When free, he was homeless, except for the 11 months between his last two imprisonments, when he was a member of an exconvict rehabilitation community. The nearest semblance to a family bond in his life began when, living in juvenile detention at age 16, he first met the woman who was the case coordinator for the Johns Hopkins genetics screening program through which he had been ascertained as being chromosomally XYY. He would see her on her frequent trips to the institution. Later, when she became case coordinator in the PRU, he maintained contact, and also adopted other PRU staff as a surrogate family. While incarcerated, he would phone as often as could be arranged, and wrote regularly. His letters from prison contained requests for stamps, candy, art supplies and other necessities, including textbooks and, once, a medical dictionary. In return he sent drawings and minor craftwork as gifts.

One letter threatening suicide was addressed to 'everyone in the PRU, and the world, except Ms. R.R.' (the coordinator). He did not declare his feelings for Ms. R.R., except in one prison letter written at age 23: 'You sometimes treat me like a mother, and at other times you appear to be like a wife or girlfriend.' He worried lest she take this letter 'as an assault.' She did not, but he did not again mention his feelings for her.

When he was not institutionalized, he was unable to hold a job. He supported himself through stealing, begging, trading his body for sex, and doing occasional odd jobs. He had no marketable job skills, and no job training except for a course in furniture upholstering, given at a state mental institution. He did not show up for interviews with prospective employers set up by counselors and probation officers. While in detention as a youth, he had completed the 8th grade. Despite his patchy academic history, he achieved a creditable level of literacy.

At age 10 he had been given the Wechsler Intelligence Scale for Children as part of a school psychiatric battery of tests (Full IQ, 106; Verbal IQ, 96; and Performance IQ, 115). When he was tested again at age 14, during a court-ordered psychiatric evaluation, the WISC IQs were 93, 82, and 107, respectively. The final testing, on the Wechsler Adult Intelligence Scale, was done at age 19 in the PRU, and the three IQs were 92, 92, and 93, respectively.

Lovemap biography

Questions about sexual imagery and practice were asked in connection with the Depo-Provera study when the patient was 19. He was uncooperative, saying 'everybody acts like police' interrogating him. Often he refused to answer at all, or gave vague answers such as, 'sex does not come to my mind at all, if at all not more than once a week.' He claimed to have a heterosexual sex life and fantasies of 'lying in bed with a girl making love.' One weekly follow-up note described the patient as 'sort of moralistic about sex. He regards as dirty any activity other than intercourse with penis in vagina. The only variation he likes is to have the woman lie on top of him.'

Later in the same year, there was contrary evidence, namely of homosexual cruising, confirmed when one of the men by whom the patient had been picked up phoned the PRU for help. He had taken the patient into his apartment and given him a place to live, only to discover that he had no way of coping with his new room-mate's violence, stealing, and refusal to leave, all of which terrorized an elderly aunt who lived in an adjoining apartment. At age 26, and no longer inhibited in revealing his sex life, the patient retrospectively explained this encounter in an audiotaped, sex-history interview: 'Since I ain't had no place but on the street, I just took the opportunity. He was showing off homosexual, but in the part of the woman, and I didn't like being in the part of the man. It wasn't me.'

The patient had, within days after his 24th birthday, declared to himself, while still in prison: 'Well, my New Year's resolution is going to be to come out.' Thereupon he wrote a letter in which, for the first time, he proclaimed himself to be gay. At the time of writing, he had spent five of the preceding six months in solitary confinement, the last four continuously. The letter was addressed to the coordinator, Ms. R.R., and dated December 31, 1978. It is reproduced, unedited, as written:

This letter here is a very personal one that I take serious and not as a joke. Because its me and I have been this away for over half of my life. I'm explainning this to you because you hold a big part in my life, cause your trustworthy, sincere, having feeling of understanding others as well as Dr. Money and Ms. H.C. but as it is said, woman understand woman better then they do a man. I'll explain later in this letter what I meant about that statement.

Well let's begin with this first, do you remember all those tests that I took at Phipps, expecially the questions and talk session for an hour a week. Well questions were like, (1) Have you ever had sex with another man, (2) Have you ever indulged in a sex act with a man, and a lot of other questions concerning sex with a man. These questions were given to me to answer, and I denied every thing concerning ever had or done or thought about having sex with the same sex.

Well let me say this first so you'll understand more better about what I saying before I go on explainning. I am GAY and that I would prefer to have sex with a man than with a woman, but I still have feeling and can have sex with a woman. I'll explain to you about my sex acts with women, teenage women and some sex acts with girls ageing from 9 yrs old to 16 yrs of age but in a fashion of oral sex.

But first I want to explain why, where, how come I have desires to have sex with men more than a woman.

MAN

Well when I was little about 7 or 8 years old I was laying naked on the floor in my bedroom coloring in my coloring book when I first expirence an nice but unusual sex act.

See when I bath at that age I put on this sweet smelling body loition of my foster-mothers, cause she said it made her skin soft and it smells good also. So after I put the lotion all over my body I lay on the floor coloring pictures.

As I was coloring I felt some soft hands on my behind, back, and legs, but mostly on my behind. I turned around and saw a foster-mother on her knees. She was saying alot of things that I didn't understand at that time, but I understood them later and that she handed me some books that I've never seen in my life, of women and men doing some weird things. At that time things like that were forbidden for kids to see and do.

My foster-mother was asking all kinds of questions, to be truthful and I'm not lying she started licking my behind. I was surprised at first, but it felt good. She took off her robe, and that the first time I had ever seen her naked or any other woman.

I saw she didn't have the same thing I had between my legs. Then I asked alot of questions, and she try to explain to me the best way I would understand. She started playing with me, then stop and took this long fat white thing out of between her legs (were her vagina was). I ask her why she had it there, and she said 'It feels good you want to try.' I didn't have no hold like she had, how was I going to get it in a hole I didn't have.

Well, she had put a lot of Vaseline on it, and put it in my ass, 'my ass.' She was slow, and it was hard for the big fat thing to get up my ass. But after a while it was up inside me. It hurt and burned, but while it was up my behind, she played with my penis.

It felt good, and we did it for the remaining years I was their, about three times a week.

Then when I went to MTSB [Maryland Training School for Boys; at age 12] I really looked more like a little girl than a boy. Long hair, soft skin, and my face was soft, smooth and my lips even looked like a girl's. And that's the first time I've ever actucally seen a Black boy. I throught history book were lieing but they weren't.

I got rape by 7 of them black boy but to me truthful their penis's felt better then a big, fat long plastic one. The reason I never said anything about them 7 what they done, was cause they said they would kill me if I said anything about it. So I never did. I stayed there for 2 years, 1 mon, and 2 days and I loved every second of it because of the male's penis's that I was getting.

And I've been doing it ever since, because I love it, I even did it in the streets at home for money. I hated selling my body for the money but I needed it to live on.

WOMAN

Everytime I've had sex with a female, I usually had oral sex and mostly more than half of them were bi-. I would love anal sex with them and they would put on or strap-on a dildo and have anal sex with me. Sometimes I had 3-some's either with another man and woman or with two females at the same time, but I wouldn't have sexual intercourse in their vagina, because the females I knew were like myself they would rather have a female than a man, but will have sex with a man but not often.

Even when I used to talk to you in your office I used to look at your side of your thighs to see how your panties fit, tight or lose or were they tight like the kind I always had on me.

See I hope I didn't afend you by the above statment about how tight your panties fited, cause I've looked at a lot of women's behind's, crotch's, and thigh's just to see how tight or lose they were and how nice they walked.

Because I've done it several times, dress up like a woman, and I had the hole works on. And truthfully speaking I looked better than most women I thought were about my age. Look Ruth ever since I've been down on this 5-years [in prison] for 25 months, I've been thinking a hell of a lot about getting my sex change so that I may become a woman. So I could stop wearing, acting, their clothes in my male life. And my male life is very miserable, that's why I'm going to work and save money so that I may have my dream come true, to be a Woman.

Ruth I hope that you could find me some of these items from your lady friend's because I'm very serious about my becomming a mature woman: see if you can get the following items: Size Blouse's 44. Pant Size, Waist, 31, don't know it in women's size, Inseam, 32, Hips, 38.

These picture's out of catalog's are the kind of clothes I like wearing. It's a younger style. The ones circle are the ones I wish you and Dr. Money could collect for me.

I'm serious about this matter so please don't take it in a joking way, because I'm not joking about this.

I have on lock-up now 390 days, I did have 210, got another 180 for what these police did to me, not me to them.

They kick my ass, thigh's, back, ribs, and head and cut me across the back proably with a razor blade.

Have a nice New Years cause I ain't.

Your's Truely,

Bobby Greenfield Better Known As Josie C.

I can only have these: 5 Blouses, 1 of them solid Red. 3 Pants, 2 of them, 1 solid Red, 1 White. 1 pr. of pajamas = 2 peice, flowers, same color. 3 socks. 3 Tee shirts (women's).

It was as a sequel to this letter that, after his release from prison two years later, the patient gave the audiotaped, sex-history interview, already mentioned. The interview began with reference to prison sex life, as follows: 'I had lovers. But sex is hard to come by in prison, because there is so much security. You hooked up in a cell together, because it was double cells. I was the part of a woman. I had a man, and he was black. He fucked me in the mouth and the ass, and he tried to masturbate me while doing it, and I hit him. He said: "Why not? You have to get your thing off as well." I said: "No, I get my own thing off." I said: "Now, if it was a cunt with a clitoris, you could do it. It would be natural then." I said: "It's not natural for a man to do it to another man." So actually, when I was in prison, I was more woman than man, because I carried myself more woman than man.'

A photograph of himself in unisex clothing and looking feminine confirmed what he said. It was accompanied by two other photographs, one of another prisoner impersonating a female perfectly in appearance, and one of his black lover, whom he called his daddy. Still in jail, his daddy had written him a love letter. 'Even right now I still think of you when I jerk off,' he wrote, 'I've had a lot of different kids since I've been here, but you are the best I ever had.' The letter ended with a reference to: 'if you still want to marry me.'

Queried as to when he first discovered the womanish part of himself, the patient retold the story of sexual seduction by a foster mother at age 7 (see above). In the retelling, he said that he now believed it had been arranged that the husband would discover what they were doing, so that he could join in and make a threesome: 'He would fuck her, and fuck her in the ass, mouth, cunt. She would strap the dildo on and fuck him, and I was to watch that. That's how they taught me. They said it was natural for children to learn sex.' That, apparently, constituted the nearest approach to formal sex education in childhood, either at home or in school.

When he was 10, his foster home was on a small farm. The neighbor was a chicken farmer, whom he would visit in the summer, when the farmer's 10 year old daughter was away at camp: 'He was talking one time about how he missed his daughter. He wanted me to dress up like a little girl, to make the same resemblance of his daughter. So I dressed up like a little girl, and he put little dabs of makeup and stuff to make me look like his daughter. That's when he did the sex. He wanted me to suck him off, to act like a little girl, daddy's little girl. He'd sit me on his lap, and get hard, and then I'd suck him off. He wanted to come in my mouth. I wouldn't, because my supposed-to-be foster father did it when I was seven, and I didn't like the taste.'

The next phase of homosexual experience began at age 12, when he was first in MTSB, the juvenile correctional institution (see above). 'As I was going through juvenile institutions and major prisons,' he said, 'I caught a lot of disrespect by being closed in. People knew that I was more of a woman than a man. And I would deny it. I would get into a lot of fights about it. And I would see the other homosexuals who did the part of a woman to a tee. They gained more respect by being their self, I guess.' So he made his New Year's resolution to come out (see above). 'And I didn't give a fuck who knew, because this is something that I want to do. It was easy for me to do there, in a male environment, because you don't have no women. But it's hard for me to get back to it, out here.'

He explained the difficulty, saying: 'See, I still have a desire for a woman, but not as strong. At times I do want to have a woman, and at times I don't. I don't know why. I can talk to her over the phone, talk about what I want to do, but when I actually get up close, I shy away. They don't appeal to me. But I do look at them, because I do want to be a woman. I look at their bodies, hoping that I'll have a body like that, one day, the dips and breasts and everything. I believe a woman is for attention, because the way a woman walks, talks, and the way her body is — a man always wants a woman. If a woman wants a man, all she has to do is walk down the street and shake her ass, and she's got one. So I feel that's where it is. A woman needs to be needed, and I need to be needed and comforted in times of depression, and bad times and hard times like now — not working, thinking about going back to prison, breaking the law because I can't get a job, can't get no funds. At nights I've been thinking if I was a woman I wouldn't have to go through that problem of being a man, because a woman can always get a man.'

The patient had, while still in prison, read about the two-year, real-life test of becoming rehabilitated in the sex of reassignment, prior to transexual surgery. The real-life test of living as a woman was, however, 'something that I'm not ready for yet,' he said. 'I couldn't stand the embarrassment of walking down the street or getting off a bus, or anything like that. But I do wear women's pants — I've got them on now. I wear women's clothing around the house. Slacks and shorts. One time I wore a negligee. I wear a nightgown, some nights.'

He believed that he would not regret a surgical sex change, because: 'I've always wanted it. When I was even younger than seven, I used to play with little girls. I used to play doll house, and doctor and nurse, and I would always want to be the girl. I wouldn't play with little boys. I always fantasized when I saw the girls with their panties on, and their dresses, and I would say wow, I wish I looked like that; because I saw, then, that girls get more attention than boys do.'

He adduced further evidence against regretting a surgical sex change: 'If I was to be a woman right now, I wouldn't get my feeling from the vagina. I fantasize a lot about getting fucked in the cunt, but that doesn't kick it off, because all my prior relationships, having sex with men, is where I really got my fulfillment by being fucked in the ass. That's where I got my sensations from. That's what really made me come.'

His masturbation imagery was also consistent with a sex change, he believed: 'I'm not thinking that I'm masturbating with a dick. I'm believing I'm masturbating with a clitoris, not a penis, even though I do know I have a dick in my hand. But that's my conscious mind telling me that. But my subconscious mind is not telling me that I have a dick in my hand because, when I'm fantasizing, it's a clitoris.'

Whereas he did not dream wet dreams, and did not have ejaculations in his sleep any longer, he did have masturbation fantasies. 'Oh, yeah,' he said, 'I fantasize about men and women as well. I fantasize about me being a woman ... the operation and stuff like that. While I'm masturbating, I'm watching them cut me open and give me an vagina, bandage it up, and then waking me in the recovery room. When the doctor takes the bandages off, he tells me to spread my legs. There's a full length mirror there, and two nurses. I can see that I've got a vagina, and I have breasts, and I look up and I have a woman's appearance, a female face and everything. ... No, I don't come then. I just keep on going further and further. I see myself getting released from the hospital. I catch a cab. I go downtown, and will be walking around. That's when

I fantasize about the men that had sex with me in prison. I bring them into the picture, and I invite them over. It's always the same three. ... It's kind of hard for me not to come before the fantasy's over with. That's why I keep repeating it and repeating it until I get up to the point where I'm sucking one, the other one's fucking me in the ass, and the other one's fucking me in the cunt. And they'll be talking about they're ready to come, and I come. And right there the fantasy disappears.'

To a warning that the feeling of orgasm might be unpredictably different following hormonal and surgical reassignment, he responded with a non sequitur in the form of a kidnaping fantasy involving 'a military M.D.' with whom he kept in contact: 'The part that I fantasize about him is kidnaping him, and making him call these different doctors that could perform a sex change. Kidnap them as well, or have them kidnaped — get three or four of my buddies to hold rifles on them to make them perform the sexual surgery, and to give me the whole complete works, the breasts and all. ... I've been dreaming all this, of making them give me a sex change. Even if I couldn't get it done legally. I've been thinking about it criminally — kidnaping and making them do this. If they don't do it, they die. It would be their choice to do it, or die.'

In concluding, his evaluation of the interview was: 'I know I feel openly about it. I don't feel scared about it. Not when someone is interested and listens and understands. Since you're into the department of sexology, you understand it real well. I don't feel as though I'm being put down about it.'

For the next interview, two weeks later, the patient was accompanied by a 19 year old woman who had recently begun having an affair with him. They had started dating after he took a telephone message from her for the director of the exconvict rehabilitation center where he was living. She was inquiring about her former husband, a convict serving time for armed robbery, whom the director had visited in a state prison some distance away. At age 19, she had a history of living on the lawless fringe of the drug culture, of being abandoned or beaten up by men friends, of supporting her lovers by working as a professional sexual escort, and of being unable to support her two children. She recounted a history of being attracted to men whom she characterized as impetuously crazy and sexually a bit weird. She knew before she married her former husband that he had served time and was actively bisexual. She accepted homosexual studs, but not effeminates. Her new boyfriend, the patient, was attractive to her because he was stereotypically macho. He had not told her of his feminine yearnings, nor betrayed any effeminate mannerisms. She rejoiced in his sexual prowess, as he could sustain intercourse for half an hour before coming and then, after a 15 minute break, could keep going with a continuous erection for up to five hours, long enough to wear her out, sometimes.

The patient's own conception of the relationship was given at the end of his own interview on the same day when he was asked if anything had been omitted. 'This won't be mentioned to her?' he asked. 'I don't like her really. She can't accept reality as it is. She lives in a whole new world of her own. When I told her I was gay, I knew she couldn't accept that I act as the woman. Since at the time I needed her money, I lied to her. I said yeah, I'm gay, but I do the man part. They suck me off and I fuck 'em. She don't interest me the least bit. She ain't nothing but a hole, and she will always be one.' He was able to perform, he explained, by having a fantasy that 'I feel somebody rubbing their dick up against my ass, and I look around, and I see who it is. And it's one of the guys at the prison, and he says, "Yep, you've been out here fucking all these women, and I still want some of that ass." And that's when I start dreaming more, making it go slow.'

Only four days after the interview in which these statements were recorded, the two were forever separated. 'I just don't know what to do,' she sobbed on the telephone. 'Something got into him. He's been throwing knives. He wants me out of his life.'

By contrast with this brutalism, he had a romantically sentimental story to tell of his first girlfriend. He met her when they were both teenagers in the same state mental hospital. Their story, as he told it, is bizarre. He was 17. She claimed to be 12, though she had the appearance of 18. With her mother's collusion, the couple went through a legally nonbinding ceremony of marriage 'in West Virginia.' The relationship was very short-lived before he was back in jail. 'I made up a whole new world for her,' he explained, 'because she OD'd on heroin, and I couldn't accept that she was supporting her habit as a whore, but she was. I was in prison. They asked me if I wanted to go to the funeral, and I said no.'

His idealization of this girl became self-idealization. 'If you've noticed,' he said, 'the name I go by as a girl is hers. See, I try to relive the life that she did, and don't want to accept that she's gone, because I can't accept it, and I won't accept it. I was in love with her, and I still am, actually, because I know she's not gone. She's the one that really changed me, changed my whole outlook. During the time with her, I thought I could stop being gay, and become a man, work eight hours, come home, take care of the family, everything.' The thought of stopping being gay failed to materialize.

Two weeks after the audiotaping of the foregoing information, the patient announced to trusted friends and to psychohormonal staff that it was time for him to go back to prison, because he couldn't get a job, and couldn't accept charity. A week later he impulsively engineered the 2 a.m. Sunday, closing-time theft of the cash register intake of a restaurant supervised by one of his gay buddies. The two left town taking with them the supervisor's gay boyfriend. All three headed for Nebraska.

Unable to await possible extradition and return to prison in Maryland, the patient engineered his arrest in Nebraska. He told his own story, unsolicited, in a long-distance telephone call, saying: 'I did a robbery out here. I turned myself in the next day, because next time I might kill someone. I practically knocked the guy's brains out with a stick. The guy pulled a knife. I grabbed it and stabbed his hand. I just went berserk, because it was either his life or mine he was playing with. I believe it was my fault, because I never did like bars, but I went there, because I felt like hustling.'

At the bar he was picked up by the victim, a man in his twenties, who invited him and his two friends back to his apartment. They had hoped he would 'pass out on alcohol and reefer' (marijuana), and then they would rob him. They did not need to rob, as all three had found work. It subsequently transpired that the victim had a history of picking up hustlers and being attacked and robbed, which suggests that he may have been paraphilically fixated on stage-managing his own victimization.

After the patient turned himself in, his two friends were arrested, one charged with assault and robbery, and the other with accessory after the fact. The patient said he was covering for the other two, 'because they can't take jail as well as I can.'

Having served two years and eight months of his four year sentence, he was resentenced to six and a half years in solitary confinement for assaulting two prison employees. His written explanation of the assault is reproduced, unedited:

'I'm emotionally and deeply in love with a guy here at the prison. My involvement in this affair is very serious, and this is why I assaulted two correctional employees. I've never experienced this type of love before with a man, and I sacrificed my freedom from prison just to be with him and to become a part of his life. He helps me considerably with all aspects of life and living. He's got thirty years in prison, and I can truthfully admit that I would love to help him do his time. ... I've actually shed real tears over him. I love him dearly, and I would rather stay with him in prison than be released into society where I'd not be loved and become a loner again. This segregation has been getting to me, because I can't be with the man I love.'

Segregation did, indeed, get to him. It was a year later that he hung himself.

EXPOSITION

The degree of sexological and psychopathological concordance in this pair of cases is all the more remarkable because of their chromosomal discordance. It offers no support to those who would like to attribute homosexuality or heterosexuality specifically to the genetics of the X and the Y chromosomes. It demolishes the equation of XYY with supermale, and of XXY with neutered male. If the supernumerary chromosome in the nucleus of every brain cell and nerve cell in each of these cases had anything to do with their sexological pathology, then the connection from chromosomes and genes to amphoteric gender and erotosexualism cannot be conceptualized as direct cause and effect, but only as derivative and multiplexly determined.

Like genetics, prenatal hormonal determinants of sexual dimorphism of the brain also cannot be implicated as a direct, cause-and-effect determinant of sexological amphoterism. It goes without saying that there have been no prospective studies beginning with fetal measurements of brain hormone levels. In the XYY syndrome, there is no evidence from which to construe, retrospectively, a prenatal hormonal anomaly that subjects the differentiation of the brain either to demasculinization, or to feminization, or to inadequate defeminization. In the XXY syndrome, by contrast, because pubertal masculinization is typically weak, it is possible to construe, retrospectively, that prenatal brain masculinization may also have been hormonally weak. However, this hypothesis has no predictive power, insofar as only a minority of XXY men are homosexual, bisexual, or otherwise amphoteric.

In postnatal life, the hormonal determinants of sexual dimorphism of the brain are replaced by social determinants that make their entry into the brain by way of the senses. In this phase of development, the supernumerary chromosome in all of the brain cells that process erotosexual development may, indeed, place XXY and XYY infants and children at a disadvantage.

In the XXY syndrome, there is an excessive preponderance of all degrees of mental retardation, as well as of other psychoneurological and psychopathological manifestations of anomalous brain functioning.

In the XYY syndrome, there is an excessive preponderance of behavioral manifestations that, taken as a whole, indicate dysfunction of the brain's governance of troopbonding, though not pairbonding. Newborn XYY boys, so far as is known, pairbond with their parents, and later in life are capable of lover-lover pairbonding. Their great deficiency is in becoming a member of a troop which has a dominance hierarchy, and a leader at the top who commands subservience and obedience. If not all, then most XYY males from infancy through maturity exist as nonmembers of the troop. They are troopless leaders who obey only their own rules. They are cognizant of consequences but, like a player of Russian roulette, take a gambler's chance. Think it, do it. Others judge them as impulsive.

Of course, there are many males, and females also, who are impulsive and indifferent to consequences, and who do not have a supernumerary Y chromosome as a marker of a genetic anomaly. The significance of the genetics of the supernumerary Y is that it is virtually, but not universally prophetic of impulsive conduct ahead, sexual conduct included.

Sexually dimorphic differentiation of the brain in the human species, as well as in nonhuman primate species for which data have been assembled, is intimately dependent on the proper timing and exercise of sexual rehearsal play. There are no norms pertaining to childhood sexual rehearsal play in XXY or XYY children. Thus it is not possible to know whether the sparse amount of information retrieved in the present two cases is syndrome typical, or whether additional personal information was not retrieved. In the XYY case, the first installment of information was greatly augmented later, without independent corroboration, however.

Personal memory may be confabulatory more than historically authentic. Thus, without more confirmatory evidence, it would be presumptuous to attribute causality to either patient's juvenile history of sexual rehearsal play, or lack of it, as told by the patient himself. It may or may not have been a determinant of the adult sexual history. The juvenile and the adult sexual histories both may have been to some degree secondary to a syndrome-related cognitive blurring of the distinction between fabrication and fact, the outcome being pseudologia fantastica. The latter was an incontrovertible feature of the letters and autobiography of the XXY patient, especially as related to exploits of supernatural powers of violence and long-distance homicide, and probably to some of his claimed sexual exploits also. Pseudologia fantastica has been found to be sporadically associated with both the XXY and XYY syndromes, in other cases.

It may happen, and perhaps might indeed have happened in the present XYY case, that, from an early age, a child's approach to people of all ages and either sex became hypereroticized, not in any pseudologically fantastic way, but quite literally. It is known to happen, for example, in association with a history of childhood neglect and abuse; and also in association with a history of clandestine and conspiratorial sexual relationships which entrap children into being damned if they do, and damned if they don't disclose them.

On the basis of two cases it is possible only to formulate hypotheses, not to resolve them. Thus, nothing can be proved about the origin of bisexuality and related amphoteric phenomena in these two cases. Nor can anything be proved about imprisonment as either a necessary or a sufficient cause or catalyst for the release of bisexual and related amphoteric expression and behavior. However, the two cases do illustrate very clearly the existence of amphoteric dualism, and that an ideal climate in which to incubate it as a fantasy is solitary confinement in a prison cell. A sex-segregated male prison provides an ideal climate for female fantasy to be brought into full flower, in practice. It is integral to the unpublicized philosophy of the modern American prison system that part of the punishment of imprisonment is to be heterosexually deprived and then to be punished again for not being heterosexual. The end result is a prison-induced sexological psychosis, more accurately named, perhaps, prison sexosis.

For the science of sexology, these two cases illustrate very well the coexistence in the same brain of the two erotosexual schemas, one female and the other male, each with its respective nonerotic extensions. In the XXY patient, the two schemas were power related and had an alternating periodicity. This periodicity is probably related to the well-known periodicity of manic-depressive psychosis. Being sexual, it is tabooed to research and research funding.

In the XYY patient, the two schemas were also power related, though there was no evidence of periodicity between the male and female schemas, but rather of sporadic war. The victorious female's life was primarily in prison, where there was no female to conquer, only a male by whom to be conquered, and for whom to fight and then to die in solitary confinement.

Theoretically, the conceptual key to both cases is inconstancy of dissociation. We all have, imprinted in our brain/mind, the schemas of masculine an feminine. One is mine, and the other thine. They keep themselves constantly dissociated in most people. When they periodically alternate, as in the XXY case, or when they permanently translocate as they did, at least in part, in the XYY case, then society is confronted with a problem for which it has, as yet, no scientific solution. It resorts instead to its prescientific religious and legal solution, which is to punish or to kill the evidence of its own scientific failure. The victim's guilt is his failure to have chosen to cure himself voluntarily and by will power.

These two cases are of theoretical significance to the differential diagnosis, prognosis, and treatment in transexualism. They exemplify inconstancy of amphoteric dissociation, with the male/female pendulum still swinging back and forth. In such cases, sex reassignment surgery is contraindicated until such time as there is evidence that the pendulum has quit swinging. Otherwise sex reassignment is likely to be regretted as a mistake. Both cases lend support to the thesis that transexualism, together with other gender amphoteric syndromes is, fundamentally and phenomenologically, a sexologically dissociative disorder. The full implications of sexological dissociation have yet to become identified and incorporated into the theory and practice of sexual medicine and therapy, and into sexological research.

Both of these men lived most of their lives in jail, one until his death there, because a vindictive society, through the inertia of its bureaucracies, couldn't have cared less about the humane treatment of their genetic disorders, or about the inordinate cost of their imprisonment to the taxpayer — more costly, per annum, than financing a student through college, medical school, residency, and postdoctoral specialty training. The adversarial and punitive system of the law is totally incompatible with the consensual and therapeutic system of biomedicine and science.

BIBLIOGRAPHY

Berlin, F.S., Bergey, G.K. and Money, J. (1982) Periodic psychosis of puberty: A case report. *Am. J. Psychiatry,* 139:119–120.

Money, J. and Ambinder, R. (1978) Two-year, real-life diagnostic test: Rehabilitation versus cure. In: *Controversy in Psychiatry* (J.P. Brady and H.K.H. Brodie, Eds.). Philadelphia, W.B. Saunders.

Money, J., Gaskin, R. and Hull, H. (1970) Impulse, aggression and sexuality in the XYY syndrome. *St. John's Law Rev.,* 44:220–235.

Money, J., Annecillo, C., Van Orman, B. and Borgaonkar, D.S. (1974) Cytogenetics, hormones and behavior disability: Comparison of XYY and XXY syndromes. *Clin. Genet.,* 6:370–382.

Money, J., Wiedeking, C., Walker, P., Migeon, C., Meyer, W. and Borgaonkar, D. (1975) 47, XYY and 46,XY males with antisocial and/or sex-offending behavior: Antiandrogen therapy plus counseling. *Psychoneuroendocrinology,* 1:165–178.

Nielsen, J. (1969) *Klinefelter's Syndrome and the XYY Syndrome.* Copenhagen, Munksgaard.

Nielsen, J., Sorensen, A., Theilgaard, A., Froland, A. and Johnsen, S.G. (1969) *A Psychiatric-Psychological Study of 50 Severely Hypogonadal Male Patients, Including 34 with Klinefelter's Syndrome, 47,XXY.* Copenhagen, Munksgaard.

CHAPTER 12

Male concordance prenatally, postnatally, and in tortuous transition to sexuoerotic maturity

SYNOPSIS

The matching of the two cases in this chapter diverges from that of other chapters in that they are matched only for concordance, not discordance. They are concordant for being the cases of ordinary males with no clinical diagnosis. The two are concordant also for having qualified as the cases of healthy university students belonging to that broad category that society recognizes as normal, and that friends and acquaintances recognize as sexually normal and heterosexual.

After they had given their sexual histories, it was further recognized that the two cases were concordant for having gone through a rather tortuous sexuoerotic development in childhood, through puberty and adolescence, and into maturity. Their type of experience was by no means unique in our sex-negative society that, even among the educated classes, explicitly neglects the sexual learning and sexuoerotic health of its children. Thus, the two cases illustrate the sexuoerotic difficulties and pitfalls that confront children who grow up with normal sex organs. These are the sexuoerotic difficulties and pitfalls that handicap the majority whose genitalia are normal, and on which are superimposed the handicaps specific to those with a history of birth-defective sex organs.

The two sex histories here presented are chronologically well matched to the histories of the cases in other chapters. In age, the two men are more or less contemporaries of the patients in the other chapters, for they are now in their mid to late thirties. They were interviewed in 1975. Their interviewer was an exceptionally talented undergraduate student who was doing a research elective in psychoendocrinology. Two of his fellow students elected to give a sex history. No other criterion was used in their selection. Thus these two cases may be considered to represent a norm against which to compare the other cases in this book.

CASE #1: LOIS'S BOYFRIEND

Lovemap biography

Daniel Schluer comes from an upper middle class family in eastern Pennsylvania. He has only one sibling, a brother five years his junior. His parents, although not deeply religious in the spiritual sense, adhere to their faith as an ethnic heritage and prefer, therefore, that their sons' girlfriends be Jewish. They are members of a swimming, golf, and tennis country club. The father is a successful businessman, and the mother an efficient homemaker. Both went to college. They have been married 23 years. Daniel attended a suburban public high school, excelled academically, and was accepted at a prestigious eastern university where he majored in physics.

Physically, he is tall and slim, 188 cm (6 ft. 2 in.) tall and 82 kg (180 lb) in weight. There are no unique distinguishing characteristics. He had moderate acne in high school which cleared by the end of his sophomore year in college.

Socially, he would be considered outgoing: a serious student who studies hard but takes time out to party. In high school his social outlet was his Jewish youth group. In college, a fraternity fulfilled that function. He would not be considered the 'all-American boy' type, for he did not fit the rah-rah, boisterous-jock, class-president mold. Rather he was seen mostly as a fairly introspective and thoughtful person. He always had a ready quip, which usually received mock groans and shakes of the head — the type of response received for cornball humor. Those who know him well say he has a good nature. In sum he qualified as your average nice guy.

As far as he could remember, he had not been explicitly seeking information at age 9, when his mother gave him his first sex-education instruction. He therefore judged his mother's actions as spontaneous. She may have feared his getting a warped lesson from peers at summer camp, and decided to prevent its occurrence. To accomplish this effectively, she purchased a descriptive sex-education text. With it she accurately described reproduction, intercourse, and pregnancy. Nonetheless, there was some confusion: 'She explained how children are made. It was rather informative, but I didn't understand how a penis could fit into a vagina, or what a vagina looked like. It was hard to picture the way the book described it.' He could not remember the title of the book, but did recall that its lack of pictures or their poor quality confused him. He was confused about the fine points of intercourse, insertion, and ejaculation. The anatomy lesson had been buried under an avalanche of words. Pictures would have been more powerful teachers.

In possession of his new wisdom, Daniel remarked: 'I went off to camp that summer feeling that I was a quotable source on human reproduction.' His credibility was tested immediately: 'In this book that my mother had handed me, it was explained that sperm were little tadpole-like creatures. I talked to the campers in my bunk and told them that sperm were little tadpole-like creatures. I was put down with: "You're full of crap, Schluer. Everybody knows that sperm are white. They aren't black like little tadpoles!" I was disillusioned.' Disillusionment dissolved shortly, however, for Daniel recalled that what he told his fellow campers was in the book. To a 9 year old, the power and the authenticity of the printed word were beyond reproach. His disillusionment then focused on his fellow campers for their blatant rejection of the truth.

Despite the challenge of the campers, the resultant disillusionment, and the preventive medicine of his mother's sex lesson, Daniel was given and still accepted a sex education from his peers: 'I picked up a few odds and ends — like what a Kotex was, like what a Tampax was, sexual baseball, what a blow job was. These were things that wouldn't be found in the book.'

Sexual talk and information-trading abounded at camp. One of the major pastimes of the summer was joking about blow jobs. After Daniel learned the meaning of the term, he wondered out loud if you could do that to a vagina. The group collectively scoffed, one member echoing their thought intolerantly: 'You're full of crap. You can't blow a cunt!' To which Daniel, comicly keeping his poise, answered: 'Yeah, but you can't blame a guy for trying.' Laughter was intense. His reply became popularized into a publicly uttered phrase, a slogan of the sexual underground, of which only the clique of campers knew the true significance.

With their sexual underground well established, the most daring of the campers sought to make contact with the outside world. To do this they approached their camp counselor, Billy, who was 18 and had been through it all. They said: 'All right, Billy, we're now going to give you the sex test. What's a Kotex?' The counselor, pretending ignorance, replied: 'That's your partner from Texas.' They all laughed at his response. Daniel did not realize until five years later that Billy was only feigning ignorance.

Sexuality at camp was not limited to discussion and random queries of 'outsiders.' Fieldwork was also in order, in the form of group masturbation. It filled the idle hours on a rainy afternoon: 'One day when it was raining and our outdoor baseball game had been cancelled, Steven, the ringleader in the bunk house, said, "All right, turn off the light. Everybody take off your clothes. We're all going to sit back and beat off." We all proceeded to take off our clothes and sit on a bed. I remember leaving on my socks. He said, "Schluer, you can't beat off with your socks on." I said, "Why not?" "Just cause you can't," he said.'

The nine campers, including Daniel with his socks still on, sat there. 'I think we all got erections. The ringleader yelled out, "Hey, look at me, I have an erection." To obtain an erection he had rubbed his penis with his hand. He gave no more orders verbally, only by example from here on in: 'I saw him playing with his dong — so I did too — stroking it gently. After a while, I went limp. There was no ejaculation. The novelty of sitting around having a group beat-off wore off. So we put our clothes back on and went outside and did the things a camper was supposed to do, like boating and swimming.'

Asked about his feelings, he said, 'I felt a little embarrassed at first and then all of a sudden I said, "Oh, hell, everybody else is doing it, so I'm going to do it, too," and that was that.' There were no deep guilt feelings, but a 'cloud of embarrassment hung overhead. ... After putting my clothes back on, it hit me: "Do you know what you did? You just took off all your clothes and played with yourself."'

The camping experience ended, and Daniel returned home. He played with himself off and on over the next several years, never ejaculating. One such incident at age 10 involved sharing a bed for a night with his 9 year old boy cousin: 'The next morning we were playing with each other's penises. It was just something we did because it felt neat.' Again, there was no ejaculation, only mutual erections. In retrospect, the episode seemed a little peculiar to him, but at the time he expressed no reservations or second thoughts, and no feeling guilty.

At camp, five summers after the first beat-off, there was another beat-off. For the first time, someone ejaculated — not himself, but another camper. Also, 'One of the campers down the way had had a wet dream one night, and he was the big hero for having come all over his sheets.' The counselor said that wet dreams were necessary to expel sperm that, if left in the body, would sap a man's strength.

Daniel himself had had no experience of a wet dream. As he was, at 14, a couple of years older than most other campers, he began to worry that something might be wrong. It was at this time, as fate decreed, that he should be given a damaging piece of false information: 'Somebody had told me that the owner of the camp slept in bed with his underwear on, and now he was sterile because his underwear kept the temperature around his testicles so warm that they could not function and now were inoperative. I was thinking — wow — I'm fourteen years old, and I haven't had a wet dream yet, and I've been wearing my underwear ever since I've been going to bed and — oh, wow! I think maybe that I made myself sterile.'

Daniel carried this weight for five and a half years, during which he thought he was abnormal. The thought of sterility did not exactly torture or torment him. It was a mildly dull ache in the back of his mind. He did not ask for help or information. 'I didn't think it was important enough to ask my mother. ... To ask your mother something like that, it takes a little courage. I guess I felt that it wasn't — I couldn't get up the courage for a matter that wasn't all that important. It was something in the back of my mind that, wearing my underwear to bed, I might be making myself sterile. It's a problem I'll have to worry about when I'm a married man, I told myself. I thought it would be very sad if I'd wake up to find myself sterile, but I didn't worry about it so much that I'd ask my mother if I was indeed making myself sterile.'

It was not until age 19 that he discovered he could ejaculate. During a session with a girlfriend, Julia, he was lying on his back while she was on top of him. They achieved a good rhythm, rubbing crotches together and simulating intercourse, fully clothed. He felt a wet feeling in his pants and at first chance went into the bathroom to check. Noticing that it was wet down there, he figured that something was working right. Then it hit him like a flash: what he had several times experienced in his sleep must have been incomplete wet dreams, for the feelings were similar. He described those feelings experienced while dreaming: 'I felt like I had this really great sensation, this really neat sensation between my legs, and then all of a sudden I felt myself taking a leak. Ah, Christ! I was dreaming of balling somebody. Oh, wow! This is really great! Then all of a sudden I felt myself taking a leak, and I said, "Ah, dammit! Spoiled another night!" But when I look back, that was the wet dream I supposed I never had!'

In making the connection between ejaculation during petting and nocturnal emissions, Daniel concluded that he was normal and had been all along.

A common deficiency of all the sex information which he received was its failure to establish the place of erotic sexuality in the lives of human beings. From his peers he had received the word that it was a taboo subject not to be discussed in adult society. From his mother he learned it was a reproductive procedure. He never knew for sure about his parent's sexual life. At age 15, still wondering if they had intercourse, he tried snooping around the house to find an answer to his question — in vain. No one told him that intercourse was pleasurable. Even the subject of nudity, a source of embarrassment and uncertainty, was ignored. He recalled an incident at age 6 when he found his mother and father taking a shower together. His intrusion into the bathroom was greeted with screams from his mother and a quick slam of the door in his face.

In his 9th-grade sex-education class, the emphasis had been on reproductive physiology. There had been no mention of the recreational aspects of sex. Topics such as oral sexuality, homosexuality, and masturbation had been totally neglected. The course did not help clear up his misconceptions about ejaculation, nor his nagging concern about sterility.

Whereas at age 16, an estimated 95% of his age-mates masturbated, Daniel had not the slightest idea of what masturbation was. Even though he had tried to masturbate years earlier at camp, he failed to draw the connection between that early incident and what he finally would achieve, a sexual climax, at age 19: 'I think that if I had known a technique of how to masturbate, of how to make myself ejaculate, I would have been doing it a long time ago.'

Once he discovered ejaculation, however, it was as if he entered a delayed adolescence: 'When I first got involved with Julia, I noticed that she and I were sort of simulating intercourse. I noticed that friction on the underside of my penis made me ejaculate. And we did something whenever we got together to make out. I tried to see to it that she was on top of me in such a position that she could create friction on the underside of my penis, and thus I would ejaculate. It's funny, the first time I tried to do it myself, that's the exact same technique that I used. I massaged the underside of my penis, while lying on my back in bed. I was rubbing there, and it really got sore, and I said: "This is no good. I've got to find a lubricant." And I thought, and then I brought out a jar of Vaseline. I was in hog heaven! I think I masturbated for the next four nights in a row. I was so ecstatic — the idea of, hey, I've found a new toy to play with!'

Once the discovery was made, Daniel couldn't wait for bedtime. He elaborated: 'I was just sort of thrilled with the idea of making myself come. It felt really neat. It was like a kid with a new toy; it got a bit tiresome. So I slacked off for a couple of days. I would say, though, from that point on I never went more than five days without masturbating, even up to now.' The average was five times a week.

During the next few months of masturbating, he experimented with different techniques. He used no devices to achieve orgasm, but once pretended that his mattress was a woman with whom he was having intercourse. It proved too strenuous to maintain, so he switched back to his hand. With his hand, he also experimented: 'My first night I applied a hand to the underside of my penis. On the second night that wasn't working fast enough for me, so I cupped my hand around, I guess like everybody else does, and just stroked up and down. I found that to be much more pleasing. It really drove me wild at the time. It's quite a sensation.' For more pleasure he would rock the shaft of his penis back and forth while tickling the hairs on his scrotum. He reported a synergistic effect, the whole being greater than the sum of its parts. He used Vaseline about one out of every 10 times for variety.

One of the various aspects of masturbation he was attuned to was the visual effect: 'One of the thrills of masturbating is watching yourself come. It's sort of fun to watch things spurt out. It's really a visual sensation, for me. ... The more consecutive nights you do it, the less you come. It's like trying to tap a dry well. It was worth stopping and waiting for a couple of days. It was like watching a geyser go off. ... After a while there was a trade-off between waiting and doing it. After a couple of days, you just don't feel like waiting. It seems that after you slack off a couple of days, you get a real nice display, so to speak. ... Just before I ejaculate, I can more or less feel a dam getting ready to break, having stored up semen over the past couple of days. Yeah! If I haven't masturbated for five days, it really feels — I seem to be able to feel more pleasure behind it. In fact, one of the other pleasures is that I actually can feel the semen traverse the urethra. And if I wait a few days, I can feel the stuff coming up the urethra longer. That's a neat feeling too.'

After he had had intercourse for the first time, he said: 'There was a period that was really strange. We got it on four nights in a row, and during that time I had no desire to masturbate whatsoever. In fact I think for even a couple of days after when we weren't sleeping together, I still didn't want to. The feeling of intercourse is just different. Screwing is a superior feeling to masturbating. Well, no, masturbating is fun, but it's different when you're fucking because you actually have a chance to see your craft, your skill, actually see someone else have an orgasm. And the sensation — you can try to get the same feeling with Vaseline, but it just isn't quite the same as being in a vaginal wall. It's different. There's something animated underneath. God, that sounds trite! Just the fact that there's another human being underneath you, or beside you, or on top of you, whatever it is, and that she's going wild.'

Responding to a query about erotic imagery, he said: 'I just had pictures in my mind — women spreading their thighs. I seem to remember a picture in *Playboy*,

Playmate of the Year, 1973. She was Swedish. I remember they had a neat little picture of her with her legs spread. That was a very stimulating picture. It had such a vivid image. I still remember it today.'

He hardly ever used *Playboy* pictures to masturbate to. It was the memory of the picture that he incorporated into a masturbation fantasy: 'I never buy *Playboy*. If you can believe it, I have this feeling whenever I am looking at the pictures that somebody is looking over my back. Really, I mean no matter where I crack it open, I've got this feeling: Uh-oh, someone's going to look at me! And it's really — it's stupid, because I just have this feeling, you know, every time I crack open a piece of pornography, I get that damn feeling.'

His fantasy imagery during masturbation was split between recall of pictures such as those in *Playboy* and thoughts of his present girlfriend. He reported no masturbation fantasies about a stranger, for example a sensually dressed woman with whom he may have come in contact during the day. A large part of his masturbation did not include fantasy, as he found enough stimulation in the frictional act alone. He had no guilt feelings over any of his fantasies: 'When I wasn't concentrating on the Playmate of the Year, I might have been picturing myself making love to Emily, my girlfriend at that time. I really wanted to do that, imagining what it would be like. I would be just sort of lying on top of a woman with her legs spread — just sort of undulating back and forth. At that time, I still had no idea of what it would be like to have your penis in a vagina. That sensation I couldn't imagine, just the idea of lying on top of a naked woman, just waiting for me to hop on top of her.'

By age 20, after he had become able to appreciate masturbation, he had a reservation: 'It felt nice, but I might have thought, "If only it were the real thing."' This feeling stemmed from a desire to be in the club — to be one of those who were having intercourse.

Masturbation itself had had no profound stigma attached to it. There had been no religious or family condemnations in his childhood. The first and only pejorative comments were heard in high school and college, where peers would use the phrases 'jerking off' or the label 'a jerk-off' derogatively as in, 'What are you doing tonight, jerking off?' In Daniel's words, it was 'as if it were some inferior activity to be involved in.' He himself did not think that there was something more healthy or more normal in having intercourse: 'Not really,' he said.

As compared with his atypical masturbation history, his early dating history was in conformity with the expected sexual stereotypes, influenced no doubt by peer example and discussion. His earliest bit of explicit heterosexual schooling had been at summer camp where he was taught the ubiquitous 'baseball system' in which kissing is first base, fondling is second base, hands in pants third base, and intercourse is a home run. Adoption of this outlook made sexuality a ritual. As he described the ritual, his girlfriend was not a partner but rather an altar on whom the rites were performed. In most instances the girl, having been made an object, became virtually inanimate. There was no talking between the pair, no response of pleasure. In fact, the only responses from his early partners that Daniel could recall were admonitions to go no further.

In his 13th year, at summer camp, he had his first dating-like experience. It was the first time that he kissed a girl: 'Summer camp is boys and girls thrown together. The idea is — it's really emphasized to pick up a girlfriend and get somewhere with her.'

Two years passed before he had his first authentic date. This two year period was a period of observation, a time of girl watching in which his tastes began to develop. It was at this time that 'breast worship' began to occur. His most prominent recollection of the first girl he dated was: 'She was voluptuous from the neck down. God, what a set of knockers.' He expanded on his philosophy: 'Back then (and still today), physical attraction drew me to a woman. And, you know, I'd see someone that I really liked a lot. Perhaps if things worked out I'd like to go to bed with her. I'd like to get things good, you know, just to get together and know her as a friend. Maybe cultivate enough interest between the two, enough mutual interest. Maybe things could get better — get in bed together.'

His amorous accomplishments, however, fell far short of his goal. A goodnight kiss alone was the sexual substance of his first date. Sexual frustration? No, on the contrary, he was satisfied. He was breaking into the adult world. The kiss was viewed as a start on the sexual road: 'The way to a girl's underwear is by taking her out on a date and getting to know her.' He was rearing to go, and had an almost perpetual erection whenever they 'made out,' kissing: 'I think I got her to French kiss maybe twice, and that was it. I could not interest her. She didn't like it. I think she was worried about bigger and for her, worse things. For me they were a lot bigger and better.'

Through his membership in a Jewish youth group, he met another girl, Sandy, shortly after he stopped dating his first girl. With Sandy, he had his initial experience with a female who wasn't totally passive. They met when she 'cornered him' at a dance and put her arms around him. Immediately he had visions of a romp between the sheets, but it was not forthcoming. They kissed quite a bit. His erection was ever present. He blamed it for scaring her away. He concluded that erections scare off Jewish girls. Gentile girls, he thought, might be more promiscuous: 'Well, it just seems to me that when a girl feels a guy with an erection, suddenly her training, you know, the ideals of being a virgin are called to light. I don't know why Jewish girls — but if you talk to any married Jewish girls, they'd say that gentile girls are a lot looser than Jewish girls. And, as my cousin said, "Boy, I like dating goyisha girls. Man, you can screw 'em first night you take 'em out." And they really seem to be a more promiscuous lot. For the life of me I can't figure out why.' This was his sexual myth to rationalize his lack of sexual success.

During the six months that he continued to see Sandy, his policy was to date only her. Then he dated different partners. 'I'd take people out, spending all kinds of bread on them, hoping to get somewhere sexually,' he said. 'It suddenly occurred to me now that a lot of girls will go along, they'll like a guy who's a little bit taller. It's an ego trip and to some extent they get to use a guy a bit.'

Then came a flash in the pan in the form of an affair with Rita. He met her at a dance and made out for the remainder of the evening: 'Wow, I thought, here's a chance. I'm really going to screw someone. Here's the chance.' The chance never came. All they did was kiss.

It was at this point, aged 17, that his feeling that intercourse was an absolute necessity began peaking: 'It was something I wanted to do, I really wanted to do, but I didn't go berserk for lack of it.' Nor did he consider a call girl: 'It didn't occur to me, but the means and the know-how I probably didn't possess. Just the idea, though, of paying for a lay when all your peers are talking about the free ass they've been getting — No.'

During his senior year in high school, he visited his cousin Fred at the State University where 'One of my cousin's roommates had found some chick he really liked, and he wanted to get it on with her. So he fed her a couple of Sopors (downers) and he was really getting it on. Then he left her briefly and told us, "Wow, she's really zonked." Though my cousin and I were leaving to pick up someone, I responded, "Gee, I'd really like to screw that girl." the roomate replied, "Why don't you go in there and do it?" So I went in.

'I remember taking off my shirt. She was really stoned at the time and thought that I was the former guy that she had been screwing. We got into bed. We Frenched a little. I played with her tits. I seem to remember massaging her vagina. Not very much because I couldn't see where it was. I just felt something soft and furry. All this time I had my T-shirt on, and my pants and belt. She asked me, "Are you ready?" I really didn't know what was coming off, and I said, "Huh?" She just sort of reached down and grabbed for my crotch and felt my belt. She said, "Wait a minute!" Suddenly her senses came to, and she felt my hair. The guy she was screwing before me had very straight hair. She felt my curly hair and said, "Wait a minute, you're not him — get out!" I blew it. I really felt, "God damn, there was my chance!" It was my only chance for the next four years. I don't know what would have happened if she grabbed my pecker, and we started screwing. I was thinking in the back of my mind, "What are we going to do for birth control?" Perhaps it was good, then, the way it happened.'

In the following autumn, he entered college. There, through a friend, he met a girl, Jane. She attended the local woman's college. So one day he made a phone call to her, explaining who he was and asking if she would like to get together with him. She said she would. Their first meeting involved putting down a carpet in Jane's room. They talked a good deal and agreed to see each other again. They soon had a date at the zoo which involved some hand holding. At the end of the week there was a Pete Seeger concert: 'I really got bold. Not having transportation, I realized that the only way we could see the concert would be if she slept over. So I asked her to sleep over.

'Many people find this hard to believe — my roommate for one and others in the dorm — that I really had no intentions of screwing this girl. And to this day I swear that when we were going to sleep over that night, I was going to sleep in my roommate's bed, as he would be away that weekend, and she would take my bed. Well, it didn't work out that way. She really got a little uptight when I took off my pants to sleep in my underwear, because she was a nice virgin at the time. It really started to worry her. We made out for a while. There was no tit action. I was about to move to the other bed, and she gave me a funny look and said, "Come to bed." So I did, though she had no intention of being screwed. During the time we made out, I had a hard-on, but she kept away from it."

There was another concert the next weekend, and they got together before it: 'Just to show what a nice guy I was, we had a long discussion, about a five minute discussion, trying to decide whether I should feel her up. And I guess she was feeling bold that night because she said, "Yeah, go ahead." My attitude had been, "I'd like to, but I don't want to do anything you don't want to do." And I guess in that light it seemed all right to her.'

It was the first time that Daniel had ever taken such an open, communicative approach towards sexuality: 'She took off her brassiere and I peeled her shirt up. I started to suck on her breast, but she made a reference as to how she didn't like that. I said: "What's the matter? Don't you like the idea of motherhood?" That really turned her off. It was a tactical error on my part.'

The next night she was very unreceptive, resisting all advances when they were alone together. They met again after doing some Christmas shopping: 'I was getting a little tit action once again, and I remember reaching for her belt buckle. She said, "No, you don't!" And then I made one of the all-time brilliant statements. My capacity to make these statements has since been curtailed, fortunately. But at the time of being rebuffed I said, "I guess I'll never find true love." She took offense at that, and things simply fell apart. Boy did we have an argument after that!

'Several weeks later, after not dating, I received a call from her saying that she wanted to stop seeing me. I said okay, and was not too upset about it. It was not until four years later that she told me why she ended it. Her decision was the product of an anti-male attitude developed after drinking and losing her virginity to an intoxicated marine.'

Daniel's next sexual contact, with a girl named June, was minimal because, he said: 'I never made any forced moves. I still don't. I move forward very slowly. ... At this time I was into fondling girls, and if I got my hand inside of her shirt, and she didn't immediately roll on her back pinning my hand or turn the other way, I took that as a sign that all was well.'

As there was no coital sexuality coming out of his relationship with June, he moved on to a girl named Jackie: 'After about a month of seeing her, I used the hand-in-shirt maneuver. I think she was trying to roll over, but I was insistent. I guess after a few minutes she just sort of gave up the ghost at that point. She seemed to get into it, though. I started kissing her as I massaged her breasts, deep kissing, embracing her tighter. I could tell she seemed to enjoy it. I remember that I started sticking my hand down her underwear and began playing with her pubic hairs. About fifteen seconds after hitting third base, the record which we were listening to ended. I was trying to inch my hand down deeper. It's hard when you've got a belt buckle in the way. I didn't unbuckle it because I figured that she would probably resist at that point. What I was trying to do was find her clitoris. I figured that if I can get my hand in a girl's pants and stimulate her clitoris, she would be all mine.

'I didn't find it. The record ended, and she got up to flip it over. That was the end of that. I just sensed the attitude that she wasn't really getting into it. And I figured that if she were really into it and wanted me to go ahead, she wouldn't have flipped the record.'

As it turns out, his knowledge of the clitoris was a remnant of peer-delivered sex education. He had looked in an anatomy book to find the proper location. But his initial knowledge and inspiration came from talk with his cousin: 'He was telling me about a girl he was seeing. She was pretty aggressive, to say the least. He had screwed her at least ten times. And he was explaining the experience to me, and you can tell when a girl is really turned on. I said, "How is that?" "Well, there's something called a clitoris, and when she's real excited, it sticks out like a little cigarette butt." He snuffed out a cigarette and held it up. So I was expecting something about as long as a cigarette butt. Later on, when I finally got the chance to see, that wasn't the way things were.

'Two weeks later Jackie and I went out. I could tell she was pissed off at me because she wasn't even interested in kissing that night. She just wanted to sit on my bed, just staring off into space. She didn't say anything. Whenever I made a move to kiss her, she moved away. ... I didn't ask her what was wrong. I take a very passive attitude when someone lodges a complaint. It's just the way I am.'

Within six weeks the relationship was terminated by mutual consent. His next experience was summed up as another night of frustration: 'I invited this girl, Margaret, up to my apartment. This is a lesson I had learned: you can be passive, but there comes a time to take the offensive. We were sitting in my room talking, lying back on my bed. She was giving me stories about how a 46 year old man back in Chicago was in love with her, and how she'd been out with older men, and all that. I realized that I really needed a piece of ass. But I couldn't bring myself to put my arm around her and begin making out, because I knew what I wanted. And I felt that my intentions weren't as noble as this other guy's who really loved her. I just couldn't. It was like my good side saying, "Don't do it, don't do it!" I felt that I didn't have enough feeling for the girl, so that I had no business being in her pants. I knew that if I started kissing her, I would end up going for her brassiere. I guess after sitting and lying on the bed for two hours, she just got tired of me doing nothing, and just got up to walk out, and said, "I'm going back to my dorm." I said, "Wait a minute, I'll get my coat and walk you back." "That won't be necessary," she replied, and left.

'I felt that I didn't have a deep enough feeling to do anything. I liked her. I liked talking to her and all that. But I felt that there was the character back in her hometown, Chicago, who really loved her and would be willing to go out on a limb for her. I wasn't at this stage yet. I wanted a piece of ass. ... It really pissed me off. Goddamn bitch walked out of my room because I wouldn't feel her up, and the reason I wouldn't feel her up is because some guy really loved her. It was frustrating for a while.'

An encounter of which the outcome was quite different took place with the return of Julia. He had begun dating her again during the summer between his sophomore and junior years in college: 'I was playing cards at the swim club when one of the girls in the card game asked me if I would like to take out her sister, a sophomore at Temple University. I said sure. There was no sense in automatically writing off something which you know nothing about.

'That night we went out. I brought some dope (marijuana), just in case. It turned out the case was right. We found a pipe and started smoking. We began making out. I went through the usual ritual. I unbuttoned her shirt and took off her brassiere — no protest whatsoever! She really got into it. First I just started cupping her breasts and massaging them. I started to draw circles around her breasts and she did the same to me. We both had our shirts off now. Then I really got bold. I unbuttoned her pants and pulled down her underwear and started playing with her vagina with my hand. I pulled her pants down below her knees and the same with her underwear. I pulled myself up closer so that I could reach down and in. I started playing with her pubic hairs. Then I reached down, feeling for her clitoris — still looking for that thing the size of a cigarette butt,' he laughed. 'I got inside the two sets of lips, majora and minora. I guess that's the vagina proper. So I was inside the vaginal opening, and she really started to lubricate. I remember that. She was good and wet. It doesn't take long to get wet once you get your finger inside. She was good and wet, and I was still looking around for the clitoris. Anyway I had my finger inside, and she started making some panting noises, like she was getting really turned on by the excitement. What was really strange was when I pulled out my finger for an instant, I started hearing sucking noises going on in her vagina as if the walls were contracting. At least this was what I imagined.

'I pulled my face level with her vagina. I was about to start eating her out and for some reason I thought: "Now if you start doing this, and you stimulate her clitoris, you're going to end up taking off your pants and screwing this girl. You don't know the first thing about birth control. You certainly don't have any, and you don't know if she has an I.U.D. or is taking the pill." So I stopped. I sat back and thought a little bit, and at that point she drew up her pants and that was that. I was kind of grateful at the time. I didn't know whether she had any birth control. I just didn't feel like taking that step. Because I knew that if I took off my pants, it would have been just like the head resident counselor at college said, "A stiff dick has no conscience." For a first date, I was more than content. As a matter of fact the next day I had some guilt feelings about it, of taking advantage of someone. Even though she enjoyed it, I still had goddamn guilt feelings.'

Apart from 'pants and grunts,' Julia had not talked while they were together. Nor had Daniel. He took it as a sign that he had gone too far when she pulled up her pants. It did not occur to him that she may have thought he was turned off, maybe by her vaginal odor. 'Christ, I never thought of it from that angle,' he said. 'The way I looked at it, I was always an intruder, and I'd gone too far. ... The next time I took her out I started to play with her brassiere. She said, "Oh, not that," so I relented.'

Once again there was no discussion to elaborate their feelings. When she indicated discontent, Daniel desisted: 'I was very obedient.' They spent the rest of the evening doing nothing more than making out, which disturbed and frustrated him. He tried to 'get her hot' so that he could go further, but it was no go. This was not the way it was going to be forever, for on their next date the sexual involvement was more like that of the first.

'This time it was more or less the same thing. I got her shirt off, and this time I pulled her pants all the way off. And while she had her underwear on, I stuck my hand in her pants and started feeling around the cleft. Once again she got lubricated, really quick. I stopped for a minute, pulled my hand out and massaged her breasts just for a little bit, and leaned back. And you know, she got off the bed, bent down, picked up her pants, and put them back on! I never asked why.'

In closing the account of the relationship with Julia, he recalled one more incident. Again the situation was set for a sexual encounter. Daniel was visiting Julia in her home state. He had rented a motel room. They returned there after an evening out. In his words: 'There is where it would be done, if it was going to be done at all.' As soon as he started to make his moves, she protested. Once again he spent the evening making out until, when the hour got late, he took her home. Daniel departed the next morning. With the exception of a few letters to be exchanged, the relationship had ended.

When the fall semester started at college, so did a new relationship. Perhaps it could be called a romance. The partner was Sue, a freshman at the local women's college: 'We were in the movie theater. I had this feeling — I guess at that time it was the urge to conquer, to find a piece of female flesh, to get some ass, love or no love. I thought, "Well, here's the point of no return." Her hand was on the armrest, and I reached for it. At that point she put her arm around my neck. Hot dog! Off to a good start! When we walked towards her dorm, I noticed a nice grassy spot in the shadows. I started to head in that direction, and she said, "That looks like a good idea." We went to the spot and started making out.

'I was kissing for about 45 minutes before I got bold. We were blowing in each other's ears, and french-kissing in the ears during that time. She was wearing a halter top. I thought it would be easy to take advantage of it, and it was. Anyway, I stuck my hand up inside her halter, and she really didn't seem to mind. I cupped her breast and moved it around in a circle. She really got into it. It was enjoyable. It was enjoyable to watch somebody get into it. I stopped at second base though.

'I remember sticking my hand down in her ass, and each time I did she promptly pulled it out. We had a movie date two nights later and did much the same thing. She seemed to enjoy, but after that she took to wearing body shirts on dates — and we all know the difficulty because of where they unsnap.

'I think she got into my cupping her breast more than my sucking on them. But when I sucked them I would sort of place my mouth around the circumference of the areola and tickle the nipple with my tongue. I bit or nibbled a little. She would respond facetiously with "My, aren't you hungry. Please don't eat me up." I wish I had, but that was as far as I ever went with Sue.'

The relationship appeared as if it could really have developed and progressed, and as if he had found what many look for — someone with whom things really clicked, affectionately as well as passionately. And then something happened. Sue was raped by a stranger one evening in late October. Luckily there was no physical injury. It was pretty much a sexual assault. A knife was used as persuasion. Otherwise the only force used was that necessary to consummate the act. The emotional impact was traumatic. As much as it affected Sue, it affected Daniel also: 'It really cramped my style. I never discussed this with her, but I sort of assumed every advance I made she was assuming I was trying to rape her, and that it would bring her distasteful memories. It still is a puzzle. She said it was legally rape, but also that it wasn't rape. I think what she was subtly trying to say was the guy had interfemoral intercourse, but that he didn't penetrate her. I saw her a week after it happened. I felt that in the name of good taste I really didn't want to discuss it graphically. I never discussed it afterwards either. But I wanted to pound the son of a bitch, and I really felt sorry for her.'

The two continued their relationship for 10 months longer, suspended in a nondeveloping void of kissing and breast fondling, frozen in place by an extraneous intrusion they could not resolve together. Sue was the first person Daniel had ever contemplated marrying, but even that degree of feeling was not enough to keep them together after Sue transferred to another college.

The loss of his virginity was a milestone in Daniel's life, the end of sexual deprivation for him, the end of denial, and a ceremony of entrance into adulthood. With virginity shed, he entered into that place from which life takes on a new perspective: 'That morning was something different, really nice. The sun breaks through the window. "This is a really great way to wake up in the morning," I said.'

The great event took place in his senior year at college. His partner was his old friend Jane. Here is the story as Daniel recalled it: 'I was sitting in my office one Friday night and, after poring over fluid mechanics and not comprehending it for the umpteenth time, I said, "I've had it. I'm going out for a beer." So I went over to the campus club. I saw a couple of people whom I knew, and then I saw Jane sitting one table over. I said, "Hi, Jane!" She didn't pay very much attention. So I repeated, "Hi, Jane! How are you?" She acknowledged my presence. I saw that she was talking to a guy, so I said, "All right, she must really be into this guy."

'Anyway, she came up to me later and said, "I really want to talk to you about something." She bought me a beer, and we began to talk. She explained why she had dropped me as a freshman — that she was pissed off at men in general. It was after she relinquished her virginity to the marine. She was really sorry and asked if she was forgiven. Being the guy that I am, I said yes.

'Her response was: "I'd like you to come back to my apartment, and we could talk things over." I said to myself, "Oh, boy, this does have its possibilities." This looked like the chance where I would get to screw someone. Because I had been pimped once before, if I could get to screw someone, more power to me.

'Anyway, I followed her back in my car. We smoked a little dope and were sitting there talking. We eventually got around to where I should sleep, and she asked me about my present girlfriend, Lois. I said, "Well, she's her own boss when she's at home, and I'm my own boss when I'm away from her." Jane mumbled, "Uh-huh, I see." At this point I was really trying. I was beginning to think that I wanted to spend the night with her. "You can spend the night on the couch," she said.

'I was feeling uncomfortable lying there for about fifteen minutes when she reappeared from the bedroom. "This isn't too comfortable," I said. "I think I'm going to, you know, take off for home." Then I changed my mind: "Why don't we just share the same bed together?" I got my memory operating and added, "Remember what we were doing together this night, three years ago? You were spending the night over in my dorm, and we shared a bed together." "All right," she said. "I won't sleep with you, but we can do that." I knew she meant we wouldn't have intercourse.

'I think we were under the covers for five minutes when I started kissing her. That really turned her on. She rolled over on me, and we started to fake screwing. We had our clothes on.' He chuckled at the absurdity of being in the same bed, under the covers, fully clothed, fake-screwing. 'I was on top of her, and I rolled away, as I wanted her on top. "I always lose the rhythm when I get on top." she said. "That's okay," I replied.

'Another five minutes of this was like being in a sauna. "Can we take our clothes off?" I asked. "It's awfully hot." You know, I only expected to take off my shirt, and leave my underwear on. I was perfectly content to leave my underwear on. I left my T-shirt on, but she took her pants off, her shirt off and her bra off and crawled into bed.

'So her breasts were exposed to me, and naturally I started playing with them and kissing them. Then we started simulating screwing again. She insisted that I take off my T-shirt. I asked her as we resumed, "What does it feel like?" "It feels like we're screwing," was her answer. ... What did I have on? My shorts. And did I ever have a hard-on!' Daniel laughed. 'Jane had on her underpants,' he continued. 'The only thing that separated my cock and her cunt were the two thicknesses of our underwear. We just started going at it, pants or no pants. As I got her excited, we started picking up the rhythm. I kept up this breakneck pace until finally she asked me to pause for a minute. I was really wearing her out. Then we picked up the pace again.

'By that time I was really wearing the skin off my pecker, but I figured, "Wow! This is really turning her on," so I kept it up. Finally, I stopped. I figured, "I'll try to eat her out. I'll find her clitoris — that big thing shaped like a cigarette butt — and then maybe we'll screw."'

Responding to an inquiry, he talked about oral sex: 'I had heard it didn't smell too neat down there. I had fingered Julia and it smelled kind of, not really pleasant, but kind of neat — a novel smell. It wasn't really pleasant, but it didn't absolutely turn me off.' Nor did it turn him on, he said: 'But I figured if I got my tongue inside, it would turn her on. That's why I persisted. I started hinting around, dropping hints, that I wanted to eat her out. So I asked her, "Could I sample your vaginal walls with my tongue?" "What?" she exclaimed. I stammered, saying, "Could I eat you out?" Jane looked at me and said, "Why don't we just screw?" I replied, "Well, I'd say okay, except that I'm a firm believer in birth control." To which she countered, "You ever heard of an I.U.D.?" So I said to myself, "Ah, home safe. So here comes the runner into home plate!" To her I said, "Yeah!"

'So she slipped off her underwear, and I slipped off mine. I was immediately getting ready to penetrate when she said, "Wait a minute! You've got to make sure I'm wet first." I thought to myself, "How stupid can I get?" I had read that I was supposed to get a girl wet, but there I was in the excitement of it all, getting ready to plunge. So I stuck my finger inside her vagina, swished it around a couple of times, and in a couple of minutes she was suitably wet, so I went in.

'I think to the day I die I'll remember this experience. Once I got inside, it was very warm. I pumped a couple of times, and it felt like I immersed my pecker in a warm bath. I lost all sensation of the vaginal walls, I think that just inside three minutes I had come. She knew that I had come, because I made a lot of panting noises. It sounded like a locomotive. Well, I wasn't sure whether she had an orgasm, so I kept on pumping. Having beforehand simulated screwing in my underwear, I'd worn some flesh loose on my pecker, and it just got to be too much for me. In fact I found a scab two days later. I just had to pull out. I said, "Wow! That's really nice." She asked, "How was I?" I said, "Gee, you were really nice. That was great — my first time." "What do you mean?" she asked, disbelievingly. "That's the first time I'd ever done that," I replied. "Really, you're kidding me!" she said in continued disbelief. I assured her, "Yes, I'm no longer a virgin as of this minute." She found it hard to believe for she had heard stories to the contrary.'

Beauty it is said lies in the eye of the beholder. Truth is sometimes there too! In Daniel's case that's where his standard of sexual performance lay. On his first night of coitus, he worried that he was impotent! Eventually, he had five different erections, five intromissions, five orgasms, and five ejaculations in the course of nine hours, but in the refractory period after the first orgasm, his sexual self-confidence began to weaken: 'I had just come with Jane for the first time and pulled out. We had talked for a while. So while I was talking, I noticed I was going soft, and I said to myself:

"My God, I sure am glad I'm not some superstud, he-man, because if I was, right now, seeing myself go soft, I'd probably go shoot myself." Even though it was after we had fucked, I didn't know — I thought you were supposed to remain hard. So I saw myself going soft, and I thought, "Oh, well, it looks like I'm impotent." Well, about ten minutes later, which is the time it takes to recycle — I knew nothing about this recycling period. As I remember, I later asked a friend about it, and he replied that it was normal. Anyway, the system came back up again. I said, "It looks like I'm not impotent anymore." So we screwed again. I didn't know, though, at that time that guys go soft between comes.'

He had failed to draw a parallel between the after-effects of ejaculating via masturbation versus the after-effects of ejaculating via intercourse. If he observed detumescence after masturbating, why shouldn't he have expected it after copulating? 'I figured it was different circumstances, that there were two different physiological processes operating. I wonder why I didn't put two and two together?' He conjectured that his misconception may have come partly by way of the media: 'I was at a movie, *Young Frankenstein*, when the monster climbed on top of Frankenstein's fiancée. After having his fun, he got off and grumbled that he was going to town. The woman yelled at him: "You men are all alike — seven or eight quickies, and then it's back to the boys." I turned to my date in the movie and exclaimed, "Seven or eight quickies — that's a night's work for me!"' A second source of misconception had involved the braggadocio of a fellow worker: 'One clown I was working with over the summer said that he came twenty times in one night.'

He hadn't swallowed the braggart's story, but it had made an impression on him. Twenty times was out of the question. Maybe he came 10 times. A five-orgasm bed session seemed pale by comparison. Even to recall the title of Jacqueline Susann's book, *Once is Not Enough*, evoked feelings that maybe five were not enough. His doubts were subsequently revived by the words of another girlfriend who measured his love for her in numbers of ejaculations per evening: 'I made a reference to the fact that I'd screwed Jane five times in one night. The girl I was screwing at the time got really upset and said: "Goddammit, you came with her five times one night, all you ever give me is three or four times." She was really upset, so I promised her I'd do seven times.'

Daniel and Jane were friends, but they were not in love when she was his partner in the first sexual intercourse of his life. They were able to share their eroticism, mutually, without becoming entangled in the game of sex and love. They did not have to persuade themselves that they were really in love, and thereby could sanction sex without marriage. They did not have to obsess over, "What's this going to do to our relationship?" And they did not have to commit themselves to an extended, long-lasting relationship with multiple, reciprocal responsibilities that they were not yet ready to sustain. Daniel agreed that he was lucky to have had it this way.

Daniel and Jane had intercourse on about 25 occasions. He referred to Jane as a very good instructor: 'She taught me how to screw. I learned about different positions, as when she got on top of me for the first time. She reminded me that girls have to be wet when you go inside. But the most important thing was what she called picking up bed time. Once, after we had been to a movie, she said, "Let's go to my place. I have a surprise for you." We came into the room, and she began pulling cushions off her sofa. Then she said, "Help me pull," and out pops a double bed. "This is your surprise." We hadn't had a double bed before. So I said: "I've got an idea. I'll take off my shirt, and you take off your shirt. I'll take off my pants, and you take off your pants." She answered, "I like the idea." That's how we got started one night. Another night she said, "Let's take all of our clothes off and see how close we can stand together." You know, standing naked, pressed against a woman, is almost as good a feeling as orgasm. On a different night, she said, "Why don't you try to seduce me?" So I reached up inside her shirt and started cupping her breast with my hand. Then I unbuckled her pants and started to stick my finger up her vagina when she said, "You know, I'm horny for you, and you're horny for me, why don't you take off your pants and let's screw?" I was having fun gradually working toward it. It gave me the impression that I was moving too slow. I did not feel secure enough to say anything.

'Once we were naked, I would try to lubricate her. I'd stick my finger inside and swish it around. One night she pointed out her clitoris. I had finally found out it really isn't like a cigarette butt! One night I really got bold and ate her out. I said, "I have a surprise for you. Spread your legs." I pressed her vaginal lips apart and started peering around and saw something sticking up which turned out to be her clitoris. She really enjoyed that and asked, "My! Where did you learn that?" I answered, "I just picked it up in my travels." First I started licking her clitoris, then I gradually moved my tongue into her vagina and started swishing it around. The odor I found somewhat repulsive, but I really enjoyed seeing her jump around.'

They got into cunnilingus about two out of every three times they had sex. Sometimes it was a part of postorgasmic play. Daniel was not repulsed by the smell or taste of his own ejaculate. He laughed slightly as he referred to it: 'I might have been so horny one night that we'd screwed first and before the second time around I ate her out. It really doesn't taste bad.'

Jane was similarly uninhibited with respect to fellatio: 'The first blow job she gave me was excellent. Once or twice she swallowed the load. Sometimes, also, if we weren't fucking, she'd jerk me off. I liked it. She had a really steady grip.'

The sunny pleasure of their relationship passed behind a cloud when Jane developed the yeast-like vaginal infection, trichomoniasis vaginalis. Knowing that trichomoniasis can pass from vagina to penis and back again, so that the infection fails to clear up unless both partners are treated, Jane asked Daniel to do as she had done, and go to a doctor: 'For some reason or another, I thought things would pass, so I didn't go. Two weeks later we went out again, and we were in bed, and we screwed. I had figured that if she were going to ask me at all, it would have been beforehand, not afterwards, When she did ask me, and I said no, she replied, "I thought I was more than a piece of ass to you!" I had had no symptoms at all. My pisser didn't even hurt. So I had let it ride, but two days later she came down with it again. I went to the doctor and took my pills like a good boy. When we were next in bed, she asked about the doctor again, whereupon I whipped out the prescription. She was happy. But several days later she came down with it again, so she asked me to start using condoms.

'There's a distinct difference of feeling with a condom on, in my opinion. I screwed her a few more times after that, but soon the relationship ended.' He offered no explanation, but added: 'At this time, though, I was starting to get into another relationship. While I'd been screwing Jane, I had been seeing Lois. I had seen her at a party with another fraternity brother. She had a really fine set of knockers. One day I saw her in the library with five other guys at her table. I said, "Well, dammit, you've been meaning to ask her out." It was damn the torpedoes, full speed ahead! I was really edgy about asking her out, because there were five guys sitting around her, and I figured they were all after her. But I asked her out anyway.

'We got in some heavy kissing for the first couple of months. Then one night we were at my place, lying on the rug in front of the fireplace. She had quite a short skirt on, and I began to tickle the underside of her thigh. She really responded to that. She kind of jumped and said, "Please don't do that. You don't know what you're doing to me!" We were kissing around, and I began pulling her blouse off when I stopped because she registered a protest.

'At that time, though, she said, "One of these nights I'm going to attack you!" "What do you mean?" I questioned. She answered, "Well, you'll find out." No sooner had I replied with, 'Why don't we find out tonight?" than I exclaimed, "Front and center! It's Gary, my roommate!" That was the end of that for the evening.'

It took about three months for them both to progess from the stage of Daniel's sexual exploration and resistance-testing to intercourse. Daniel considered that during this three-month period, their relationship developed a fair degree of emotional breadth. But it did not follow a soap-opera script in which love continuously intensified to where intercourse, a solemn, sanctified event, took place. There was endearment, but at the same time having intercourse was not till death do us part.

That bothered Lois: 'She has a problem about screwing guys just for the sake of the sensuousness of it. She really feels that she is sinning if she does. If she saw a nice-looking guy and wanted to screw him, maybe she did. But afterwards she was really sorry for she felt just having sex with a guy was wrong. She had a lot of guilt. It turns out that the night before I was with her when we had intercourse for the first time, she was with another guy and had intercourse with him. She felt that she had sinned. She was upset and prayed for forgiveness. To me it didn't matter heads or tails.'

Actually it mattered to Daniel more than he admitted. He shared Lois's double standard covertly, as when he referred not to her fucking or being screwed, as he did when talking of Jane, but to her 'being with' someone. Also, he lowered his voice and reluctantly, almost apologetically disclosed that she had been with 'no more than four' other partners.

The double standard did not prevent their eventually having intercourse: 'We were both lying down talking, and I started unbuttoning her shirt. She didn't say no, so I continued and took off her shirt and brassiere — a really nice set of breasts. I licked the nipples a couple of times — sucked on them. She seemed to like that. We kissed a little. I rubbed on her breasts in a circular motion. She really liked that. Having had my experience with Jane, I was feeling a bit bold. So I asked, "Should we make love tonight?" "That's a good question," was her reply.

'It seems that at one point before a guy screws a girl, there's a credentials check. What have you done? Have you screwed before? What do you know about birth control? Will you think of me as a slut after we do it? I thought something like this might be in the offing, so I had brought along a half-dozen condoms.' He spoke with a snicker of embarrassment. 'We agreed to have intercourse, but she said, "My heart is really not in it, but I'd like to, so why don't we?" So I take off my underwear. She has on a nightgown. I pull off the covers and start to pull up her nightgown and look at her. She protests, "You're not doing that!" whips the nightgown back down, and pulls over the covers. Mocking, I say to myself, "Oh, boy, this looks like a night of fun."

'I guess the idea of my viewing her vagina was repulsive to her. Well, anyway, I had been talking to her about my sexual experiences with Jane. She was upset with me, and it came through. She said something to the effect of "I just want to screw. I don't want to sit and analyze things clinically. Let's just get it on." "All right," I said, "We don't have to say anything. We'll just make love and make animal noises." That really sent her up the wall. "That's it," I said to myself. "Turn on the light. Not tonight. Ah, Christ, look what I did — talked myself into a dead end."

'So I handed her her underwear. Her response was, "Gee, I certainly do lubricate." I thought to myself, "I bet if I act like a nice guy and don't press matters, and ask her why she was so upset, I bet we'll screw tonight." What we did still strikes me as very strange. She was very difficult to screw that night. She wouldn't spread her legs enough. I had her insert me, because frankly it was hard to do. I just didn't set in her right. I asked her to spread her legs some more, and she just said in a loud voice, "No!" It didn't bother me much, but I thought that maybe she hadn't had as much sexual experience as I had, or that my technique was strange to her, and she didn't like it. I immediately thought after I had come, "Boy, I know what's going to be the pitfall of this relationship," meaning her sexual inexperience. All the while I was thinking what a lousy screw it was.

'I put the condoms away and said, "I'll have to save these for later." She exclaimed in a disturbed tone, "Are you serious?" So I figured here's a good time to back off and said, "No, I was just kidding." She rolled over to go to sleep. Before I did the same, I kissed her goodnight. "Gee, thanks," she said, "I needed that." It was subsequently established that Lois's surly mood and lack of cooperation had been the products of Daniel's being excessively late for their evening together, coupled with his failure to apologize. Her tactic was to deny sex to him to teach him a lesson.

'A couple of weeks later we managed to get in bed together again. This time we made a lot of jokes about how uptight we were the first time. We even joked about the time she said "No!" when I asked her to spread her legs. This time it was really nice.'

Prior to having sex with Daniel, Lois was anorgasmic: 'She said that she didn't have orgasms. She said that in all her other affairs, she may have had an orgasm once or twice, but she wasn't sure.' If it was bad for her, it was bad for Daniel also: 'After a while, I got sick and tired of getting my rocks off, and she not getting her rocks off. That bugged me. It felt like I was just sort of using her as a vehicle to get my rocks off, and she wasn't getting any pleasure out of it. To me, I equated pleasure with orgasm. I would feel great when I came. After five minutes, it's not so great when the other party is still staring at you. It gets annoying. After a while, I got tired of entertaining myself.

'I didn't try to assign blame,' Daniel said, but he gave himself credit when Lois finally joined him in being orgasmic: 'I fixed that! I hate to sound like a wise-ass stud, but I fixed that! She comes very close to crying with delight. She sometimes has more than one orgasm in an evening. She says I really trigger her. She triggers me too. Once I was just at the verge, and she contracted a couple of times and that just, "spt!" triggered me.

'I usually have my orgasm first, and that sends her off about ten seconds afterwards. The first time it doesn't take me long at all, but afterwards, the second time when I start going into my long distance running, 20 to 35 minutes, she'll have a couple of little panting sessions, and then I might come. It's hard to tell when she has an orgasm, but we talk about it afterwards: "Did you come?" "Yes," she says with feeling. I have trouble feeling her contractions through a condom. One time, though, I really noticed it, in a position that Jane taught me. You put the girl's legs up straight in the air while she's on her back, and you enter while on your hands and knees and rock your can back and forth. In this position she says she feels like I'm screwing her brain. It's that sensitive!

'At any rate I noticed this one time she was really hotter than a firecracker. I mean she was rolling her eyes and panting and moaning and groaning, and finally she asked me to stop. She thought she would start crying. So I chose a less sensual position. She later said that I really satisfy her. It's really gratifying to me that I can inspire an orgasm. First of all I think it was a matter of getting acclimated to each other. Now I gradually work up speed in pumping, and then I just sort of stop, and then I try to sort of move my penis around inside. It's sort of like sticking your finger in a wad of clay and reaming it into a cone. I do that for a little bit, and — maybe this sounds a little ridiculous — but after a while I reverse directions.' He laughed a bit, embarrassed: 'It works better than just straight thrusting. I just do this for variation. If you just keep thrusting away, it gets to be just a little bit boring. You look for variation. It keeps her on her toes — what to expect, you know. ... She really gets stimulated when I start licking her clitoris with my tongue. We have a lot of oral sex, but not with the intent of orgasm. We got into sixty-nine occasionally, and really got into it. She got into it, especially my sucking-action around her vagina.' Only once did he come in her mouth: 'It was the first time she ever swallowed. She pulled back, came up to me and kissed me, and said, "How does that taste?" I said, "A little salty." She answered, "I've never done that before. It tastes kind of funny." She didn't seem to be grossed out by it.'

Among the recollections of all the times that he has experienced orgasm, there was one episode which he recalled as quite remarkable: 'There was one time when I had spent all night studying and had been up for about 40 hours, before we were in bed together. It's really strange, because I'd come and was still pumping around. I hadn't gone soft. About four minutes later, or about five minutes later, I came yet again. This was really strange having one orgasm after another. I thought something was wrong with me. It was ridiculous because it usually takes me thirty-five minutes to recycle. It seems though that Lois experienced the same phenomenon, remarking that "It seems like lack of sleep does the same thing for you as it does for me."'

At the outset of their relationship, Daniel used a condom as a contraceptive, as a matter of convenience. 'Now,' he said, 'Lois uses a diaphragm. She used to use birth control pills. The pill really fucked up her system. It made her gain all sorts of water weight and gave her excessively long periods. So now we use the diaphragm, alternating sometimes with a condom.' He is not pleased with either method. For the present, though, he views them as a satisfactory compromise, as their future plans call for them to be separated by a thousand miles for several months at a time. Eventually, he would like Lois to get a second opinion on an I.U.D. The one gynecologist she consulted turned thumbs down on the I.U.D. method. Before Daniel, Lois's only contraceptive precaution was coitus interruptus — pulling away.

Despite their contraceptive caution, they did have a pregnancy scare. It was probably a false alarm, but a three-week trauma, nonetheless: 'First she was a week late, then she was two weeks late, and then she went home for spring break. She said she was going to go to her gynecologist. Somewhere between going home and going to her doctor, when an estimated two weeks and four days late with her period, she discussed the problem with her mother.

'She started out diplomatically by saying to her mother that today is not like yesterday, and that kids are a lot freer with sex. I don't know, but I don't think her mother bought that, for when Lois told her mother she was sleeping with me, she received a lecture to the effect that you don't have to sleep with a boy to keep him.

'Anyway, she went to the doctor and had a test which proved inconclusive. It was negative but given too early to tell absolutely. And then it came. She was really fortunate that the day she left for school her period started.' They both breathed a sigh of relief. What would they have done if pregnancy had been confirmed? 'I would have chipped in for an abortion. I had no intention of jumping ship, but I was only going to pay half, because I was only going to take half the blame. My feeling was "Just my luck."

'She said aborting was something she really didn't want to do. She didn't want to go through with it, but it sounded like she was going to. But it would be an experience she didn't like. It would be like being prepared to have something ripped out from inside of her. It really made her feel better that I stood by her and didn't jump ship. ... I thought it would disappear, clear itself up if it were ignored.' They are not sure about planning a pregnancy. In fact they are not sure about how well they might live with one another in the roles of husband and wife.

CASE #2: SALLY'S BOYFRIEND

Lovemap biography

By age 13, if my parochial school training had taught me nothing else, it had taught me that if I had faith in something, it would happen. Yes, the nuns said it was so. And so, with the effects of their tutelage unbeknown to these sisters of the Catholic church, I prayed for a woman. She would be my salvation from the addiction of masturbation.

At that time I was into talking to God personally, and with each night went my monologue, 'God, give me a woman so I can have real sex and stop harming myself. Give me a woman right now. I know you can do it. I want to feel it. I need to feel what it's like.' I would continue further explaining to the Lord that I had to have the experience so that in later life I wouldn't be a fumbler. But alas, my woman was not delivered.

I grew up in an environment in which the atmosphere was permeated with sexual repression. My mother was the sole source of sex education in the family. She was hardly a well-spring of knowledge, but rather a shallow well from which I could draw the answers to questions if I had them. I remember one such question at about age 12. I had had my first seminal emission and I thought there might be something wrong. 'No, there's nothing wrong,' was her answer and with it I received an accurate lesson in the functioning of my anatomy. That was the extent of her teaching on sex. From my father I never heard a word, directly or indirectly. I asked nothing, and he told nothing. Now that my parents are separated, I have gotten to the point of contemplating their sexuality. What were things like behind their bedroom door?

Peers were my salvation. I remember the first time I learned what the word fuck meant. I was in the fourth grade, and standing in the lunch line. Until then, my idea of conception required only the sharing of a marital bed as prerequisite to the generation of new human beings. I was aghast to find this to be untrue. When I learned of the workings of penises and vaginas, my first reaction to the new truth was: 'My parents *did* that?' The emphasis was on *did*, for I was absolutely sure that if by chance they *did* do it at one point, they no longer were *doing* so. I reasoned that only unmarried people had to resort to such an unesthetic deed. Married people wouldn't have to, for they would acquire the pristine powers of immaculate conception.

I can't recall how long I was satisfied with my dichotomy, but I do recall that I consulted a great many books and found with dismay that my theory was wrong. My lunch-line informant had been correct. Married people, my parents, did have to fuck to have children.

I know for certain that I did not start out being guilty about masturbation. In fact my introduction was very benign. I was 12 that summer and was on a camping trip with my 16 year old cousin. As a way of telling someone to stop doing something, my cousin used the expression, 'quit jacking off.' He had to explain what he meant. It was right before turning in one night, and we were lying down in the tent. Along with his verbal explanation, he gave a demonstration. He grabbed his penis and stroked it a few times. I watched in the light of a flashlight and was somewhat surprised. The next day I tried it myself and, lo and behold, I ejaculated. The feeling was like nothing I had ever before experienced. Later I asked my cousin if he still did it. Avoiding self-incrimination, he said no, and claimed that he had gotten tired of it and had quit a long time ago. Like me, he's probably still jerking off, and maybe feeling guilty. When I didn't quit, I remember the guilt began.

I would masturbate every day, and worry about what I was doing to myself. The antimasturbation propaganda of my Boy Scout Handbook, with its definitive condemnation, sticks out in my memory. When the Boy Scout Handbook said something, it was to be believed! Its subversive influence continued until I was 18 and for the first time came across a positive statement on masturbation. It was in an article in *Playboy* magazine. For nearly seven years, I had been drowning in a sea of pious, righteous falsehood, with the dry land of fact nowhere in sight.

I masturbated mostly in the bathroom, but often in bed at night. Neither the bedroom or bathroom had a lock on the door, so I was always worried about being caught. I never was. The closest call was when my mother came into my bedroom one night to look for something. Anxiety over being caught in the act dictated my masturbation practices. I didn't have much variety. It was get it over with as quickly as possible. I didn't want my father asking why I was in the bathroom too long. The shower provided the greatest degree of privacy, and I used it well. There was no equipment other than my penis and my right hand. I had no visual stimulation. No *Playboy* centerfolds. I didn't have the guts to go into a store and buy the magazine, and have it around as incriminating evidence.

I resorted to storing erotic stimuli in my head. When, during the day at school, I would see a well-turned thigh or the outline of a pretty girl's breast, I would file away the mental snapshot which my eyes photographed. I would do the same whenever I was able to surreptitiously scan *Playboy* on a newsstand. If I were more fortunate, my friend's father would have a copy, and I would see it at his house. I would devour it. To hell with the articles, I wanted the pictures. Like a spy photographing top secret documents, I would record them in my brain in intimate detail. It was pretty easy, as the poses and seductive looks were all the same. Only the faces changed.

One day the ultimate bonanza came my way. Basically it was indirectly made possible by my father. An Air Force buddy of his came to visit us. He was one of these super-macho types who referred to himself as a 'chopper pilot in 'Nam.' Along with his uniform and assorted paraphernalia, he brought a copy of *Playboy*. What a windfall! It was God sending me a substitute for the woman he couldn't send in the flesh. I knew that, sooner or later, my moment would come. The magazine was unguarded in the guest's bedroom. The adults were celebrating, drinks in hands. Miss December and I had a date. We met in the bathroom — the ultimate Christmas gift. After being satisfied, I returned her to her proper owner.

Satisfying myself was my goal, and I was more concerned with the goal than the getting there. Orgasm was my destination and to hell with the scenery along the way. One might say it's a crude approach, like sitting down to a seven course meal and eating everything in 15 minutes just to have a long, satisfying belch at the conclusion, but there was no savoring while I masturbated. My fear was of getting caught.

There was a co-factor to my fear of being caught, however. Why should I be enjoying myself? I was doing something wrong, and to take pleasure on the way would have been an even greater sin. I compromised. For a few moments of ecstacy, I would forego enjoying myself beforehand. I would not make a pig of myself. By ejaculating as quickly as possible, I would satisfy all parties involved. I wish I had known back then that *I* was the one and only party involved.

It bothered me very much that I was masturbating every day. To deal with my guilt, I vowed to limit myself. I began by setting three masturbations a month as my quota. I firmly resolved that masturbation could not control me. I had to control it. I had to break the addiction. I knew that God didn't like people masturbating. Amidst all this trouble, I was beginning to think that I was some sort of freak. Oh, for sure I knew that other people masturbated — at one time. But they would quit, just like my cousin had said he had quit.

I struck another pact with God. My father had been given a job transfer, and I had left all my friends several thousand miles away. It was summertime, with almost no chance to make new friends. I proposed a trade-off with the Lord. I offered to stop masturbating for one month, if we could move back to Arizona. As a gesture of good faith, I would pay in advance. We never did get to return, but I did stop masturbating for a month. That was quite a feat. When my pact was not honored, I resumed my habit. It didn't much matter anymore if we did move back, for school was starting up, and I was making new friends.

As the year passed I didn't 'get tired of it' as my cousin had said he had done. Masturbation has continued to serve as my foremost sexual release. At the present time, I masturbate about four times a week. It serves as a tension release from the rigors of premedical academic life. I find that it increases in frequency as exams begin and the work load increases. I'm fortunate these days in having *Playboy* and *Penthouse* readily available in the house in which I live, but I still don't purchase them myself.

Though I still file away the mental imagery of that 'well-turned thigh,' my fantasies have a regrettable dichotomy between the girls I would fuck and the girls that I would marry. I fail to be able to have my steady girlfriend in a fantasy. I would feel guilty if I did. I try to find ways to shed this unwanted guilt and the dichotomy it imposes.

At around age 15, sexuality with a partner began to enter my life. It was at this age that I had my first date. I went to a dance with Judy our first time out. She had a history which included an attempt at suicide. I don't know the details or its seriousness. It could have been either a theatrical display or an almost successful attempt. I do know the circumstances surrounding the attempt. It was a case of dishonor for her. Three of my friends had been passing her around, each seeing who could score the most. Later they compared notes. She found out what they were doing. For me it is tragic that attempted suicide would be the price when the highest score was fondling, not intercourse.

I don't know why I took Judy out, except that she was just another person with whom the social rituals could be practised. I don't think I took her out for sex alone, for at the time I was scared of sex. But maybe deep inside I was attracted by the illicit possibilities of going out with a girl who 'had been around' and 'knew what she was doing.' One of my masturbation fantasies which now comes to mind involved such a situation in which the female would take the initiator's role and sexually initiate me. Perhaps the other side of the coin was the 'knight in shining armor,' a sort of sterile masculinity — I would date this girl with sex being unimportant and far-removed. Perhaps I would be the one who would show that there were some clean individuals to whom sex was not the number one objective.

On another date, we went to the movies. Again there was no sex. I can remember a sort of primeval lust which existed deep inside me. When we held hands in the movies, I had an erection. At that age, I had erections from everything. I spent much time adjusting my trousers, with hand in pocket, to provide a tent for my troublesome, engorged penis.

By the time I was 16 and a junior in high school, I was always on the lookout for girls. Joan was the female lead in the drama club. As if it had been ordained in fantasyland, I was the lead male. The stresses and hectic pace of the production and frequent rehearsals drew us together. The Harvest Ball was coming up. It was a Sadie Hawkins deal, in which the girls asked the guys. I was standing at my book-locker at school when she asked me. I was thrilled, but tried not to show it. Instead, I was cool and replied, 'I thought you'd never ask.'

At the end of the dance, we were standing outside waiting for her mother to pick us up. We were all bundled up against the cold, scarves, sweaters, hats. I moved to kiss her. The dark and the clothing caused me to miss her mouth, but even more embarrassing was that it all occurred as her mother drove up with headlights shining on us. I was only mildly embarrassed. We were both laughing about it in a fun sort of way. We parted with my saying, 'It will get better.'

It took a while for both me and my erection to subside after having kissed a girl for the first time. My feelings gradually became feelings of stupidity and uncoolness. I was aware of my lack of sophistication, and it bothered me. Eventually I felt my sociosexual behavior to be very ritualistic. That also upset me. It was a case of going through the motions expected of me, but with no real feeling behind them.

In the ensuing five months, my sexual experience broadened. Up to that time, it had amounted to clutching, kissing, and hugging on the dance floor. But soon we moved on. I got my hands under her shirt and finally into her pants. When I first did that, she began to cry. I intended to relieve her inhibition with more stimulation, and eventually I did, but I apologized first, before moving on. She did place limits, nonetheless, as did the Minnesota winter. The cold kept our clothes not only on, but almost fully buttoned, as most of our contact was in a parked car. The lack of a driver's license forced us to double-date, and the company further inhibited us. In placing the limit on how far we would go, all Joan would say was, 'No, I don't want to.' She gave no reason. When she said she loved me, I doubted it, if she would not show it in a physical way. When I said that I loved her, I felt sincere.

We engaged in activities like actors you might see in the movies or in commercials on television — going sled riding on a winter's day, she would fall in the snow, and I would fall into her arms, and then we would kiss. I knew the script well. We both played our parts. I felt really great at the time, for I was finally doing what was supposed to be done.

When the spring thaw came, her feelings remained wintery. No longer protected from my sexual advances by the Minnesota cold, perhaps she considered it best to exit to safer territory. She did, and thus our relationship ended. In retrospect, I was pretty happy with what had transpired. I was getting into the mainstream of sexuality, and hopefully leaving masturbation behind. I had, however, always wished that she would be more aggressive. I grew tired of being the initiator, the doer, as if manipulating a puppet with a mind of her own only when it came to resistance.

I was destined not to be lonely, for Harriet from Australia floated briefly into and out of my life. She was a girl in the drama club. She had the biggest boobs, I thought, that I had ever laid eyes on. When we were hugging and kissing backstage, I tried to feel them up, but couldn't as they were squeezed up against me. There was no room for a hand. We got into heavy petting pretty quickly, too quickly, according to my protocol at the time. When I said to her that sex shouldn't be just physical, she laughed. Later she got knocked up, and went back to Australia to have her baby.

Soon, Sally and I began dating. She avoided all physical contact, even holding hands, though there was plenty of flirtation on her part — seductive looks and suggestive double entendre. Over a period of six months, she would not let me kiss her. Summer came, school ended, and I didn't see her until the fall. I worked on the beach near my home, played with the dog, and masturbated.

Getting things together sexually between Sally and me was laboriously slow. For months, closed-mouth kissing was the rule. We discussed our sexual relationship, and I was permitted to fondle her breasts and rub her crotch through her clothing. We had been dating a year and a half when I went away to college. The high water mark at that time had been fondling her breasts under her shirt. No matter what stimulation I might try, she never caught fire. She was always passive.

When Christmas vacation rolled around in my freshman year, we progressed by leaps and bounds. We got as far as shirts off in her bedroom, and were able to lie chest to chest. We discussed having sexual intercourse, but it didn't materialize on this visit home. But, when May rolled around, to my amazement, after two and a half years, we were finally able to make it. We did some talking about it, and finally did it.

It was a total mess of fumbling and bumbling, but it had one important benefit. No matter how half-assed and awkward it was, I was still able to rise from the bed and say to myself, 'You finally did it. You're not a virgin anymore,' even though it had been a frenzied, crude encounter. We were on the bed in her room. There was no contraception. Our clothes were still on. I didn't even know where the hole was, but it wasn't even necessary for, before I could get anywhere near intromission, I ejaculated. I apologized for coming so quickly. She said that it was all right. That was roughly the extent of our post-coital conversation. I do distinctly remember her saying 'I love you,' and my answering in kind. We agreed that if we both loved each other, it was all right.

But here was where my intense inner conflict erupted. It tortured and tormented me. I developed pregnancy fears. My feeling was that pregnancy was the most disastrous thing that could happen, and if it happened, it would be all my fault. I asked myself, 'How could you subject someone you love to such a dangerous situation? Why would you persist in subjecting her to the possibility of pregnancy? If you really loved her, you wouldn't. Do you really love her?' Around and around I went in this obsessional circle, trying to rationalize sexual intercourse.

The conflict pushed me to the breaking point on her 18th birthday. I had planned for a big evening. My parents were out, and I had the house to myself. Things couldn't have been better: plenty of time, soft lights, music, wine. We then had sex. I had planned ahead and bought lubricated condoms. She was fully protected, or so I thought. Upon pulling out, I went into the bathroom to remove the condom. I glanced at it. It had broken. It was as if my insides were being ripped apart. I said, over and over, 'You stupid son of a bitch,' in between my tears and sobbing. I sat on the edge of the bed, my head hanging down, staring at my bare feet, shaking my head from side to side, saying, 'You stupid, inconsiderate son of a bitch.' Her reaction astonished me. She was absolutely cool, with about as much concern as if she had a book overdue at the library. It was a good thing. I don't know how I would have reacted had she been frantic. She just said, 'It's okay, don't worry about it.'

I wonder what her feelings inside really were. My feelings came back to haunt me as her period was late. I was torn, and didn't know what to do. I knew that she did not want to be tied down with a child, but at the same time that she did not want an abortion. Furthermore, I didn't want a child, and I didn't want her to have an abortion. We also did not want to marry. Not yet. We had never talked about this contingency of pregnancy ahead of time.

Her next move was to her gynecologist, and with this her parents found out. The gynecologist prescribed a treatment which brought her period around. I was rescued from my nightmare. We continued to have intercourse, but it was much different from my standpoint. It turned into an 'I'll satisfy myself as quickly as I can' routine. With this attitude, I would pull out and draw away from her as soon as I'd finished. It was almost as if I were afraid to be near her.

That self-centered reaction changed after Sally went on the pill. I was reassured that contraception would be effective. My endeavors were directed more to satisfying her than satisfying myself. Before that, I was inanely looking for any expedient to prevent the possibility of a pregnancy. I even remarked one day, 'Wouldn't it be lovely if I were sterile? There would be no problem then.' My pregnancy fear has disappeared, but in its place came two new problems. We were at the point of almost breaking up our relationship when I realized that it was I who had taken her virginity. What would it mean for her in the future? Might it cause a break-up in a future relationship if she told her boyfriend that she was not a virgin? I was beginning to blame myself for something which had not yet happened and might never happen. I brought the situation to her attention, and asked for her thoughts. Her reaction, to my surprise, was one of no concern. She couldn't see any problems from it. If she wasn't concerned, I guessed I shouldn't be. I put the matter out of my mind and dealt with my other, more serious dilemma.

My efforts, once the pregnancy fear abated, had turned to sexually satisfying my partner. With each act of intercourse, though, I was realizing no success. I felt I was using her. She was a repository for my penis, a sex object which I was using for my pleasure only, for she never had an orgasm. My failures resonated through my person, as I felt less of a man for them. The lack of her orgasm, my failure to induct it, was eating me away. I had always wanted to be a good sexual performer. It had been one of my earliest desires, reflected in masturbation to 'Nick Carter' detective stories. I saw myself as one day being as proficient as the hero himself, the master of sexual expertise. No way! I tried many things. It got to the point where I was so concerned with pleasing her that I focused exlusively on that goal. I became impotent on several occasions.

We did not ever discuss the situation, to any extent. That's the way in all of our sexual discussions. We talk about it, but not in specifics, and never graphically. Euphemisms are pretty common. The verbal initiation of sexuality is cloaked in such language as her saying, 'I want you.' I don't know what she would say if I said, 'Let's go fuck!' It probably would blow her mind. She probably would be turned off, if I said it in a serious manner. I don't really have any tremendous urge to make such a statement, though.

I do have some desires that do go unfulfilled with Sally, by her choice. Many things turn her off. She is pretty conventional and wants nothing to do with fellatio. I finally got her to try it once, and her immediate reaction was 'yech' as she pulled her head away. When it comes to avoidance, she usually is pretty good with using only sounds. She never has told me exactly what was wrong with blowing me. Furthermore, she objects to having me ask her to do something, or to suggest something new to her. I would like to try different positions, but she doesn't. She is always insisting on the missionary position and won't stray from it. I would like to investigate what different penetrations felt like. It would be nice just to spend an afternoon screwing in different ways. I'd like to fuck until we were raw, as they say.

'Yech!' was her original response to our having intercourse during her period. The blood didn't bother me. Things felt the same, so I didn't see any reason for not doing it. What's more, the amount of time that we are together is strictly limited, so I say make the best of what time you have. Once again, we never sat down and talked it out. I guess I always presumed that it was some kind of carry-over of the tribal and biblical prohibitions surrounding menstruating women, so I didn't question it.

Cunnilingus also is prohibited in her book, or perhaps I should say semi-prohibited. She offers a groan of displeasure when I first start to go down on her, but this all changes when I really begin my oral caressing, If her initial groans aren't sufficient to deter me, they soon melt into utterances of pleasure. I suspect that her initial aversion is more for my sake than hers. I think that she enjoys the feelings, but I also think that she considers the performance demeaning to me. I don't insist on cunnilingus at those times when she is adamantly opposed.

While she tolerates fucking on the rag, and my eating her, Sally really objects to semen. 'Yech' is what I hear if she gets semen on her skin. She gets very annoyed, chastising me, like she did the first time I ejaculated on her during coitus interruptus. So I apologize.

I do desire her to be more aggressive in our lovemaking, but it's difficult to ask her to be so. I suppose that it's not in her nature. During sex she is very playful and prefers to keep sex easy-going, lots of little jokes and tickling. It really isn't a somber scene, and that's really nice.

At the present I am at a crossroads. It bothers me to say it, but it seems the more I am with her, the less I like her. This presents a substantial problem, for her philosophy is: 'If there isn't love, then I feel like a whore.' I am at the point where I don't think love is a prerequisite for sex. In other words, I could enjoy going out and having a good fuck, whereas that would be impossible for her. Consequently, I could carry on our relationship as just something sexual, but for her it would be impossible. So my problem is what to do. I really don't know exactly where my feelings lie, and I really don't feel like starting a new relationship after so many years. Furthermore, I don't wish to hurt her. It's a difficult predicament from which I don't know where to turn.

As far as the pure sexuality of the relationship is concerned, I have always felt that two people must compromise and reach a median of sexual compatibility. In other words, I have no right to expect her to engage in fellatio if she doesn't want to, even if I want to, but she should attempt to compromise to please me.

I have always been taught that sex is secondary in a relationship. It's other things like compatibility, mutual tastes, and philosophies that are more important — sex isn't everything. I can even remember the guy who married my cousin talking in those same terms. Maybe he was right, but I've been doing a lot of talking to a friend about it lately.

He suggested to me that maybe society has it upside down and turned inside out. He said that sex is the most important aspect of a relationship and that the bedroom is the place for compatibility, not compromise. Compromises are made over which color the drapes in the living room will be. It was his contention that society has made the living room drapes the most important aspect, and has tried to discount sexuality in an attempt to preserve its taboo.

The summer is fast approaching, and I'll be with Sally more. Neither of us knows much about creating compatibility and compromise, but I have derived from the layout of this sexual biography a more astute conceptualization of their place in the sexuality of our relationship. Revising the relationship will be on the summer agenda.

EXPOSITION

The biographies of the preceding chapters of this book do not exist in a historical vacuum. They belong to people who attained adulthood in the era that is now historically labeled as the sexual revolution, though its more precise label would be the sexual reformation. It was also the era of the anti-Vietnam war protests, and the civil rights protests. Its antecedents as an era of sexual emancipation were, in the 1940s, the newly discovered efficacy of penicillin and other antibiotics for the prevention and cure of syphilis and gonorrhea; and in the 1950s, the newly discovered efficacy of the recently synthesized and commercially produced steroidal sex hormones as an oral contraceptive that gave women personal control over their own fecundity.

The lovemap biographies of the two men in this chapter, together with the glimpses of those of their girlfriends, are historical documents of what it was like to come of age in the era of the sexual revolution. They brought to their college years the handicaps of an unemancipated childhood, and the difficulties engendered from that period of their development. As compared with the genitally birth defective, they showed that being anatomically normal has its hazards, too.

They gave their sexual histories at a time, 1975, when sexual emancipation was sufficiently widespread to allow people to make intimate disclosures without too much fear of self-incrimination or of social reprisal. The militant forces of counterreformation from the new right were not yet ready to mount their attack. 1975, one sees in retrospect, was the peak period of the sexual reformation. The sunny window of sexual emancipation that looked out on the antibiotic victory over syphilis and gonorrhea was still wide open. Herpes simplex was barely yet in view. It would not be until after 1981 that AIDS, unvanquished still, would slam the window shut.

In the two cases of this chapter, and in others with similar sexuoerotic histories, it is not possible to ascertain any genomic, neurobiological, or neurochemical factor that might be identified as a vulnerability factor, independent of the conditions of sexual rearing. One assumes that the course of development is, in fact, contingent on the conditions of rearing. Bad rearing superimposed on a vulnerability factor is still bad rearing. It has a negative effect on the vulnerable, even if the nonvulnerable have a resilience with which to surmount its bad effects.

An adventitious feature of these two biographies is that they are concordant in portraying male sexuoerotic development as a personal, interoceptive experience totally at variance with the stereotype exteroceptively projected onto males in the currently fashionable, militantly feminist, male-bashing literature. Both men were familiar with the macho stereotype of the male as a relentless and promiscuous fucking machine. It did not fit. They were intimidated by the shibboleth that made them, alone, responsible for the totality of the woman's erotic response and pleasure. They yearned for the reciprocality of lust to accompany the reciprocality of love. They were lovers, affectionate and caring. They knew the personal meaning of limerence – of being pairbonded in love – and of love requited and unrequited. They desperately longed for success as both lovers and lusters. Their search for success was by trial and error. There were no courses they could enroll in. They had no mentors — only the remarks of their peers, and the practicums with a sophisticated partner. They bear living witness to the incontrovertible truth that, beyond the stereotypes that block the view, pairbonded love combined with pairbonded lust are not gender differentiated. They are the same for men and women. When two partners are not symmetrically matched on both, then the relationship is as much impaired for the one as for the other partner.

A second adventitious feature of these two cases is that they were discordant for ethnic and religious tradition of rearing: a Jewish ethnic heritage in one case, and an Irish-Catholic ethnic heritage in the other. It goes without saying that some religions and sects are more sexually puritanical and prudish than others, but it is society as a whole that sustains a sex-negative climate. This climate affects those with a history of birth defects of the sex organs, as well as those without such a history. It is a climate that is as diabolically resistant to change now as it was during the long and hideous history of the Inquisition. Its effect on sexology and the practice of sexual medicine is a catastrophe.